MATERIALS SCIENCE AND TECHNOLOGIES

NEW RESEARCH TRENDS OF FLUORITE-BASED OXIDE MATERIALS

FROM BASIC CHEMISTRY AND MATERIALS SCIENCE TO ENGINEERING APPLICATIONS

MATERIALS SCIENCE AND TECHNOLOGIES

Additional books in this series can be found on Nova's website
under the Series tab.

Additional e-books in this series can be found on Nova's website
under the e-book tab.

MATERIALS SCIENCE AND TECHNOLOGIES

NEW RESEARCH TRENDS OF FLUORITE-BASED OXIDE MATERIALS

FROM BASIC CHEMISTRY AND MATERIALS SCIENCE TO ENGINEERING APPLICATIONS

AKIO NAKAMURA
AND
JUNICHIRO MIZUSAKI
EDITORS

nova publishers
New York

Additional color graphics may be available in the e-book version of this book.

Library of Congress Cataloging-in-Publication Data

New research trends of fluorite-based oxide materials : from basic chemistry and materials science to engineering applications / editor, Akio Nakamura (Advanced Science Research Center, Japan Atomic Energy Agency, Ibaraki, Japan) and Junichiro Mizusaki (Inst. Multidisciplinary Research for Advanced Materials, Tohoku Univ., Sendai, Japan).

pages cm

Includes bibliographical references and index.

ISBN 978-1-63117-350-9 (hardcover)

1. Fluorides--Industrial applications. 2. Fluorspar--Industrial applications. 3. Zirconium oxide--Industrial applications. I. Nakamura, Akio (Materials scientist) II. Mizusaki, Junichiro.

TA455.F54N49 2014

620.1'93--dc23

2014003594

Published by Nova Science Publishers, Inc. † New York

CONTENTS

PREFACE

Fluorite-based oxide materials such as stabilized zirconias, doped cerias and urania represent a group of the most important key engineering materials in our modern society, with their well-known expanding electrochemical, ceramic and nuclear etc. applications, which is made possible by their multi-lateral combined excellent properties such as high chemical/ structure stability, high oxide ion conductivity, superior mechanical strength, and unique nuclear properties, and so on. Concurrently with such expanding technological applications, the rapid progress of various theoretic, computational and experimental research techniques in recent years has brought a great advance and new evolution in our understanding of their key basic properties, such as thermal, thermodynamic, mechanical, defects and transport properties, grain boundary structure, gas/solid interfacial electrochemical, catalytic and local-structural properties, and radiation tolerance, etc. This book represents a first attempt to compile such as wide as possible range of basics of these fascinating oxide materials toward their diverse applications in a single book. To attain this goal, total eleven experts of these materials from various regions of the world (two from Europe, one from India and eight from Japan) are invited to contribute each one chapter and intend to convey a timely up-to-date comprehensive review of their latest research trends. Chapters 1 to 8 mainly deal with various bulk and interface properties of stabilized zirconias and doped cerias (and thorias, etc) in connection with their electrochemical (solid state ionic) and ceramic applications. The last three Chapters 9 to 11 mainly deal with those of stabilized zirconias and urania (and its mixed oxide (MOX) with plutonia) in connection with their nuclear applications.

First in Chapter 1 a comprehensive introductory review of solid state ionics of defect-fluorite oxides, primarily of stabilized zirconias and doped cerias, is given, focusing both on bulk defect chemistry of oxygen-vacancy diffusion behaviour and kinetics of interfacial gas/ electrode reactions crucial for their solid oxide fuel (SOFC) applications. Then in Chapter 2 an another-type global review of ceria-based solid solutions (doped cerias) as to their structure, phase-diagram behavior, ionic-conductive, catalytic and magnetic etc properties is presented in connection with their various electrochemical, catalytic and nuclear etc applications. Chapter 3 describes many new results of crystal structure, structural disorder and oxide-ion diffusion pathway of a number of important fluorite-type oxides such as doped ceria, bismuth oxide, mixed zirconia-ceria solid solution, etc. firstly clarified by advanced high-temperature neutron diffraction techniques. Chapter 4 gives a comprehensive detailed review of new defect-crystal-chemistry approach to coupled macroscopic non-Vegardian lattice-parameter behavior and microscopic non-random defect structure recently proposed for doped cerias and thorias, providing a new intriguing interpretation for controversial ionic-conductivity maximum behavior these systems exhibit. In Chapter 5 difference in solid oxide fuel cell (SOFC) electrode reaction mechanisms between yttria-stabilized zirconia (YSZ) and doped ceria is described and discussed based on various electrochemical and spectroscopic data of gas/solid interfacial and bulk transport properties. Further in Chapter 6 a practically very important but least-studied and not-well understood in SOFC technology, namely, mechanical properties of stabilized zirconias and doped cerias are comparatively discussed, clarifying some important problems in long-term high-temperature operation of these complex ceramic oxide materials. Chapter 7 gives a timely latest review of recent remarkable achievements on grain-boundary structures of yttria stabilized zirconia (YSZ) as a vital property-determinant of poly-crystalline ceramic materials based on advanced analytical electron-microscopic techniques in combination with computer grain-boundary structure simulation techniques. Then, In Chapter 8, as one such rapidly growing practical application example, a compact review is presented on the current research and development status of tetragonal-zirconia polycrystalline (TZP) as a new popular dental material. It highlights the crucial role of such interfacial grain-boundary structure and property in zirconia ceramics.

The last three Chapters 9-11 deal with stabilized zirconias and urania (plutonia and mixed oxide etc) systems in connection with their nuclear applications. Chapter 9 describes and discusses results of computer-simulation study of some important transport (both of oxygen and actinide ions), thermal (thermal conductivity, melting, etc.) properties of urania, urania-plutonia mixed oxide (MOX) and stabilized zirconias. This enables us to know the current status and future direction of such silicon-based various computational numerical-simulation techniques and their feasibility in reproducing and/or predicting various key properties of these generally ‘hard-to-measure’ radioactive nuclear-fuel materials. Then Chapter 10 focuses on an attractive urania (UO_2) image or negative crystal, i.e., a well-faceted single-crystalline jewelry-like void in bulk crystal, formed in helium atmosphere at high temperature, and gives a first successful detailed explanation for its formation and morphology-change mechanism; It also clarifies that such image-crystal formation is not only of basic-scientific interest but also of significant practical nuclear-technological problem closely related to minor-actinide (MA)-bearing target nuclear-fuel fabrication and its operation. Lastly in Chapter 11 results of persistent research effort in Europe for realizing stabilized zirconia as inert-matrix (IF) nuclear fuel for actinide transmutation and conditioning are extensively and in detail reviewed, illustrating rich potential applicability of stabilized zirconia in nuclear-energy area.

As above delineated, subject areas of these fluorite-based oxide materials are quite widespread and diversified. Yet, it should be mentioned that their full range of basics and applications could hardly be covered by this book. For example, one interesting basic physical subject of electronic and magnetic properties of these 4f- and 5f-electron fluorite oxide materials and their ab-initio calculations, and other directly-application connected high-priority mechanical-property subject of monoclinic/cubic transformation-toughening mechanism in partially-stabilized zirconias, and alumina/zirconia super-plastic deformation mechanism, are not at all covered in this book.

Future version of this book may be able to cover such subjects. Anyway, we hope that this book will not only further stimulate the interest of concerned chemists, physicists and materials scientists in these materials but also serve as an unique guide for young undergraduate and graduate students who would owe their future.

Lastly but not the least, we express our heartfelt thanks to all the chapter contributors, and also to Nova Science Publishers staff for their persistent support and patience for the repeated delay of the initial publication schedule of the book.

In September, 2014

Akio Nakamura
(Tokai, Japan)
and
Junichiro Mizusaki
(Funabashi, Japan)

In: New Research Trends of Fluorite-Based Oxide Materials ISBN: 978-1-63117-350-9
Editors: Akio Nakamura and Junichiro Mizusaki

Chapter 1

SOLID STATE IONICS OF DEFECT-FLUORITE OXIDES: DEFECT CHEMISTRY OF OXYGEN VACANCY DIFFUSION AND KINETICS OF GAS ELECTRODE REACTIONS

Junichiro Mizusaki*

Professor Emeritus, Tohoku University, Katahira, Aoba-ku, Sendai, Japan

ABSTRACT

In this review, two topics are selected from the application of fluorite-type defective oxides in the field of solid state ionics. One is on the nature of oxygen ionic conductivity maximum with dopant concentration. Oxygen vacancy diffusion coefficients of various systems were calculated from the reported ionic conductivity data and were displayed as a function of oxygen vacancy concentration and dopant i.e. impurity cation concentration. It is revealed that the logarithm of oxygen vacancy diffusion coefficients monotonically decrease with defect i.e. oxygen vacancy concentration. The phenomenon can be explained if we assume that the activation energy for vacancy diffusion is a linearly increasing function of vacancy concentration. Yet, a kind of mixed host cation effect upon homovalent cation substitution (Zr ↔ Ce) is not understood well on this basis, and remains as a challenging unresolved problem.

The other is on the nature of gas electrode reaction of solid oxide electrolyte cells. Conventionally, it has been accepted that the charge transfer reaction is rate determining when the Butler-Volmer equation holds for the overall current - potential relationship of the electrode reaction. It is clarified here that this is not necessarily the case, demonstrating that the Butler-Volmer equation also represents the chemical reaction controlled kinetics. Some typical gas/electrode reaction kinetics are picked up and reviewed to illustrate that in many or most cases they are indeed controlled by the chemical reaction processes, but not by the electrochemical one. It should however be also mentioned that again here controversy is still going on as to what the Butler-Volmer equation actually means, awaiting more in-depth and more extensive future research.

*Corresponding author: home address: 2-20-7-204 Nishifuna Funabashi-shi, Chiba, 273-0031 Japan. E-mail address: mizusaki@tagen.tohoku.ac.jp.

1. INTRODUCTION

It is widely known that fluorite-type compounds MX_2 are good X ion conductors. Utilizing the ionic conduction of solid inorganic ceramics, typically fluorite type compounds such as stabilized zirconia, variety of the fields of solid state ionics have been developed. Representative application fields of solid state ionics are solid oxide fuel cells, zirconia oxygen sensors for automobile and blast furnace applications, NaS batteries and lithium secondary batteries. Here, fluorite-type oxides play important roles in the oxygen sensor and fuel cell applications.

In early 1970's, already at the infant stage of such technological applications, several comprehensive review books were published on the basic concepts and fundamental theories of the electrolytic and mixed electronic-ionic nature of materials, and their applications [1-5]. These books are still now recognized as classic of this area. The title of this area, "solid state ionics" was named by T. Takahashi and O. Yamamoto. They called this area "Solid Ionics" in 1960's in Japan and the name appeared on the international journal in 1970 [6]. It was renamed in 1971 as Solid State Ionics [7]. The journal entitled "Solid State Ionics" has been issued since 1980. The series of International Conference of Solid State Ionics, ICSSI, has been held almost every two years since 1976. The journal "Solid State Ionics" and the series of ICSSI have reinforced and led the development of the field of solid state ionics. There are many related journals and series of conferences in the field of electrochemistry, ceramics and materials science. However, one can find the current status as well as history and future direction of solid state ionics best by surveying the journal "Solid State Ionics" and the proceedings issue of ICSSI which are also published as special issues of "Solid State Ionics."

The largest application area of fluorite oxides and also one of the largest areas of solid state ionics is the technology of solid oxide fuel cell, SOFC. Most of the recent research and development on fluorite oxides in the field of solid state ionics are closely related to the research and development of SOFC. The rapid developments of SOFC technology are led strongly by the series of International SOFC Symposium started since 1989. This symposium has been held every two years and the proceeding volumes [8] have been covering whole area of science and technology of SOFC. Also, due to the popularity of SOFC technology, tremendous amount of books and reviews have been published on SOFC technologies and the related materials, as can be easily seen by internet surveys. The author considers two books [9, 10] and a review on Ceria [11] as the representative ones.

As described above, tremendous studies have been made on fluorite-type oxides as solid electrolyte. However, there remain varieties of issues to be clarified. In this chapter, two topics are selected.

One is on the well-known phenomena of the dopant concentration dependence of oxide ion conductivity. In most of the fluorite-type dioxides, MO_2, the doping of the oxides with the lower valent oxides such as rare-earth sesquioxides ($ReO_{1.5}$) and alkali-earth monoxides (MgO, CaO) result in the formation of oxygen vacancies, giving rise to oxygen vacancy diffusion or oxide-ion conduction. However, in either doped CeO_2 or stabilized zirconia (ZrO_2), the ionic conductivity increases with dopant oxide concentration up to roughly 10 mole percent and then decrease drastically with further increase of the dopant concentration by more than one order of magnitude [12-14].

In spite of various models proposed for this intriguing phenomenon, two viewpoints seem to have been overlooked. They are the mixing effect of host and guest homo-valent cations observed in ZrO_2-CeO_2 system [15-17] and the effect of large oxygen nonstoichiometry typically observed in oxygen-deficit nonstoichiometric $CeO_{2\text{-}d}$. The latter can be regarded as the mixing effect of Ce^{3+} in Ce^{4+} lattice [18]. This topic will be treated in the next section.

The other is the kinetics of the gas electrode reaction of solid oxide electrolyte cells. This process is vitally important in the response mechanism of zirconia oxygen gas sensor and also reducing gas sensing, because the electrode kinetics determine the response rate of the sensors. They are also very important in fuel cell application because the electrode overpotential or the energy loss due to electrode reaction is determined by the electrode kinetics of oxygen gas electrode and fuel gas electrode.

One of the most important electrode reaction kinetics is the charge transfer process described in many cases by Butler-Volmer equation [19, 20]. Since the gas electrode reactions of zirconia cells often obey Butler-Volmer equation, people tended to conclude that the rate of the gas electrode reaction is controlled by such charge transfer process [21-23] as described by Butler-Volmer equation.

The author and the co-investigators examined in detail the relationships between the response of gas sensors and various electrochemical kinetic data, and concluded that the gas electrode reaction of solid electrolyte cells is mostly controlled by chemical reaction process [24-35] and the observed electrode capacitance is so-called chemical capacitances arising from gas adsorption on electrode materials [24,27,29], gaseous component dissolution into electrode materials [33], nonstoichiometric composition variation of electrode materials [34, 35] and so on, even if the overall relationship between the rate of electrode reaction and the over potential apparently obeys a Butler-Volmer type equation.

However, arguments are still continuing as to the physical meanings of so-called charge transfer resistance in the impedance analysis or the degree of the contribution of charge transfer kinetics to the rate determining reaction process [36-40]. In the last section, discussion will be made for the conventional belief on Butler-Volmer equation and its-associated misleading interpretation on the gas electrode reaction of zirconia cells.

2. Oxide-Ion Conductivity and Oxygen Vacancy Diffusion Coefficient

2.1. Phenomenology of Conductivity Maximum and Valence State of Dopant

Ionic conductivity maximum behavior observed in various ceria and zirconia-based defect-fluorite oxides is depicted in Figure 1 as conductivity–oxygen vacancy concentration ($[V_O^{\bullet\bullet}]$) plot, where $[V_O^{\bullet\bullet}]$ formed by the doping of lower valence cation oxides such as AO and A_2O_3 into CeO_2 or ZrO_2 is calculated assuming the electro-neutrality condition of the system: That is, using the Kröger-Vink notation for the effective charge state of defects, when a M^{4+} cation in the host fluorite-type dioxide MO_2 is replaced with a A^{2+} ion in AO to from A_M", two positive charges will be missing on the cation sublattice, so that to maintain the charge balance, one O^{2-} ion is to be removed from the anion sublattice, leaving an oxygen vacancy, $V_O^{\bullet\bullet}$. Likewise, when a M^{4+} cation is replaced witha A^{3+}cation in $AO_{1.5}$ (= (1/2) A_2O_3) to form A_M', 1/2 of $V_O^{\bullet\bullet}$ is formed.

In Figure 1 the conductivities of ZrO_2 based solid solution is shown with filled triangles and circles while those of CeO_2 based solid solutions are shown in open or lightly filled triangle and circles. Reported majority data [12-14,41-44,46] show that the conductivity maximum is observed for the oxygen vacancy concentration at ~0.1 mole $V_O^{\bullet\bullet}$ / 1 mole MO_2, i.e., $[V_O^{\bullet\bullet}]$ = ~0.1 per fluorite chemical formula MO_2 or d in the chemical formula MO_{2-d}.

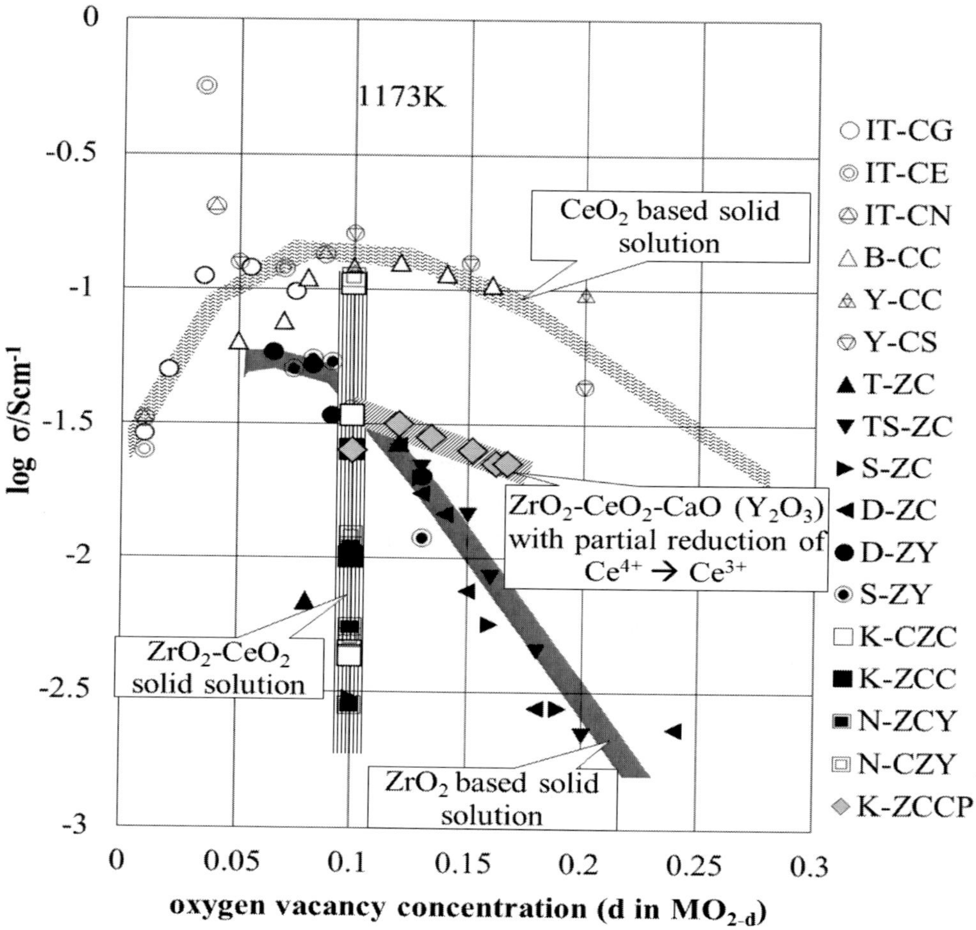

Figure 1. Oxide ion conductivity of some fluorite-type oxides as a function of oxide-ion vacancy concentration in mole / one mole oxide, $[V_O^{\bullet\bullet}]$, at 1173K.All the data except for those for K-ZCCP are those taken in air or in oxygen. The composition of materials and the source of respective data are as follows. IT-CG: CeO_2-Gd_2O_3, Ivers-Tiffee and Oel [41], IT-CE: CeO_2-Er_2O_3 Ivers-Tiffee and Oel [41], IT-CN: CeO_2-Nd_2O_3 Ivers-Tiffee and Oel [41], B-CC: CeO_2-CaO Blumenthal et al. [14], Y-CC: CeO_2-CaO Yahiro et al. [42], Y-CS: CeO_2-Sm_2O_3 Yahiro et al. [42], T-ZC: ZrO_2-CaO Tien [43], TS-ZC: ZrO_2-CaO Tien and Subbarao [13], S-ZC: ZrO_2-CaO Strickler and Carison [44], D-ZC: ZrO_2-CaO Dixon et al. [12], D-ZY: ZrO_2-Y_2O_3 Dixon et al. [12], S-ZY: ZrO_2-Y_2O_3 Strickler and Carison [44], K-CZC: CeO_2-ZrO_2-CaO Kawamura et al. [16], K-ZCC: ZrO_2-CeO_2-CaO Kawamura et al. [16], N-ZCY: ZrO_2-CeO_2-Y_2O_3 Naito et al. [15] and Otake [45], N-CZY: CeO_2-ZrO_2-Y_2O_3 Naito et al. [15] and Otake [45], K-ZCCP: ZrO_2-CeO_2-CaO Kawamura et al. [18].

So far as aliovalent cation doping is concerned, conductivity is essentially determined by the host oxide, either ZrO_2 or CeO_2. The different dopant gives somewhat different conductivity but the difference seems not so large compared with the difference in the host oxide. σ(ion)of CeO_2-based systems being much larger than that of ZrO_2-based systems. In contrast, the most attractive data in Figure 1 are those of co-host (or mixed) CeO_2-ZrO_2 solid solutions shown by square and diamond symbols at d=0.1.

These data are given in the open, filled-black and double square-symbols in Figure 1. The authors studied the conductivity variation of these two kinds of mixed CeO_2-ZrO_2 solid solutions, $Zr_{0.9(1-x)}Ce_{0.9x}Ca_{0.1}O_{1.9}$ (0.2<x<1) [16] and $Zr_{0.8(1-x)}Ce_{0.8x}Y_{0.2}O_{1.9}$ (0.2<x<1) [15,45]. The more detailed conductivity data of these solid solutions are shown in Figure 2. In these systems, Zr^{4+} is replaced with Ce^{4+} (filled black squares) or vice versa shown by open square symbols. That is, charge balance is kept constant and vacancy concentration is not affected, being fixed at d=0.10. Simple expectation would be the conductivity of the solid solution to vary with the increase in Ce concentration from the ZrO_2 based lower curve to the CeO_2 based higher curve. In reality, as shown in Figure 2, the conductivity of the ZrO_2-CeO_2 solid solution is generally much lower than that of the both end members, and the conductivity for the nearly 1 : 1 ZrO_2-CeO_2 composition is more than one order of the magnitude lower than those of the either end members [15,16,45], exhibiting an unexpected pronounced interferent mixed host-cation effect.

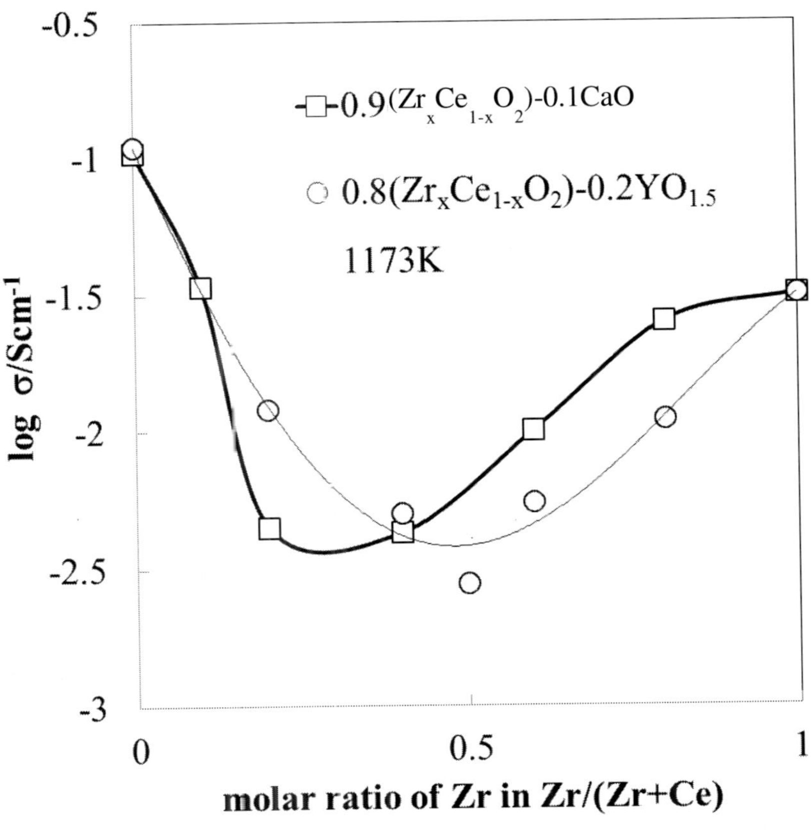

Figure 2. Oxide ion conductivity of ZrO_2-CeO_2 solid solutions in 1bar O_2(g) at 1173K [15,16].

The conductivity of CeO_2 varies by replacing Ce^{4+} with trivalent cations such as Y^{3+} or Gd^{3+} as shown in Figure 1. Then, what will happen when we further replace Ce^{4+} with Ce^{3+}? Such a replacement is possible by partial reduction of CeO_2, in which a part of Ce^{4+} is replaced by Ce^{3+} and extra oxygen vacancies are formed to compensate the charge imbalance.

We have tried such study by using ZrO_2-CeO_2-CaO solid solution, $Zr_{0.72}Ce_{0.18}Ca_{0.1}O_{1.9-d}$, where d is half of the Ce^{3+} concentration formed by such partial reduction [18]. The results are shown in Figure 3 as log d and log σ vs. log $P(O_2)$. The results are also shown by the diamond shaped symbols in Figure 1. The absolute value of the conductivity is more than one order smaller than those of the CeO_2 based solid solution. It is natural because this solid solution is based on ZrO_2. However, the variation of the conductivity with the Ce^{4+} reduction to Ce^{3+} is essentially parallel to the conductivity variation of CeO_2 based solid solutions. That is, Ce^{3+} in CeO_2 indeed behaves as if it were trivalent impurity such as Y^{3+} and Gd^{3+}.

We thus see that the conductivity variation with the partial replacement of host cation with dopant takes place not only by aliovalent cation doping associated with oxygen vacancy formation but also by host homo-valent cation substitution and also by partial reduction of host cation such as $Ce^{4+} \rightarrow Ce^{3+}$.

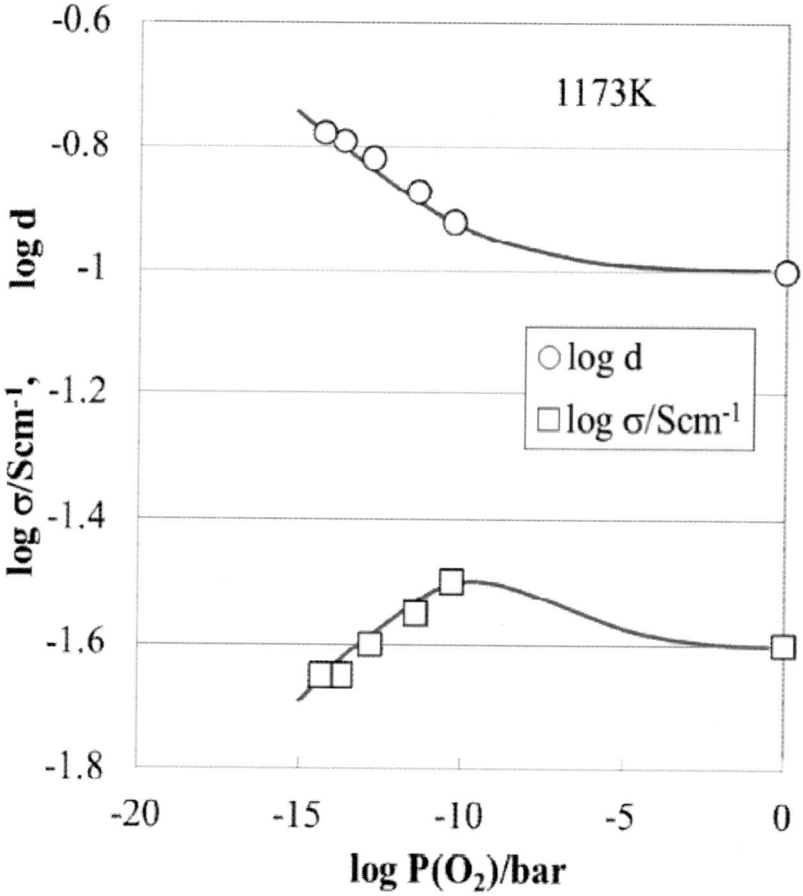

Figure 3. Nonstoichiometric deviation d. of oxygen content and oxide-ion conductivity σ(ion), in $Zr_{0.72}Ce_{0.18}Ca_{0.1}O_{1.9-d}$, as a function of oxygen partial pressure, $P(O_2)$ [18].

2.2. Diffusion Model in Fluorite-Type Oxides and Concept of Vacancy Diffusion

The conductivity maximum and the conductivity decrease for heavy doping of aliovalent cations as above revealed in Figure 1 have already been known since 1960's. Various interpretations for this phenomenon have been attempted. They can be categorized largely in two groups: One is the approach from defect chemistry or statistical thermodynamics to emphasize the role of lower valence cation - oxygen vacancy interaction. The vacancies tend to be associated with the lower valence dopants. The effective mobile vacancy concentration becomes reduced because a large number of vacancies are associated with the immobile dopant cations. Modeling based on this type of approach has been tried since 1960's [47] and one of the most thorough models is proposed recently by Nakamura [48,49]. The more recent other is classical or first principle calculations based on molecular dynamics and/or monte-carlo simulations [50-57]. The calculated results largely agree to suggest that the mobility of oxygen (oxide ion) vacancy tends to be reduced when two or more of the four nearest neighbor cations are the lower valence dopants, also suggesting the important role of vacancy-vacancy repulsion in reducing the mobility of the vacancies. Vacancy mobility is discussed in terms of vacancy diffusion coefficient, D_v [54,55].

In fluorite- and perovskite-type oxides, the transport of oxide ions proceeds by vacancy mechanism. Since oxide ion is mobile only when it places next to the vacancy, the average oxygen diffusion coefficient, which is called oxygen self-diffusion coefficient, D_O, increases with increase in vacancy concentration, C_V, which may vary in several orders of magnitude, e.g., minimum 10^{-6} to maximum 0.1. When C_V is not extremely high, vacancy is considered to be always surrounded by the lattice oxide-ions as shown in Figure 4, being able to jump to any of the nearest neighboring oxygen sites. The vacancies migrate as if they were free carriers and the vacancy diffusion coefficient D_V, is essentially constant independent of C_V [58].

The concept of 'constant D_V' works well for the perovskite-type oxides, ABO_{3-d} [59]. That is, the values of D_V calculated both from the oxide-ion conductivity and the oxygen isotope diffusion coefficient are found to be almost independent of oxygen vacancy concentration, also generally showing good agreement with each other.

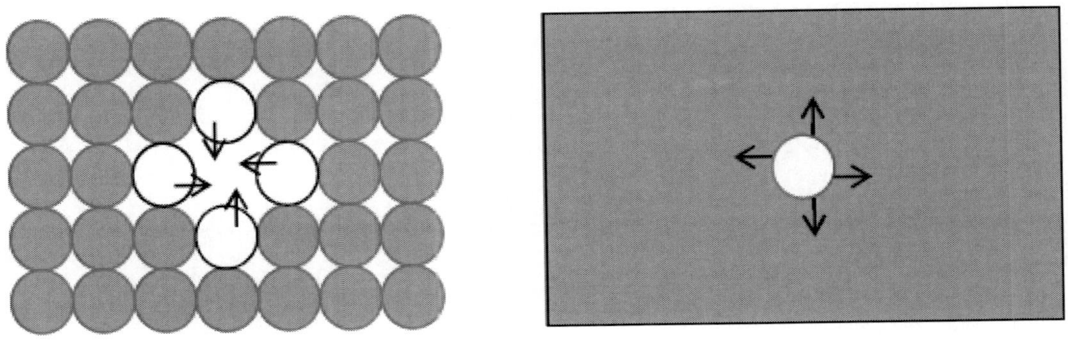

Figure 4. Quasi-free motion of vacancy: Ions cannot move unless there exist vacant sites next to them. The mobility of ions varies with the concentration of vacancy. Dilute vacancies behave as if they are free particles because the entire lattice around vacancy is occupied and it can move to any direction.

As to the fluorite-type oxides, the conductivity decreases obviously with increase in defect concentration as shown in Figure 1. It is clear that the hypothesis of 'constant D_V independent of defect concentration' does not hold. It would be worthwhile to examine in detail D_V in fluorite-type oxides [60].

The value of D_V is calculated from the ionic conductivity, σ(ion), in Figure 1 using the Nernst-Einstein equation,

$$D_V = \sigma V_m RT/(4F^2[V_O^{\bullet\bullet}]) \quad (1),$$

where V_m is the molar volume of oxides, R the gas constant, F the Faraday constant, and T temperature in Kelvin (K). For the calculation of V_m, room temperature lattice parameters are used, which are available in the cited literatures and/or Powder Diffraction File (PDF) data [61]. The conductivity data are those at 1173 K and some of them are in reducing atmosphere. However, the difference in the lattice parameters between room temperature and 1173 K or upon partial oxidation/reduction is less than one percent (< 0.01) and is negligible considering the large variation of σ(ion) and D_V in orders of magnitude.

The calculated D_V values for σ data in Figure 1 are shown in Figure 5 as log D_V vs. oxygen vacancy concentration $[V_O^{\bullet\bullet}]$ plot. It is clear that the maximum observed in the log σ vs. $[V_O^{\bullet\bullet}]$ plot at around $[V_O^{\bullet\bullet}] = 0.10$ in Figure 1 is no more observed in the log D_V vs. $[V_O^{\bullet\bullet}]$ plot in Figure 5. Rather, we find a smooth near linear log D_V vs. $[V_O^{\bullet\bullet}]$ relationship, which is mutually significantly different between CeO_2-based and ZrO_2-based solid solutions. The partial reduction of Ce^{4+} to Ce^{3+} forms another line different from the above two. These results indicate that log D_v of all these three kinds of solid solutions near linearly decreases with increase in $[V_O^{\bullet\bullet}]$. On the other hand, the conductivity variation caused by homo-valent substitution of $Ce^{4+} \leftrightarrow Zr^{4+}$ is on the vertical line because this causes no variation of $[V_O^{\bullet\bullet}]$.

A primitive physico-chemical interpretation is that the former relationship results from the linear increase of activation energy for D_V, E_a, with $[V_O^{\bullet\bullet}]$. That is,

$$D_V = D_V^0 \exp\{-(E_a + \varepsilon[V_O^{\bullet\bullet}])/(RT)\} \quad (2),$$

Where D_V^0 is the pre-exponential factor of D_V and ε is a proportional constant for the variation of activation energy with the vacancy concentration change. Equation (2) can be written in the logarismic form as

$$\log D_V = \log D_V^0 - (E_a/(2.303RT) - \{\varepsilon/(2.303RT)\}[V_O^{\bullet\bullet}] \quad (3)$$

which accords with each linear line in Figure 5, excluding the vertical line for mixed ZrO_2-CeO_2 solid solutions.

2.3. Vacancy Diffusion Coefficient as a Function of Dopant Concentration

As shown in Figures 2 and 3, the composition variation of ZrO_2-CeO_2 solid solution, or the partial substitution of Zr^{4+} with Ce^{4+} and vise-versa, results in a drastic decrease in the oxide ion conductivity or oxygen vacancy diffusion coefficient, in spite of the number of oxygen vacancy being kept constant. This strongly suggests that there exists other factors than the number of oxygen vacancy that control the vacancy diffusion coefficient. One possibility is the disturbance of homogeneity in the distribution of periodic potential due to the largely different ionic radii of both cations, resulting in vast potential variations somewhat similar to that of glassy state.

We selected the total of dopant cation concentrations, $[M_A'']$ + $[M_A']$ + $[M_A^x]$ as the independent variable instead of oxygen vacancy concentration, $[V_O^{\bullet\bullet}]$, disregarding the difference in the effective charge. According to this formalism, the effect of partial reduction of Ce^{4+} to Ce^{3+} cannot be taken into account. The plot of log D_V against total dopant concentrations is shown in Figure 6.

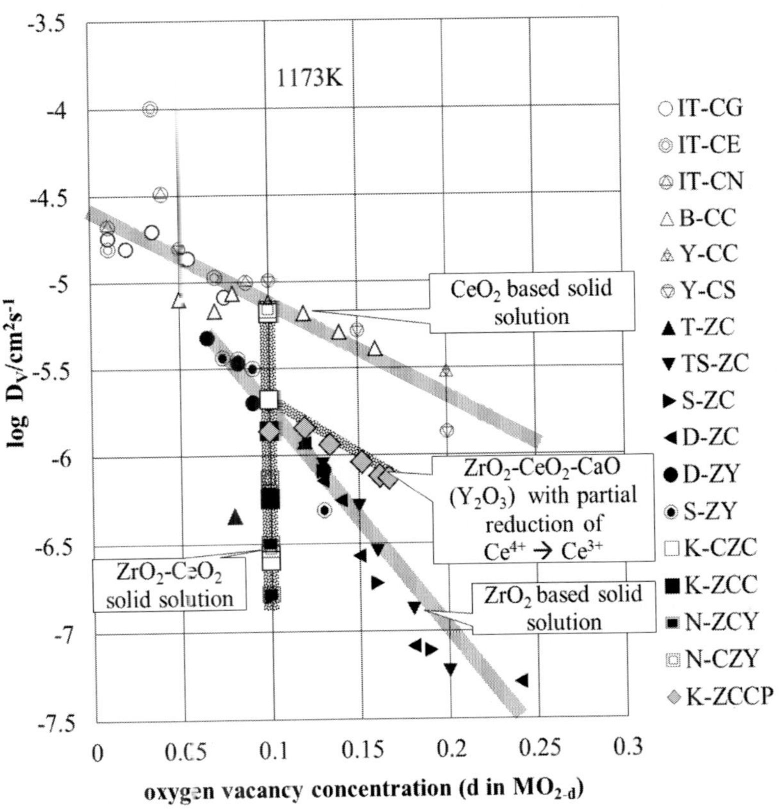

Figure 5. Oxygen vacancy diffusion coefficient, D_v, of some fluorite-type oxides as a function of oxygen vacancy concentration in mole / one mole oxide, $[V_O^{\bullet\bullet}]$, at 1173 K. All the data except for those for K-ZCCP are in air or in oxygen. The composition of materials and the source of data are as follows. IT-CG: CeO_2-Gd_2O_3, Ivers-Tiffee and Oel [41], IT-CE: CeO_2-Er_2O_3 Ivers-Tiffee and Oel [41], IT-CN: CeO_2-Nd_2O_3 Ivers-Tiffee and Oel [41], B-CC: CeO_2-CaO Blumenthal et al. [14], Y-CC: CeO_2-CaO Yahiro et al. [42], Y-CS: CeO_2-Sm_2O_3 Yahiro et al. [42], T-ZC: ZrO_2-CaO Tien [43], TS-ZC: ZrO_2-CaO Tien and Subbarao [13], S-ZC: ZrO_2-CaO Strickler and Carison [44], D-ZC: ZrO_2-CaO Dixon et al. [12], D-ZY: ZrO_2-Y_2O_3 Dixon et al. [12], S-ZY: ZrO_2-Y_2O_3 Strickler and Carison [44], K-CZC: CeO_2-ZrO_2-CaO Kawamura et al [16], K-ZCC: ZrO_2-CeO_2-CaO Kawamura et al. [16], N-ZCY: ZrO_2-CeO_2-Y_2O_3 Naito et al. [15] and Otake [45], N-CZY: CeO_2-ZrO_2-Y_2O_3 Naito et al. [15] and Otake [45], K-ZCCP: ZrO_2-CeO_2-CaO Kawamura et al. [18].

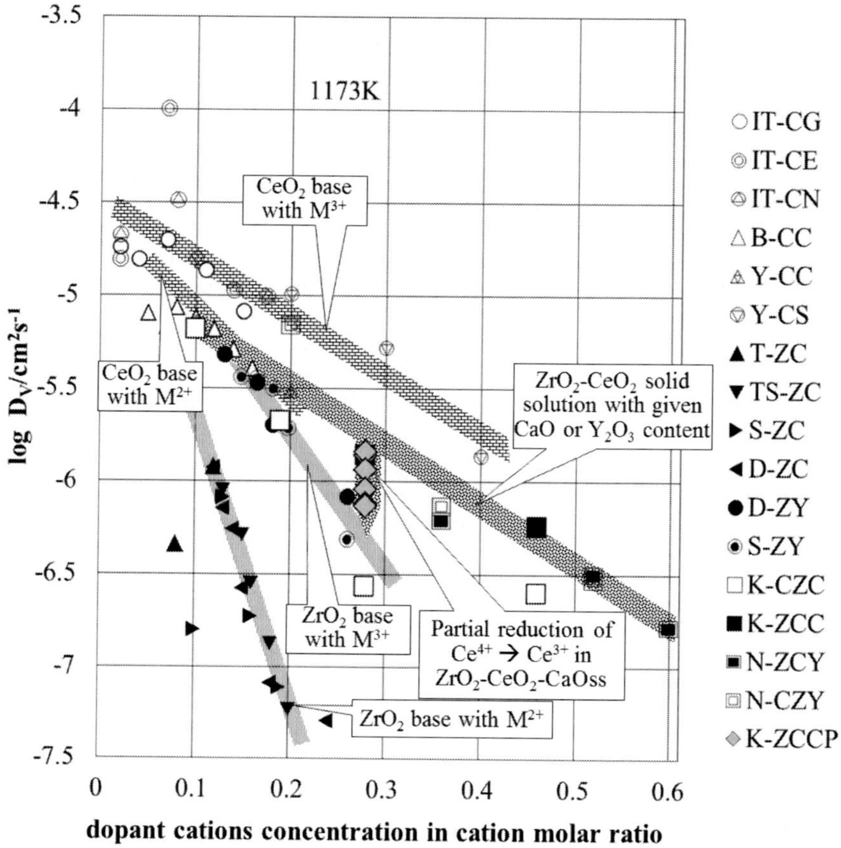

Figure 6. Oxide ion vacancy diffusion coefficient, D_V, of some fluorite-type oxides as a function of the concentration of doped aliovalent impurity cations in mole / one mole oxideat 1173K. All the data except for those for K-ZCCP are in air or in oxygen. The composition of materials and the source of respective data are as follows. IT-CG: CeO_2-Gd_2O_3, Ivers-Tiffee and Oel [41], IT-CE: CeO_2-Er_2O_3 Ivers-Tiffee and Oel [41], IT-CN: CeO_2-Nd_2O_3 Ivers-Tiffee and Oel [41], B-CC: CeO_2-CaO Blumenthal et al. [14], Y-CC: CeO_2-CaO Yahiro et al. [42], Y-CS: CeO_2-Sm_2O_3 Yahiro et al. [42], T-ZC: ZrO_2-CaO Tien [43], TS-ZC: ZrO_2-CaO Tien and Subbarao [13], S-ZC: ZrO_2-CaO Strickler and Carison [44], D-ZC: ZrO_2-CaO Dixon et al. [12], D-ZY: ZrO_2-Y_2O_3 Dixon et al. [12], S-ZY: ZrO_2-Y_2O_3Strickler and Carison [44], K-CZC: CeO_2-ZrO_2-CaO Kawamura et al. [16], K-ZCC: ZrO_2-CeO_2-CaO Kawamura et al. [16], N-ZCY: ZrO_2-CeO_2-Y_2O_3 Naito et al. [15] and Otake [45], N-CZY: CeO_2-ZrO_2-Y_2O_3 Naito et al. [15] and Otake [45], K-ZCCP: ZrO_2-CeO_2-CaO Kawamura et al. [18].

In this figure, ZrO_2-based solid solutions form two lines, being one that of ZrO_2-MO solid solutions and the other that of ZrO_2-M_2O_3 solid solutions. The CeO_2-based solid solutions also forms two lines, being those of CeO_2-MO and M_2O_3 solid solutions. The line slopes of CeO_2-based solid solutions are smaller than those of corresponding ZrO_2-based ones.

In the figure, the ZrO_2-CeO_2 solid solutions indicated by square symbols form a line lying in between those of CeO_2-MO and CeO_2-M_2O_3. Among the square symbols, the filled symbols are those for the solid solutions with ZrO_2 as the majority component while the open symbols represents those of the solid solutions with CeO_2 as the majority one. Black symbols represent the data with 10mol percent CaO while double square symbols show the composition with 20mol percent Y_2O_3. At present, we find no clear systematic relationship between ZrO_2 and CeO_2 rich solid solutions and between Y_2O_3 and CaO doped solid solutions. The diamond symbols in the figure forms a vertical line which shows the Dv variation caused by the partial reduction of Ce^{4+} to Ce^{3+}. That is, the vertical line indicates D_V of the system varies with valence state of cation dopant in these iso total-dopant concentration systems.

Figure 6 shows very clearly that D_V of the fluorite-type oxides is affected by both the oxygen vacancy concentration and the dopant cations' concentrations. The drastic decrease in D_V caused by the solid solution formation between ZrO_2 and CeO_2 suggests that the decrease in D_V by the increase in oxygen vacancies and that by increase in doped alio- and homo-valent cations are not only due to the well-known electrostatic interaction effects but also due to the local lattice distortion and anisotropy caused by vacancy formation and ionic-size difference of constituent cations.

2.4. Summary of This Section

The conductivity maxima observed in the conductivity vs. dopant concentration plots for fluorite type oxides were reconsidered and the use of the oxygen vacancy-diffusion coefficient was proposed. The reported conductivity decrease by the formation of ZrO_2-CeO_2 solution was discussed in terms of the decrease in the oxygen vacancy diffusion coefficient. Oxygen vacancy diffusion coefficient is shown to decrease with the increase not only in its concentration but also in dopant impurities. For the $ZrO_2 - CeO_2 - MO_x$ solid solutions with varying Zr / Ce ratio, the oxide with the smaller concentration can be treated as a dopant.

3. Gas Electrode Reaction; Kinetics and Mechanism

3.1. Characteristic Reaction Process and Rate Determining Process of Oxygen Gas Electrode on Solid Oxide Electrolyte

In electrode reactions, reactants come to the surface of an electrode from electrolyte or other phases such as gas phase in contact with the electrode and electrolyte. After some chemical processes to convert the reactants to the suitable chemical forms for redox reactions, the reaction takes place, exchanging electrons between the reformed reactants and the electrode to form primal products, and then, after some chemical processes to convert the primal products to more stable ones, they leave from the electrode into the electrolyte bulk and/or other coexisting phases.

This is the scheme of general electrode reaction process whatever the state of the electrolyte phase is. However, apparent electrode reaction process is often very different among different states of electrolytes. Typical example is the difference of the electrode reactions between the aqueous electrolyte redox system and the gas electrode / solid oxide electrolyte system, such as O_2(g), Pt / YSZ described in reference [62] that is our prime concern here.

In aqueous electrolyte, ions exist as solvated forms. To vary the valence state of solvated ions, charge transfer must take place between ions and surface of electronic conductors, either electrodes or electrode catalysts. However, this charge transfer process is simultaneously accompanied by desolvation process prior to and re-solvation process after the charge transfer.

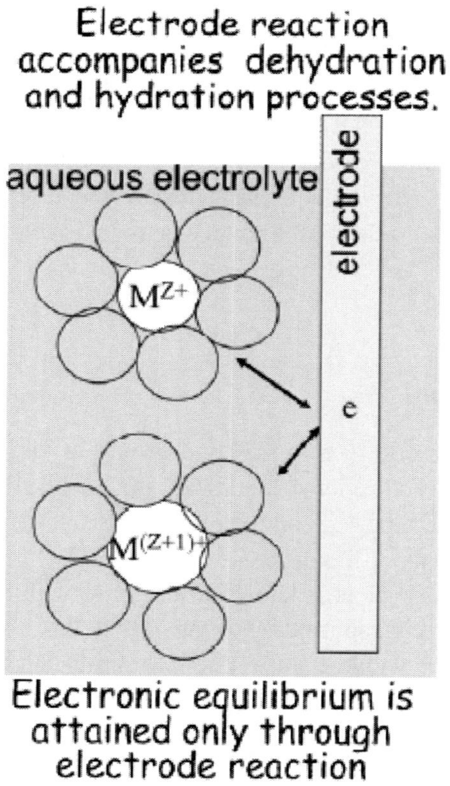

Figure 7. Charge transfer controlled electrode reaction kinetics in aqueous solutions.

That is, the charge transfer process in aqueous electrolyte involves a chemical reaction process which requires activation processes, as shown schematically in Figure 7. The electronic equilibrium between solid electronic conductor and aqueous solution is attained only by the charge transfer process. In other words, the electronic equilibrium at the electrode / aqueous electrolyte interface is attained only through the ionic equilibrium with the relaxation process involving electrochemical reaction.

In solid electrolyte / electrode systems, the situation is quite different from such aqueous-solution systems. The characteristic of oxygen gas electrode reaction on solid oxide electrolyte is summarized in Figure 8. When we consider the electrode process on solid oxide electrolyte, the electrolyte oxides are often treated as if electronic insulators. However, they are actually semi-conductors with very wide band gap, as illustrated in Figure 8(a). That is, purely electronic equilibrium between the electrolyte and electrode is considered to be established much faster than many other ionic processes, usually in less than the order of pico -second.

Oxygen gas molecules may adsorb on the electrode surface and dissociate to form adsorbed O atoms, O_{ad}, as shown in Figure 8(b). There are arguments whether the adsorbed oxygen is ionized or in the atomic state. These arguments are, however, considered meaningless. Adsorbed oxygen atoms are vibrating irregularly on the surface, similar to any other chemical species bonded each other, with the vibration frequency of the order of giga to tera hertz. With this vibration, the bond length between the surface metallic element and the adsorbed oxygen varies.

When the bond length is shorter and the adsorbed oxygen is close to the surface element, the exchange interaction becomes stronger and larger charge transfer may take place between them. The adsorbed oxygen can be said ionized. When the bond length is larger and the adsorbed oxygen is rather far from the surface element in the vibration cycle, the exchange interaction becomes weaker and the adsorbed oxygen may be only weakly charged.

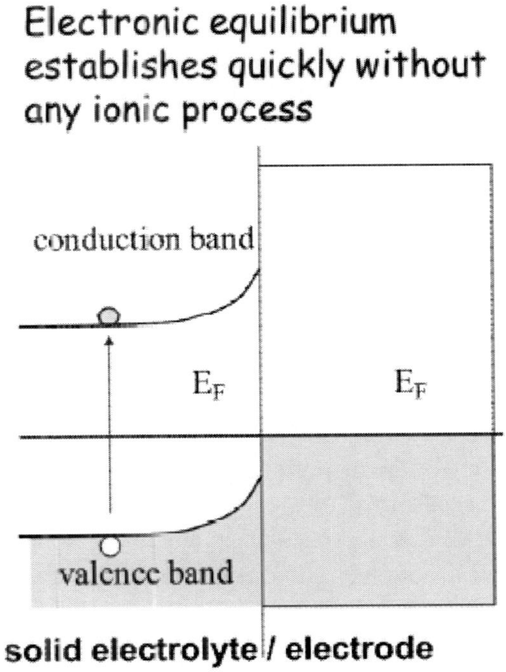

a

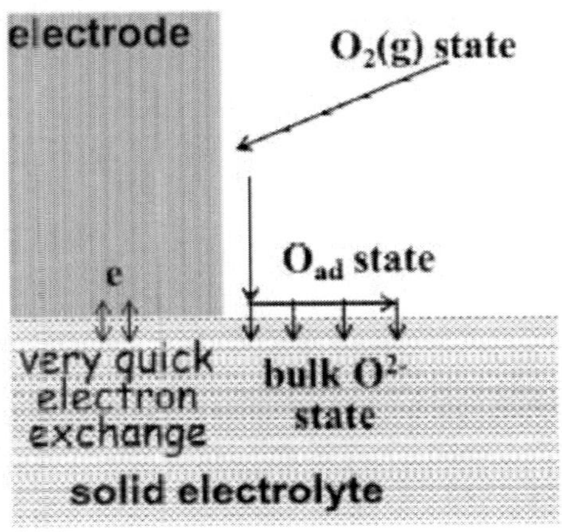

b

Figure 8. (Continued).

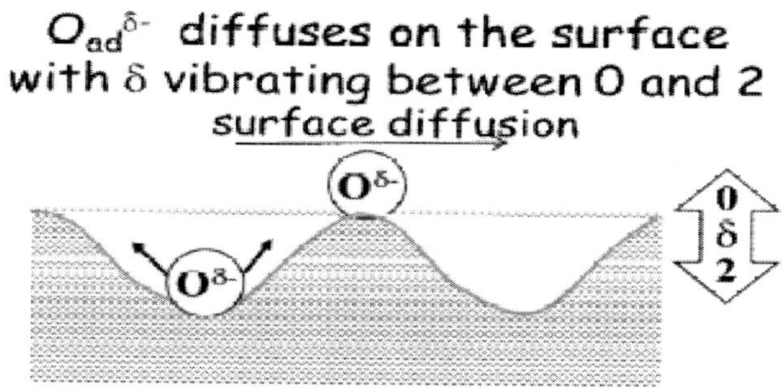

c

Figure 8. Schematic of reaction process at solid oxide electrolyte gas electrode. (A): When the electrode and solid electrolyte are placed in contact with each other, electronic equilibrium is established spontaneously at the interface.(B): $O_2(g)$ molecules adsorb on the electrode surface and dissociate to form adsorbed oxygen atoms, O_{ad}. Then, O_{ad} atoms diffuse on the surface of electrode metal, spill-over on the electrolyte surface, and diffuse on the surface of the electrolyte. O_{ad} atoms are incorporated into oxide bulk during the surface diffusion process. (C): During the surface diffusion process, the interaction or bonding state between O_{ad} and oxide surface atoms (oxide ion or cations) may differ almost periodically between atomic oxygen state (O_{ad}) and oxide ion state (O_{ad}^{2-}).

When the bonding is weak enough to allow the adsorbed oxygen to migrate to the neighboring surface site, surface diffusion may take place, as schematically shown in Figure 8(c). Such migration may take place as a probabilistic event of the vibration of the adsorbed species. The order of the frequency of surface diffusion is considered less than giga hertz. The degree of ionization of the migrating oxygen species is considered low, rather close to atomic state. That is, the adsorbed oxygen is considered to experience between nearly atomic state during migration and ionic state during bonding state with the surface atom.

When the O_{ad} atoms on metal electrode diffuse on the metal surface to the triple phase boundary, TPB, of metal / oxide / gas phase, they may be spillover from metal to the oxide surface and diffuses onto the oxide surface. On the surface of oxide near the TPB, the O_{ad} atoms are likely to proceed preferentially to the strongly chemi-sorbed $O^{2-}{}_{ad}$ state and further to the bulk O^{2-} or $O_O{}^X$ species [24-26]. Thus, in this reaction process, there exists no large electrostatic potential step and that there are no such clear charge transfer process associated with chemical reaction as the one we see in aqueous systems.

However, as mentioned in the first section, a few papers claim still now that charge transfer reaction is the rate determining step of the gas electrode process of zirconia cells. The main reason for this is that they observe the electrode polarization of gas electrode on solid oxide electrolyte, which appears to obey the Butler-Volmer equation expressed by

$$i = i_0 [\exp\{\alpha nF\Delta E/(RT)\} - \exp\{-(1-\alpha)nF\Delta E/(RT)\}] \qquad (4),$$

where, i is the current density, i_0 the exchange current density, α is the transfer coefficient, n is the number of exchanged charge in the rate determining reaction being considered, and ΔE is the applied overpotential at the electrode. Text books often mention that the reaction rate is considered to be controlled by the charge transfer process if this equation holds.

When the overpotential is very large, i.e., FΔE >> RT, Butler-Volmer equation is simply expressed by

$$\log i = A + B\Delta E \qquad (5),$$

where A and B are empirical constants. The equation is called Tafel equation. Not a few people conclude the electrode reaction process of solid electrolyte / gas electrode to be controlled by the charge transfer kinetics, when the observed current vs. overpotential behavior for the electrode can be expressed by Tafel equation. Such misleading conclusion seems to arise from the lack of detailed understandings of the complex electrode reaction processes involved here, as delineated in the following.

3.2. Reaction Rate Expressed in the Form of Butlar-Volmer Equation: Charge-Transfer Reaction Controlled Kinetics and Chemical Reaction Controlled Kinetics

3.2.1. Derivation of Butler-Volmer Equation from Charge Transfer Controlled Reaction Kinetics

Figure 9 shows the schematic potential variation in the electrochemical redox reaction,

$$Ox + ne \leftrightarrow Red \qquad (6),$$

where Ox and Red are oxidant and reductant, respectively, and n is the number of electrons involved in this reaction. The height of potential barrier between the oxidant and reductant or the energy of transient state is assigned as ΔG^T. Under equilibrium, the potential of (Ox + ne) and Red are the same as shown in Figure 9(a). At this stage, the rate of forward reaction, $i_{\rightarrow}$ is the same to that of the backward reaction, $i_{\leftarrow}$. The forward current and backward current under equilibrium is called the exchange current, i_0.

$$i_{\rightarrow} = i_{\leftarrow} = k_0 \exp\{-\Delta G^T/(RT)\} = i_0 \qquad (7),$$

where k_0 is the rate constant.

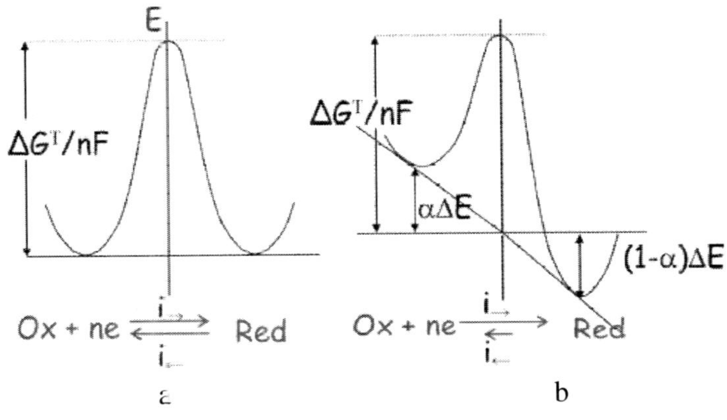

Figure 9. Potential diagram for the charge transfer controlled electrode reaction kinetics in aqueous solutions, based on the absolute rate theory [20]. (a) Equilibrium condition (b) After potential difference ΔE is applied.

When a potential difference, ΔE, is applied to the system as shown in Figure 9(b), the forward and backward current can be expressed as follows.

$$i_{\rightarrow} = k_0\exp(-(\Delta G^T - n\alpha\Delta EF)/(RT)) = i_0\exp((n\alpha\Delta E)/(RT)) \quad (8),$$

$$i_{\leftarrow} = k_0\exp(-(\Delta G^T + n(1-\alpha)\Delta EF)/(RT)) = i_0\exp(-(n(1-\alpha)\Delta E)/(RT)) \quad (9),$$

where, α and $(1-\alpha)$ is the divided part of ΔE used to raise the potential of reactant, Ox + ne, and that used to reduce the potential of the products, Red, respectively. Then, the total net current is obtained as

$$i = i_{\rightarrow} - i_{\leftarrow} = i_0[\exp\{(n\alpha\Delta EF)/(RT)\} - \exp\{(-n(1-\alpha)\Delta EF)/(RT)\}] \quad (10)$$

Equation (10) is exactly the same as Equation (4).

When the charge transfer process typically expressed in Figure 9 is the rate controlling process of the whole reaction, Equation (10) and hence Equation (4), the Butler-Volmer equation, holds. However, it does not always mean that the rate determining process is the charge transfer reaction process even when the reaction rate is expressed by Equation (4).

3.2.2. Derivation of Butler-Volmer Equation from Chemical Reaction Controlled Kinetics

In this section, we will see that the rate equation for usual chemical reactions can also be expressed by a Butler-Volmer type equation. Since the absolute reaction rate theory and the concept of activated complex [20] describes not only the charge transfer kinetics of electrode reaction but also chemical reactions and diffusion processes in general, it is natural that the chemical reaction rate can be expressed in a similar style as that of Butler-Volmer equation.

Here, we show briefly the relationship between conventional chemical rate equation and Gibbs free energy based expression similar to Butler-Volmer type equation. When the reaction takes place between neutral chemical species A and B to form the products of neutral chemical species C and D, the reaction and the reaction rate can be expressed in terms of activity of the chemical species involved here;

$$A + B \rightarrow C + D \quad (11),$$

$$v = k a_A a_B - k' a_C a_D \quad (12),$$

where, k and k' in Equation (12) are the forward and backward reaction rate constants, respectively, a_A, a_B, a_C and a_D are the activities of the chemical species A, B, C and D, respectively. Reactivity of a neutral chemical species, A, can be expressed using activity, a_A, or the chemical potential, μ_A, of the species A given by the equation

$$a_A = \exp\{(\mu_A - \mu_A^o)/(RT)\} \quad (13)$$

where μ_A^o is the standard state of μ_A. Similarly, a_B, a_C and a_D are expressed in terms of chemical potential μ_B, μ_C, μ_D and theirs standard states μ_B^o, μ_C^o and μ_D^o by the equations

$$a_B = \exp\{(\mu_B - \mu_B^o)/(RT)\} \quad (14)$$

$$a_C = \exp\{(\mu_C - \mu_C^o)/(RT)\} \quad (15)$$

$$a_D = \exp\{(\mu_D - \mu_D^o)/(RT)\} \quad (16)$$

In terms of chemical potential using Equations (13)- (16), Equation (12) can be rewritten as

$$v = k\exp[\{(\mu_A + \mu_B) - (\mu_A^o + \mu_B^o)\}/(RT)] - k'\exp[\{(\mu_C + \mu_D) - \{(\mu_C^o + \mu_D^o)\}/(RT)]$$
$$= k_R \exp\{(\mu_A + \mu_B)/(RT)\} - k_R' \exp\{(\mu_C + \mu_D)/(RT)\} \quad (17),$$

where, k_R and k_R' are renormalized reaction constants including the rate constant and standard chemical potential term.

Gibbs free energy change, ΔG, of the reaction in Equation (11) can be expressed using Equations (13)-(16) as;,

$$\Delta G = (\mu_C + \mu_D) - (\mu_A + \mu_B) \quad (18)$$

Since Gibbs free energy can be converted into electrical work and the relationship is given by the Nernst equation, we have,

$$\Delta G = -ZF\Delta E_{Nernst} \quad (19).$$

where ΔE_{Nernst} is the electromotive force obtained from ΔG.

We know that the reaction proceeds or ΔG becomes more negative when the chemical potential of the products, $(\mu_C + \mu_D)$ becomes smaller or the chemical potential of the reactants, $(\mu_A + \mu_B)$, becomes larger. However, only form the change of the ΔG value, we cannot distinguish either $(\mu_C + \mu_D)$ decreases or $(\mu_A + \mu_B)$ increases. That is, ΔG in Equation (18) can be divided into two parts, $\alpha\Delta G$ as the increased chemical potentials of reactants and $(1-\alpha)\Delta G$ as the decreased ones of the products. Then, we have

$$v = k_R \exp\{\alpha\Delta G/(RT)\} - k_R' \exp[\{-(1-\alpha)\Delta G\}/(RT)] \quad (20).$$

Applying Equation (19) to Equation (20), we have the identical equation to the well-known Butler-Volmer equations. That is,

$$v = k_R \exp\{(\alpha ZF\Delta E_{Nernst})/(RT)\} - k_R' \exp[\{-(1-\alpha)ZF\Delta E_{Nernst}\}/(RT)] \quad (21).$$

Equation (21) shows that when we set the reaction to take place between two rooms, such as set-ups for fuel cells or oxygen permeation systems, we can observe the Butler-Volmer type relationship between reaction rate and driving force very similar to the Butler-Volmer type whether the reaction is controlled electrochemically or chemically.

3.2.3. How to Determine Reaction Kinetics from the Observed Current – Potential Relationship

Many text books on electrochemistry describe that you can conclude the rate determining process of the electrode reaction you are studying to be of the charge transfer type one, if the Butler-Volmer equation holds. It is not true, as shown in the previous section. That is, whether the observed current-electrode potential relationship obeys Butler-Volmer relationship or not cannot be the criterion to determine whether the rate determining reaction step is a charge transfer process or not. On the other hand, when the electrode reaction is controlled by the chemical reaction process, you may easily find, comparing Equations (12), (17) and (21), that the slope of the log v vs. ΔE_{Nernst} plot gives the order of reaction, or the number of equivalent involved in the unit reaction represented by Z.

In order to clarify the rate determining reaction process whether charge transfer one or chemical one, additional information other than simple current - potential relationship is necessary. One is the nature of the overpotential: Whether the potential difference you observed between working electrode and reference electrode is really the electrostatic potential difference between the electrolyte or not. Another is the comparison of the observed data with other data obtained in surface science, diffusion, catalytic research, and so on. This topic is, however, beyond the scope of this book on the reseach of fluorite type oxides. In the next section, an example is discussed. For further study, please refer the original publications. [25-40].

In brief, Butler-Volmer type equation represents the rate equation of the rate determining reaction process whether the rate determining process involves charge transfer electrochemical reaction or non-electrochemical reaction kinetics.

3.3. Typical Reaction Kinetics: A Case Study on Gas Electrode System with Pt or Mixed Oxide Conductor, O_2 (g) / Oxide Electrolyte

When we consider porous metal electrode in which the solubility of oxygen is negligible such as Pt, there are two possible routes for the oxygen gas molecule, $O_2(g)$, to approach near the triple phase boundary, TPB, of gas phase / oxide electrolyte / Pt or oxide electrode,as shown schematically in Figure 10 [24-31]. One route is via the surface diffusion process: $O_2(g)$ lands from the gas phase on the electrode surface in the vicinity of TPB. Upon landing, the $O_2(g)$ is to decompose there into adsorbed atoms, O_{ad}, and the adsorbed atoms diffuse to TPB on the surface. The surface diffusion length may extend typically up to the order of micrometers and in some cases, more than 10μm. Another is the direct dissociative adsorption at the TPB or in the band with the width of nanometers or a lattice-size width on both sides at the boundary. These two pathways are considered to work in parallel, as shown in the figure. For mixed conducting oxide electrode, the third route, diffusion in the bulk of electrode oxide shown in Figure 10, is also important, particularly when the oxide ion conductivity of the electrode is large enough.

Experimental results of oxygen electrode reaction on porous Pt / YSZ system are shown in Figures 11-13. In Figure 11 at 973K under $P(O_2)$ = 1 bar, anode polarization current, or oxygen evolution reaction, tends to saturate with the increase in the electrode potential.

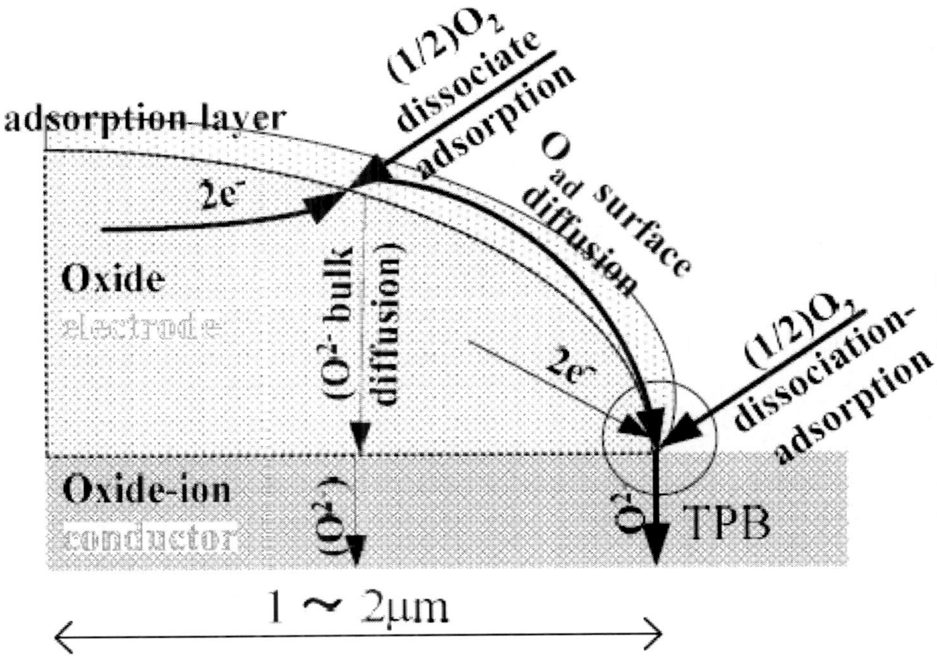

Figure 10. The cathode reaction kinetics in the vicinity (1~2μm area) of TPB in porous cathode with 10~100μm thickness.

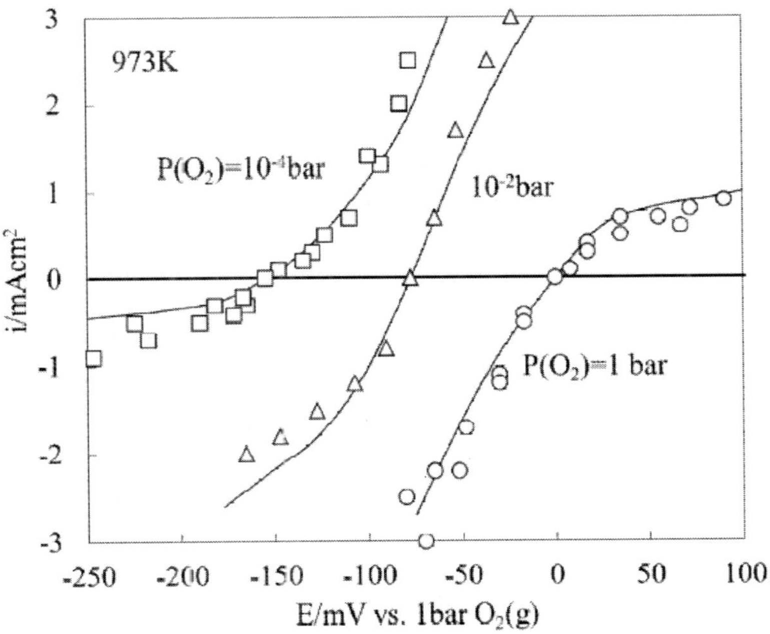

Figure 11. Polarization curve (current – electrode potential relationship) of O_2(g), porous Pt / YSZ electrode. Solid curves are calculated by the model with the assumption for the rate determining reaction process as oxygen surface diffusion on Pt [26].

On the other hand, at $P(O_2)=10^{-4}$ bar, cathode reaction, or the oxygen incorporation reaction, tends to be saturated under low current density, or low reaction rate. These tendencies are well explained in 3.3.1 by the surface diffusion model of adsorbed oxygen on Pt electrode.

The $P(O_2)$ dependent variation of the shape of polarization curve can be evaluated by the slope of i – E relationship at i = 0. The slope is called the electrode interface conductivity, σ_E [25].

$$\sigma_E = (\partial i/\partial E)_{i=0} \qquad (22)$$

The observed σ_E for O_2(g), Pt / YSZ electrode is shown in Figure 12 as a function of $P(O_2)$ at different temperatures. The data shown by thick curves are explained by the surface diffusion controlled kinetics described below and the data displayed by dashed lines are calculated by the surface reaction controlled kinetics described in 3.3.2. The kinetics takes place in parallel. Therefore, the total reaction is the sum of the reaction of these two rate processes. In Figure 13, the Arrhenius plot for σ_E is shown to reveal the relationship of these two kinetic processes.

3.3.1. Surface Diffusion Controlled Kinetics [25]

Surface diffusion of adsorbed oxygen may take place by the successive jumps of adsorbed oxygen atom to the neighboring surface vacant site, V_{ad}. The microscopic mechanism can be expressed by the site exchange of O_{ad} and V_{ad} on the site I and II.

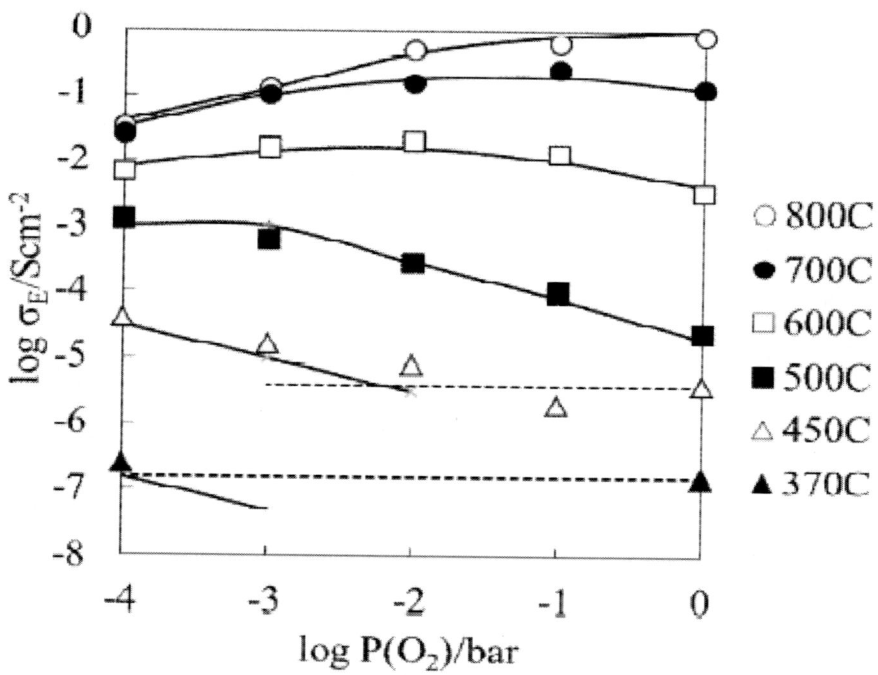

Figure 12. Electrode interface conductivity, σ_E. as a function of $P(O_2)$. Sold curves are calculated from the surface diffusion controlled kinetics in 3.3.1. The dashed lines are calculated by the surface reaction controlled kinetics in 3.3.2. Since these two kinetics take place in parallel, the observed σ_E is the sum of the reaction rates of both kinetics.

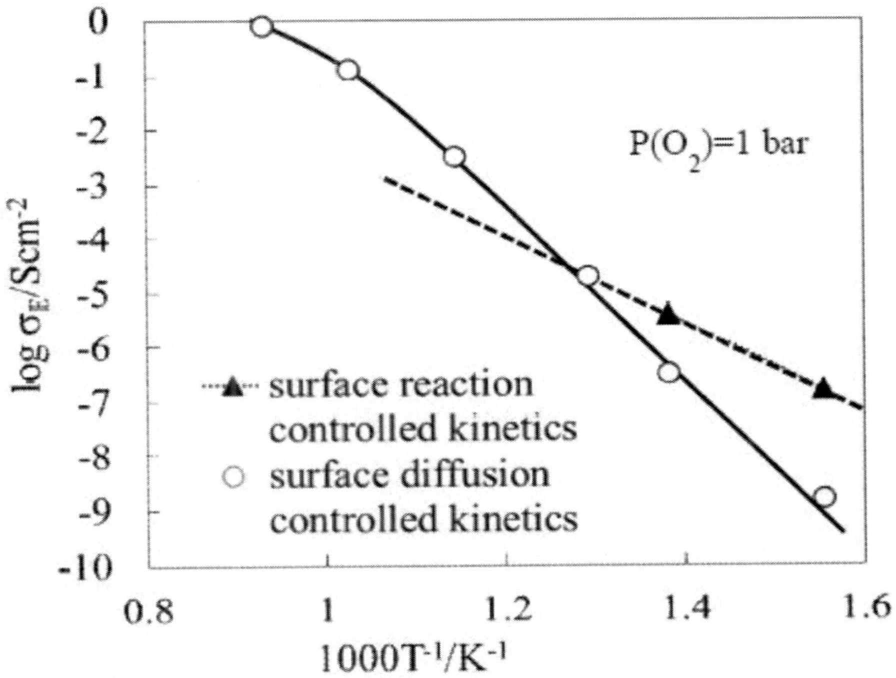

Figure 13. Arrhenius plot of σ_E for O_2(g), porous Pt / YSZ electrode [26].

That is,

$$O_{ad}(I) + V_{ad}(II) \rightarrow V_{ad}(I) + O_{ad}(II) \qquad (23).$$

When we put the surface coverage of O_{ad} by θ, the surface diffusion becomes proportional to $\theta(1-\theta)$. The driving force of the diffusion is the chemical potential gradient of oxygen on the surface. Then microscopic expression of the surface diffusion is given by

$$J = -B\theta(1 - \theta)\ \partial\mu(O) / \partial x \qquad (24),$$

where, B is the constant related to the mobility, the atomic distance and the vibration frequency of O_{ad}.

In the above expression, we used the gradient of oxygen chemical potential on the surface. That is, the chemical potential is defined in two-dimensional plane. The interaction with the gas phase and that with the bulk of the solid are not taken into account explicitly. The meaning of this thin film chemical potential is very similar to that of local equilibrium assumption. The relationship is illustrated in Figure 14.

When a certain position of the plane is equilibrated to gas phase, the oxygen chemical potential of the surface position, μ_O(surf), or the oxygen activity there, a_O(surf) can be expressed by the following equation.

$$\mu_O\,(surf) = RT \ln a_O(surf) = (RT/2)\ln P(O_2) \qquad (25).$$

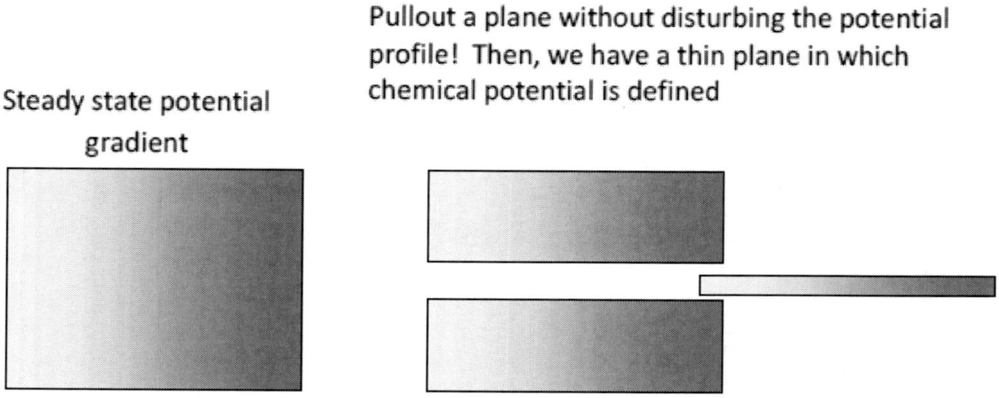

Figure 14. Definition of surface chemical potential.

In principle, we can take out the monolayer plane without disturbing the thermodynamic conditions in the plane. Then, we can define the two dimensional chemical potential or activity.

On the adsorbed surface layer, we assume a Langmuir-type adsorption isotherm. Using the surface oxygen chemical potential and activity defied above, we have

$$\theta/(1 - \theta) = (P(O_2)/P(O_2)^*)^{1/2} = a_O / a_O^* \quad (26)$$

here, $P(O_2)^*$ and a_O^* are $P(O_2)$ and a_O at $\theta=1/2$, respectively.

When we put the average length of the surface diffusion from the TBP as δ, the current produced by the oxygen flux is expressed by;

$$i = -2FJ = (1/\delta) \int_{x=0}^{\delta} J\, dx = 2FBRT\, a_O^* P(O_2)^{1/2}[1 - \exp(-2F\Delta E/(RT))]$$

$$/ [\delta(a_O^* + P(O_2)^{1/2})\{ a_O^* + P(O_2)^{1/2}\exp(-2F\Delta E/(RT))\}] \quad (27).$$

Here, ΔE corresponds to so-called cathode overpotential used in the study of electrode reaction kinetics. In this model, ΔE comes not from electrostatic potential difference, namely faradic process, but from non-faradic potential difference. That is,

$$2\Delta EF = \mu_O(TBP) - \mu_O(\text{free Pt surface}) \approx \mu_O(TBP) - (RT/2) \ln P(O_2) \quad (28).$$

From Equation (22) and Equation (27), we obtain

$$\sigma_E = 4\sigma_E^* a_O^* P(O_2)^{1/2}/(a_O^* + P(O_2)^{1/2})^2 \quad (29)$$

Here,

$$\sigma_E^* = BF^2/\delta \quad \textbf{(30)}$$

The parameters a_O^*, σ_E, δ can be determined from experimental data. For example, in Figure 12, the maximum σ_E is given at $P(O_2) = a_O^{*2}$ and the maximum σ_E is σ_E^* itself. By dividing σ_E^* with F^2, we have B/δ. Using the parameters determined from Figure 12, the fitting curves in Figure 11 were calculated.

3.3.2. Surface Reaction Controlled Kinetics

When the surface diffusion becomes slow, the oxygen from gas phase is supplied essentially via direct dissociative adsorptions on the electrode surface near the TBP. The chemicals reaction related to this dissociative adsorption can be expressed by

$$O_2(g) + 2\,V_{ad}(\text{Pt at TBP}) \rightarrow 2\,O_{ad} \qquad (31).$$

When this equation determines the oxygen supply from the gas phase, the resulting current can be expressed by the equation:

$$i = kP(O_2)(1-\theta)^2 - k'\theta^2$$

$$= kP(O_2)[a_O^* / \{a_O^* + P(O_2)^{1/2}\exp(-2F\Delta E/(RT))\}]^2 -$$

$$k'P(O_2)\exp\{4F\Delta E/(RT)\}/\{a_O^* + P(O_2)^{1/2}\exp(-2F\Delta E/(RT))\}^2 \qquad (32).$$

Here, k and k' are the forward and backward reaction rate constants in Equation (31), respectively.

In Figure 12, surface reaction controlled kinetics appears at lower temperatures under rather high $P(O_2)$, that is, at $P(O_2) >> a_O^{*2}$. Then, from the combination of Equation (22) and Equation (32), we have

$$\sigma_E = 16Fka_O^*/RT \quad (P(O_2(g)) >> a_O^*) \qquad (33)$$

As we see in Figure 12, at temperatures below 450°C, σ_E is independent of $P(O_2)$ at the higher $P(O_2)$ conditions. This is consistent with Equation (33).

It is known that Equations (27) and (32) hold for the oxygen gas electrode reaction also on $La_{1-x}Sr_xMnO_3$ / stabilized zirconia (SZ) [63,64], and the equations well explain the reported experimental results.

3.3.3. Reaction Kinetics with Mixed Conducting Oxides

When the electrode material is oxide electronic conductor, oxygen diffuses inside the oxide as oxide ion, as schematically shown in Figure 8. Then, not only can the vicinity of TBP, but also the whole electrode surface be the possible reaction pathway. Here we have essentially two types of kinetics. When the diffusion in the oxide electrode is not fast enough and the gas phase transport or the surface diffusion of oxygen is much faster, the reaction proceeds similarly to the case of metal electrodes. Typical example is $La_{1-x}Sr_xMnO_3$ electrodes mentioned above [63,64].

The other is the case in which the diffusion in the bulk oxide is fast enough to supply necessary oxygen for the desired electrochemical process to proceed. However, what actually observed so far is that, when oxygen diffusion is fast enough, the rate determining reaction step becomes not the diffusion process but the preceding surface reaction process. A schematic potential profile for oxide cathode interface is shown in Figure 15. In this case, the driving force of the surface reaction is the chemical potential difference of oxygen, μ_O, between that in the gas phase, $\mu_O(g)$, and that at the oxide electrode / electrolyte interface, μ_O(electrode/electrolyte). However, the gradient of the oxygen chemical potential in the electrode oxide is very small due to the large oxygen diffusivity in the electrode, and therefore, the difference in μ_O through the oxide is small. That is, most of the difference in μ_O is applied to the surface, as shown in Figure 15.

In recent years, many reports have been published on the experimental determination of surface reaction constant, k. [65-69], of the similar process to Equation (31)

$$O_2(g) + 2\,V_{ad}\,(\text{surface of oxide}) \rightarrow 2\,O_{ad} \qquad (31)'.$$

However, since there would be a variety of rate controlling kinetics at the electrode surface, much of detailed kinetic study of this surface process remains as a future problem.

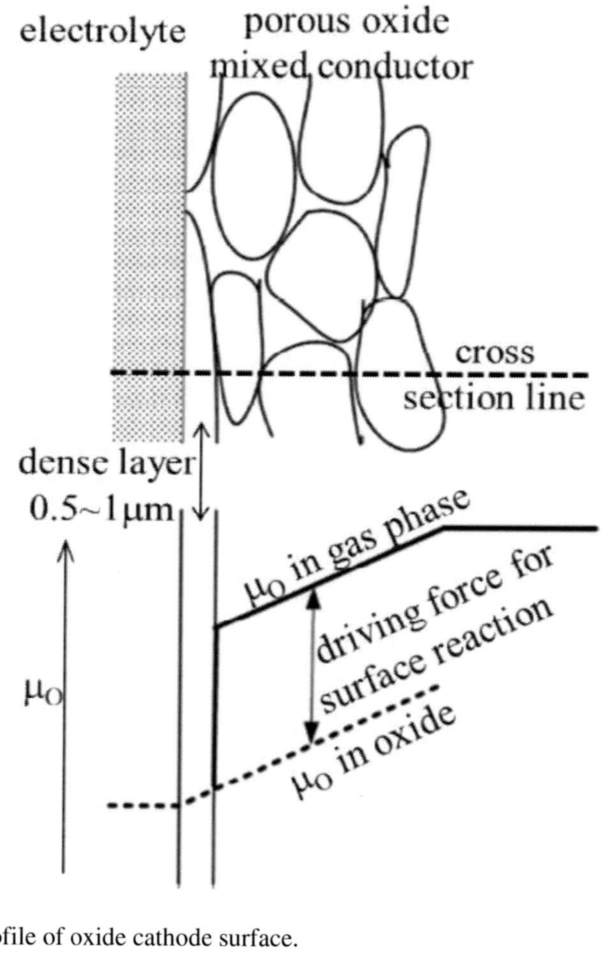

Figure 15. Potential profile of oxide cathode surface.

Recently, catalytically enhanced surface reactions are also attracting lots of interest. The first finding is by the present author and the co-investigators on $La_{2-y}Sr_yCoO_4$ dispersed $La_{1-x}Sr_xCoO_3$ [70,71]. Catalytic activity of $La_{1-x}Sr_xCoO_3$ [66,69] is not so high compared to those of the similar perovskite-type oxides such as $La_{1-x}Sr_xFeO_3$ [65,68], $La_{1-x}Sr_xMnO_3$ [68], and $La_{1-x}Sr_x(Fe,Co)O_3$ [67]. However, when we disperse $La_{2-y}Sr_yCoO_4$ on $La_{1-x}Sr_xCoO_3$, the surface reaction constant becomes orders of magnitude enhanced. The detailed investigation and analysis of this phenomenon is now being challenged by many research groups.

3.4. Summary of This Section

Electrochemical application of fluorite-type oxide ion conductors involves gas electrode reactions. The mechanism of the rate controlling reaction process was discussed. The rate of electrode reaction is often expressed by Butler-Volmer type equation. The physical meaning of the Butler-Volmer equation was discussed. It was shown that the equation represents either charge transfer controlled reaction conventionally known and also the chemical reaction controlled one. Typical electrode reaction kinetics such as Pt, O_2(g) / oxide electrolyte and oxide electrodes, O^2(g) / oxide electrolytes are discussed.

REFERENCES

[1] K. Hauffe, *"Reactionen in und an den Festen Stoffen"*. (1966), Berlin, Springer Verlag.

[2] H. Schmalzried, *"Festkörperreaktionen"* (1971), Weinheim, Verlag Chemie.

[3] J. Hladik, *"Physics of Electrolyte", Vol. 1 and 2* (1972) London, New York, Academic Press.

[4] H. Rickert, "*Elektrochemie fester Stoffe*" (1973) Berlin, Springer-Verlag.

[5] F.A. Kroger, "*The Chemistry of Imperfect Crystals*", Vol. 1-3 (1974) New York, North-Holland.

[6] T. Talahashi and O. Yamamoto, *J. Electrochemical Soc.* (1970) 117, 1-5.

[7] T. Takahashi and O. Yamamoto, *J. Electrochem. Soc.* (1971) 118, 1051-1057.

[8] *Solid Oxide Fuel Cells I* (1989), III (1993)-XII (2011), 1989-2005: ECS proceedings volume, 2007-ECS Transaction, The Electrochemical Society Inc.

[9] N.Q. Minh and T. Takahashi, *"Science and Technology of Ceramic Fuel Cell"*, (1995) Amsterdam, Elsevier ISBN 978-0-444-89568-4.

[10] S. Singhal and K. Kendall, *"High-temperature Solid Oxide Fuel Cells: Fundamentals, Design and Applications"* (2003) New York, Elsevier, ISBN 1856173879.

[11] H. Inaba and H. Tagawa, *Solid State Ionics* (1996) 83, 1-16.

[12] J.M. Dixon, L.D. LaGrange, U. Merten, C.F. Miller and J.T. Porter II, *J. Electrochem. Soc.* (1963) 110, 276-280.

[13] T.Y. Tien and E.C. Subbarao, *J. Chem. Phys.* (1963) 39, 1041-1047.

[14] R.N. Blumenthal, F.S. Brugner, J.E. Garnier, *J. Electrochem. Soc.*, (1973) 120, 1230 - 1237.

[15] H. Naito, N. Sakai, T. Otake, H. Yugami, H. Yokokawa, *Solid State Ionics* (2000) 135, 669-673.

[16] K. Kawamura, K. Watanabe, T. Hiramatsu, A. Kaimai, Y. Nigara, T. Kawada, J. Mizusaki, *Solid State Ionics* (2001) 144, 11-18.
[17] V. Ruehrup and H.D. Wiemhoefer, Z. Naturforsch., (2006) 61b, 916 – 922.
[18] K. Kawamura, K. Watanabe, Y. Nigara, A. Kaimai, T. Kawada and J. Mizusaki, *J. Electrochem. Soc.* (1998) 145, 2552-2558.
[19] J.O'M. Bockris and A.K.N. Reddy, *"Modern Aspects of Electrochemistry"* (1970) New York, Plenum Press.
[20] S. Glasstone, K. Laider and H. Eyring *"The Theory of Rate Processes"* (1941) New York, London, McGraw-Hill.
[21] D.Y. Wang and A.S. Nowick, *J. Electrochem. Soc.* (1979) 126, 1155 – 1166.
[22] H. Okamoto, G. Kawamura and T. Kudo, *Elecrochim. Acta* (1983) 28, 379 – 382.
[23] A.J.A. Winnubst, A.H.A. Scharenborg, A.J. Burggraaf., *Solid State Ionics* (1984) 28, 319 – 327.
[24] J. Mizusaki, K. Amano, S. Yamauchi and K. Fueki, *Proc. Int'l Meeting Chemical Sensors (Analytical Chem. Monogrh. 16), ed.* by T.Seiyama et al., Kodansha / Elsevier, Tokyo / Amsterdam (1983) 279 – 284.
[25] J. Mizusaki, K. Amano, S. Yamauchi and K. Fueki, *Solid State Ionics* (1987) 22, 313–322.
[26] J. Mizusaki, K. Amano, S. Yamauchi and K. Fueki, *Solid State Ionics* (1987) 22, 323–330.
[27] J. Mizusaki, M. Aoki, Y. Miyaki, S. Yamauchi and K. Fueki, Denki Kagaku (1990) 58, 1169 – 1177.
[28] J. Mizusaki, H. Tagawa, Y. Miyaki, S. Yamauchi, K. Fueki, I. Koshiro and K. Hirano, *Solid State Ionics* (1992) 53-56, 126 – 134.
[29] J. Mizusaki, H. Tagawa, K. Isobe, I. Koshiro, M. Tajika and K. Hirano, *Sensors and Actuators B* (1993) 13-14, 121 – 124.
[30] J. Mizusaki, H. Tagawa, K. Isobe, M. Tajika, I. Koshiro, H. Maruyama and K. Hirano, *J. Electrochem. Soc.* (1994) 141, 1674 – 1683.
[31] S. Onuma, A. Kaimai, K. Kawamura, Y. Nigara, T. Kawada, J. Mizusaki, H. Inaba and H. Tagawa, *J. Electrochem. Soc.* (1998) 145, 920 – 925.
[32] J. Mizusaki, H. Tagawa, K. Tsuneyoshi and A. Sawata, *J. Electrochem. Soc.* (1991) 138, 1867 – 1873.
[33] J. Sasaki, J. Mizusaki, S. Yamauchi and K. Fueki, *Solid State Ionics* (1981) 3/4, 531 – 535.
[34] T. Kawada, J. Suzuki, M. Sase, A. Kaimai, K. Yasahiro, Y. Nigara, J. Mizusaki, K. Kawamura, and H. Yugami, *J. Electrochem. Soc.* (2002) 149, E252 - E259.
[35] T. Nakamura, K. Yashiro, A. Kaimai, T. Otake, K. Sato, T. Kawada, and J. Mizusaki, *J. Electrochem. Soc.* (2008) 155, B1244 - B1250.
[36] S.B. Adler, *Solid State Ionics* (2000) 135, 603 – 612.
[37] S.B. Adler, *Chem. Rev.* (2004) 104, 4791 – 4843.
[38] S.B. Adler, X.Y. Chen, J.R. Wilson, *J. Catalysis* (2007) 245, 91 – 109.
[39] J. Fleig, A. Schintlmeister, A.K. Opitz and H. Hutter, Scripta Materialia (2011) 65, 78-83.
[40] E.C. Shin, P.A. Ahn, H.H. Seoi, J.M. Jo, S.D. Kim, S.K. Woo, J.H. Yu, J. Mizusaki and J.S. Lee. *Solid State Ionics* (2013) 232, 80 – 96.

[41] T. Ivers-Tiffee and H.J. Oel *"Electronic conductivity of ceria" In: High Tech. Ceramics*, P. Vincenzini, Ed. (1987) p.1933-1941, Elsevier Science Publishers b.V., Amsterdam.
[42] H. Yahiro, T. Ohuchi, K. Eguchi and H. Arai, *J. Mater. Sci.* (1988) 23, 1036 – 1041.
[43] T.Y. Tien, J. Amer. *Ceram. Soc.* (1964) 47, 430 – 433.
[44] D.W. Strickler and W.G. Carison, *J. Amer. Ceram. Soc.* (1964) 47, 122 – 127.
[45] T. Otake, Ph.D. Thesis *"Defect Structure and mixed conducting properties of ceria based oxides,"* 1975, Tohoku Univ.
[46] H. Tannenberger, H. Schachner and P. Kovacs, rev. Energ. Primaire. *J. Int. d'Etude Polics Combustible, Bruxwllwa* (1965) 3, 19-26.
[47] F.A. Kroeger, *J. Amer. Ceram. Soc.* (1966) 49, 215-218.
[48] A. Nakamura, *Solid State Ionics* (2010) 181, 1543-1564.
[49] A. Nakamura, *Solid State Ionics* (2010) 181, 1631-1653.
[50] A.D. Muirray, G.E. Murch and C.R.A. Catlow, *Solid State Ionics* (1986) 18-19, 196 – 202.
[51] F. Shimojo, T. Okabe, F. Tachibana, M. Kobayashi and H. Okazaki, *J. Phys. Soc. Jpn.* (1992) 61. 2848-2857.
[52] F. Shimojo and H. Okazaki, *J. Phys. Soc. Jpn.* (1992) 61. 4106-4118.
[53] R. Car and M. Parrinello, *Phys. Rev. Lett.* (1985) 55, 2471-2474.
[54] R. Krishnamurthy, Y.D. Yoon, D.J. Srolovitz and R. Car, *J. Am. Ceram. Soc.* (2004) 87, 1821-1830.
[55] R. Pornprasertsuk, P. Ramanarayanan, C.B. Musgrave and F.B. Prinz, *J. Appl. Phys.* (2005) 98, 103513 (1-8).
[56] C.H. Cheng, S.F. Lee and C.W. Hong, *J. Electrochem. Soc.* (2007) 154, E158-163.
[57] J. Hooper, A Ismail, J.B. Giorgi and T.K. Woo, *Phys. Chem. Chem. Phys.* (2010) 12, 12969 – 12972.
[58] P. Kofstad, *"Nonstoichiometry, Diffusion and Electrical Conductivity in Binary Metal Oxides",* (1972) New York, Wiley.
[59] J. Mizusaki, *Solid State Ionics* (1992) 52, 79 – 91.
[60] J. Mizusaki and K. Yashiro, Chapter Two, *"Nano-Scale Energy and Mass Transport Phenomena Involved in Fuel Cell Processes"* in *"Nano-Mega Sclae Flow Dynamics in Energy Systems", The 21st Century COE Program, International COE of Flow Dynamics, Lecture Series, Vol. 9.* Ed S. Maruyama, (2008) 9, 39-78, Sendai, Tohoku Univ. Press.
[61] *The Powder Diffraction File*, JSPDS-International Centre for Diffraction Data, Newton Square, PA.
[62] J. Mizusaki, *Proceedings of the 12th Asian Conference on Solid State Ionics*, ISBN 978-7-5629-3159-1, (2010) 12, 473-484, Wuhan, Wuhan University of Technology.
[63] H. Kamata, A. Hosaka, J. Mizusaki, H. Tagawa, *Solid State Ionics* (1998) 106, 237-245.
[64] K. Yasumoto, N. Mori, J. Mizusaki, H. Tagawa, M. Dokiya, *J. Electrochem. Soc.* (2001) 148(1) A105-A111.
[65] J. Richter, P. Holtappels, T. Graule, T. Nakamura and LJ. Gauckler, *Monatsh. Chem.* (2009) 140, 985 — 999.
[66] A.V. Berenov, A. Atkinson, J.A. Kilner, E. Bucher and W. Sitte, *Solid State Ionics* (2010) 181, 819—826.
[67] S. Diethelm and J.V. Herle, *J. Eur. Ceramic Soc.* (2004) 24, 1319—1323.

[68] G.J. la O and Y. Shao-Hornz: *J. Electrochem. Soc.* (2009) 156B816–B824.

[69] T. Kawada, K. Masuda, J. Suzuki, A. Kaimai, K. Kawamura, Y. Nigara, J. Mizusaki, H. Yugami, H. Arashi, N. Sakai and H. Yokokawa, *Solid State Ionics* (1999) 121, 271-279.

[70] M. Sase, F. Hermes, K. Yashiro, K. Sato, J. Mizusaki, T. Kawada, N. Sakai, and H. Yokokawa, *J. Electrochem. Soc.* (2008) 155(8), B793-B797.

[71] M. Sase, J. Suzuki, K. Yashiro, T. Otake, A. Kaimai, T. Kawada, J. Mizusaki, H. Yugami, *Solid State Ionics* (2006) 177, 1961-1964.

In: New Research Trends of Fluorite-Based Oxide Materials
Editors: Akio Nakamura and Junichiro Mizusaki
ISBN: 978-1-63117-350-9

Chapter 2

CERIA-BASED SOLID SOLUTIONS: STRUCTURE, PROPERTY AND APPLICATIONS

*V. Grover and A. K. Tyagi**
Chemistry Division, Bhabha Atomic Research Centre
Mumbai, India

ABSTRACT

In view of their tremendous multifaceted applications, the importance of ceria and ceria-based systems cannot be over-emphasized. There has been a lot of research done on exploring various functionalities in ceria by tuning the composition, size, structure and morphology and further attempts have been made to optimize and tune these by synthesizing solid solutions with other appropriate materials. This chapter gives an overview on the ceria and ceria-based materials describing synthesis, structure and various properties. The structure and applications of ceria-based materials followed by phase relations in technologically important solid solutions of cerium oxide with various rare earth oxides have been discussed. The phase relations, in general, consist of fluorite ceria-type phase, a C-type bixbyite type phase and in some cases biphasic phase fields consisting of any two of F-type, C-type and/or monoclinic phases. The lattice parameter variations in various phase fields were controlled by average ionic radii and presence of vacancies/interstitials. This chapter discusses various synthesis routes employed for obtaining ceria and ceria-based solid solutions with best possible powder morphologies and phase purity. Various applications of these materials as ionic conductors, catalysts, nuclear, magnetic and luminescent materials etc. have been discussed in connection with the structure, powder properties (in nano-particle or bulk form) and synthetic-route variations.

* Correponding author: A. K. Tyagi. E-mail address: aktyagi@barc.gov.in; vinita@barc.gov.in.

1. INTRODUCTION

Rare-earth oxides have attracted a lot of attention in terms of their applications in various areas of prime importance such as energy and environment. Ceria, in particular, has wide-range applications and it would not be a misnomer to call it a wonder material. It is known to crystallize in fluorite-type lattice (Space Group; Fm-3m). MO_2-type oxides of tetravalent lanthanides and actinides, e.g., CeO_2, ThO_2, UO_2 and PuO_2 etc. adopt the same fluorite-type structure, which can be defined as a cubic eutaxy of M atoms with O atoms in tetrahedral interstices [l].

A simple representation of fluorite-type structure is shown in Figure 1(a), wherein green color sphere(s) indicate oxygen and blue color sphere(s) M (e.g., Ce). Another way of describing the fluorite-type structure is to consider it to be built of OM_4 tetrahedra, i.e., M tetrahedra with oxygen in its center. A simple calculation shows that for spherical atoms in a fluorite-type compound MX_2, the contact between an M atom and its eight co-ordinated X atoms would occur when the ionic radius of M is greater than 0.73 times that of X, i.e., for $r(M) / r(X) \geq 0.73$. However, this is only an approximate relationship, for many MO_2 (including CeO_2 itself) do not necessarily fulfill this relationship.

Fluorite-type structure is one of the most versatile structures. If some oxygen atoms are removed from the structure, a homologous series of oxygen-deficient (i.e., oxygen-vacancy (V_O)-type) binary rare-earth oxides is obtained [2-4], of which the limiting lowest oxygen-content system is the so-called bixybite C-type sesquioxide (RE_2O_3) with all the oxygen six-fold coordinated RE^{3+} (Figure 1(b)). This structure is adopted by heavier rare earth binary oxides and is formed when one quarter of the oxygen atoms of the stoichiometric fluorite lattice are lost.

Thus, the C-type structure is closely related to the fluorite lattice with respect to the ordered vacancies ($a_{C\text{-type}} = 2a_{Fluorite}$). Crystal structures of these compounds are characterized by several types of oxygen vacancies' arrangements in the fluorite-like framework.

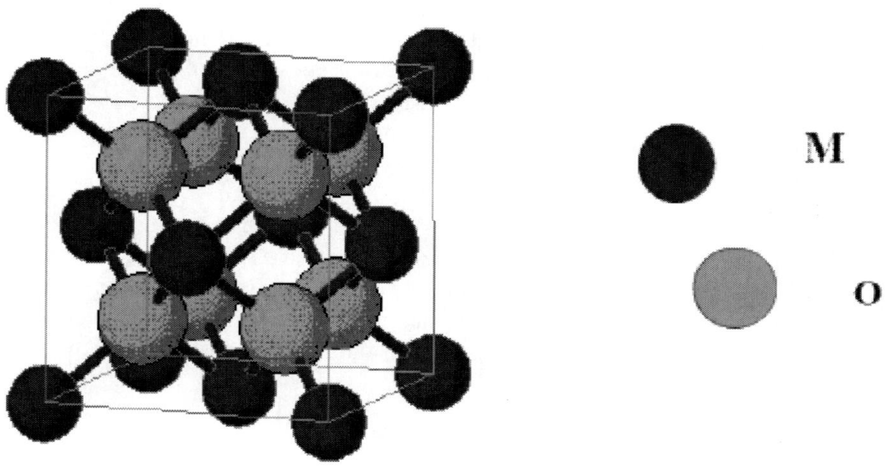

Figure 1(a). A pictorial representation of fluorite-type structure, MO_2.

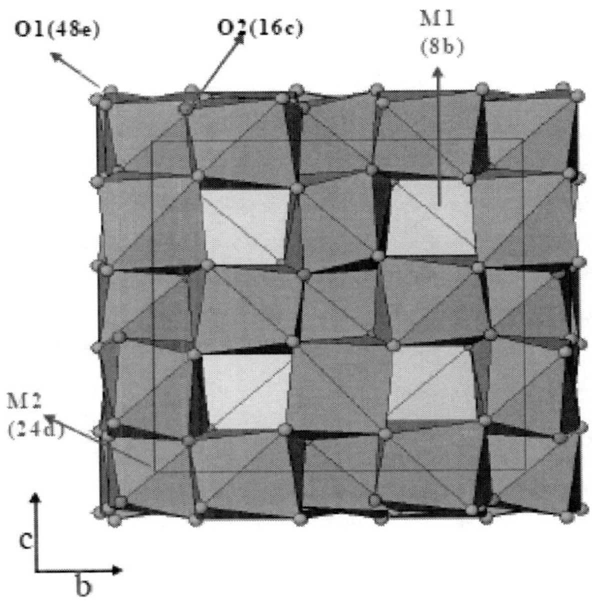

Figure 1(b). Pictorial representation of C-type structure adopted by rare earth sesquioxide (M_2O_3). M1 and M2 are two types of metal ion sites (both CN=VI).

In order to classify these structures, Kang and Eyring [2,3] proposed an elegant method based on the use of distinct fluorite modules composed from oxygen-centered metal tetrahedral, wherein they have nicely elucidated how the crystal structures of fluorite-related binary rare earth oxides may be described as constructed by the linking of these distinct fluorite modules via common faces.

In fact, the structures of fluorite-related anion or cation deficient arrays have been a subject of intense interest in crystallography.

A lucid article written by Krivovichev et al. [5] also discusses about the description of different classes of binary oxides in terms of fluorite related modules.

Most of such compounds can be described as having frameworks which may be obtained from fluorite structure by removal of certain number of anion-centered tetrahedra from a fluorite module. Also, there are many frameworks built from fluorite-related blocks rotated relative to each other.

Fluorite-type structure is also one of the most studied structure-type by material scientists for a long time and this can primarily be attributed to a lot of void space that it possesses, its ability to accommodate a wide variety of isovalent and aliovalent cations and anions and its thermal and chemical stability.

All these characteristics are the backbone feature for a wide variety of applications that are exhibited by a fluorite-type structure. Indeed, many known materials which are significant from application point of view are either based on fluorite-type structure or can be derived from it. Pyrochlore is one such example of an important class of materials.

The compound of interest here, CeO_2, which contains cerium in +IV oxidation state, is the most stable oxide of cerium at ambient temperature and pressure. It is the most abundant of the lanthanides and produced after the extraction of Th from Monazite (Ce,La,Y,Th)PO_4.

The other oxide of cerium, viz., sesquioxide Ce_2O_3 (containing Ce(III)) can be obtained only at very lower oxygen partial pressures of about 10^{-93} atm.

As the oxygen pressure is increased to about 10^{-40} atm and above, Ce_2O_3 gets oxidized to CeO_2 passing through a homologous series Ce_nO_{2n-2m} at intermediate oxygen pressures [6]. Ceria itself forms a nonstoichiometric oxide(s) of general composition CeO_{2-x} where $0<x<0.5$, at high temperature under reducing atmosphere [7], which is decomposed into such discrete homologous Ce_nO_{2n-2m} phases at lower temperatures. It has been shown by using several techniques such as temperature programmed reduction (TPR), X-ray diffraction and magnetic measurements that the fluorite structure of ceria is retained down to 900 K under reducing atmosphere [8-14].

Also, as Ce^{4+} gets reduced to Ce^{3+}, the lattice parameter is found to increase, thus showing an expansion in the face-centered cubic (FCC) lattice [14]. An increase of 1.8% in the lattice parameter has been reported after the reduction treatment at 873K with respect to stoichiometric ceria [14]. Also, when ceria is reduced at 1070-1170K, C-type Ce_2O_3 phase (Figure 1(b)) is formed which oxidizes to an expanded ceria phase gradually [14]. Interestingly, when reduced at temperatures greater than 1273 K, it gets crystallized as a hexagonal (A-type) Ce_2O_3 which is isotypic to La_2O_3. It is this capability of ceria to *easily switch between* Ce^{3+} *and* Ce^{4+} which makes it an indispensible material for applications utilizing this type of redox reaction(s), such as catalysis [11,15].

Due to the presence of highly versatile fluorite structure as well as ease of reduction-oxidation, ceria-based materials have been intensively investigated for a variety of structural and chemical properties [15,16] in connection with oxygen storage capacity [9-11], redox behavior and its-associated non-stoichiometry [8], redox catalysis [8-13,15,16], metal-ceria interactions [12,13], areas of environmental catalysis [9,11,17-23], wet catalytic oxidation of organic pollutants [23,24] etc. The prominent role of ceria has therefore been established in three-way catalysis, catalytic wet oxidation, water-gas-shift reaction, oxidation/combustion catalysis and solid oxide fuel cells [16,19,18-27]. The structural features of ceria in combination with oxygen storage and release properties are crucial for most of these applications. Ceria can be employed as an oxide carrier or a mixed oxide carrier with a transition metal oxide providing unique catalytic properties. A number of CeO_2-based systems such as CeO_2-ZrO_2, CeO_2-Al_2O_3, CeO_2-SiO_2, CeO_2-La_2O_3, Pd/CeO_2 and Au/CeO_2 have been explored for their structural, mechanical and catalytic properties [14,15, 11-12, 16-18]. Addition of ~ 12 mol % of CeO_2 to zirconia (ZrO_2) stabilizes its tetragonal form which shows exceptional mechanical toughness and strength [28]. Also ceria-doped zirconia is used in thermal barrier coatings on metal surfaces as aircraft engine parts in order to reduce the high temperature corrosions of metal substrate [29].

Another application of growing importance is of ceria-based catalysts where ceria is a catalytically active component for removal of pollutants from automobile exhaust [30]. It can also act as a stabilizer for the high surface area alumina in catalytic converters, as a promoter of the water-gas shift reaction, as an oxygen storage component and as an enhancer of the NOX reduction capability of rhodium. It is also useful as a catalyst in petroleum refining and in metallurgical applications. The substitution of ceria and to forming its solid solutions with rare-earth oxides and transition metal oxides, has been utilized and explored not only for enhancement of its properties as catalytic and fuel cell materials but also for the introduction of new properties like photocatalysis, luminescence etc. The stability range of the phase compositions varies depending on the dopant or substituent subject to their ionic radii and electronic properties relative to CeO_2.

Among all doped ceria systems studied, rare-earth (RE)-doped ceria holds an important place in view of applications in oxygen concentration cells and in solid oxide fuel cells [31]. It is also being considered as a potential catalyst for mitigating hydrocarbons, CO and NOX from automobile exhausts [32,33]. RE-doped ceria is very stable and exhibits enhanced oxygen mobility. Following electroneutrality condition aliovalent RE^{3+} substitution for Ce^{4+} ion in ceria lattice leads to the creation of oxygen vacancies in order to compensate the effective negative charge produced by the dopant cations [34,35] as has been proved by various studies.

Structure and property of these defects resulting from such cationic substitution not only depend on the nature of the dopant, i.e., its crystal structure, electronic charge, ionic radius, electronegativity, etc., but also on its mole fraction within the host matrix, because these drastically affect the oxygen diffusion within the host lattice as well as oxygen uptake and release at the surface of the system and so on [36,37]. Many research groups have also directed their efforts to modulating the structure and property of ceria-based systems by synthesizing them by different chemical routes. This in turn controls the particle size, powder properties, morphology, etc. of the resulting product and has a strong bearing on the physico-chemical properties. Of late, ceria has proved its importance in nuclear research as surrogate for plutonium dioxide, PuO_2. The hazardous radiotoxic nature of Pu largely restricts handling and use in experiments, requiring sophisticated fume chambers and hot cells but still limiting its handling to a very dilute quantity. Ceria is a material which can simulate plutonia (PuO_2) in terms of their many structural, chemical, and physical similarities, besides its non-radioactive nature [38]. Also, most of the burnable poisons used in present day nuclear reactor systems are the rare earth species, e.g., Gd^{3+}, Er^{3+} etc. [39] and hence rare earth doped ceria systems have been investigated for simulating plutonia fuel doped with burnable poisons. Thus, in the context of nuclear energy as well, study on ceria-based systems holds tremendous significance. Detailed investigations of ceria-based systems have also been performed in our group. They have been explored as nuclear materials, catalytic materials, magnetic materials and also as luminescent materials. A lot of work has also been devoted to phase analysis and structural studies on these mixed ceria based oxides, primarily ceria doped with rare-earth ions. Synthesis of ceria-based powders has been attempted by various routes in order to tailor the powder properties. This chapter attempts to summarize the research that has been carried out in the field of ceria and ceria-based solid solutions, primarily their synthesis, phase analysis, properties and various applications.

2. Phase Relations in Rare Earth (RE: La – Lu, Sc, Y) Doped Ceria

Extensive phase relations have been established in $CeO_2 – REO_{1.5}$ system (where RE: La, Sm, Eu, Gd, Dy, Ho, Er, Tm, Yb, Lu, Sc, Y). Structurally, rare-earth oxides are an interesting class in themselves that adopt different structures depending upon the radius of rare-earth ion. Adachi and Imanaka [4], in their review article on binary rare-earth oxides, have elaborated on the structure of rare-earth sesquioxides as a function of temperature and ionic size. Three types of polymorphs, designated as A (hexagonal), B (monoclinic) and C (cubic) have been reported in RE_2O_3, at temperatures below 2000°C, depending on the rare–earth ionic size.

In the rare-earth doped ceria, lattice parameters and other physical properties of the samples change with composition and exhibit an interesting interplay with relative sizes of Ce^{4+} and RE^{3+} as well as the oxygen vacancies created as the result of this aliovalent substitution. Along with the change in anion vacancy concentration, the substitution may also lead to the stabilization of high temperature phases of rare-earth oxides at ambient temperature. This makes the study of phase evolution in rare earth substituted ceria an interesting research area both from the point of view of structural evolution as well as properties. A lot of studies in this respect were carried out in our group wherein the mixed ceria - rare earth oxide samples were synthesized by a multi-step heating procedure and the entire sub-solidus studies were reported under slow-cooling conditions. The CeO_2–RE_2O_3 systems were found to exhibit interesting trends in cell parameters with compositions.

It is interesting to observe that almost all the rare earth oxides showed 30-40 mol% solubility of $REO_{1.5}$ in ceria while retaining the cubic fluorite lattice. Such high degree of solubility can be explained based on the fact that majority of rare earth oxides adopt C-type structure which is actually a superstructure of the fluorite cell. If one-fourth of oxygens are removed along the non-intersecting strings in the four <111> directions, then all metal ions will have a six-fold co-ordination (CN=VI) resulting in a C-type structure. It has two kinds of Ln^{3+} sites with Wyckoff notations, 24d and 8b, in 3:1 ratio, the former having face-diagonal Vo-Vo connection. (Figure 1b)

Hence, the C-type structure is closely related to the fluorite lattice by the way of the ordered vacancy in fluorite lattice ($a_{C\text{-type}} \sim 2a_{Fluorite}$). Such similarity in structure along with relatively similar ionic sizes of Ce^{4+} and RE^{3+} [40] ensure high mutual solubility. It is noted that the solubilities of $REO_{1.5}$ in CeO_2 obtained in our studies differ from those observed by Bevan et al. [41] on similar systems, most plausibly due to the fact that the results presented here are obtained via slow-cooling (2°C/min) of the compositions from high temperatures whereas Bevan et al. performed the studies under quenched conditions. The difference may also be attributed to different synthesis temperature employed in the two studies. The phase relations develop in an interesting way depending on the relative ionic sizes of Ce^{4+} and RE^{3+} ions. The bar diagrams representing phase relations for various $Ce_{1-x}RE_xO_{2-x/2}$ (RE: Rare earth ions, Sc^{3+}, Y^{3+}) are shown in Figure 2. Various nominal compositions were synthesized by a three step heating protocol at 1200°C, 1300°C and 1400°C with intermittent grindings. Heating and cooling rates were maintained at 2°C/min. It is noted that RE_2O_3 have been represented as $REO_{1.5}$ in the following discussion of individual systems for easy comparison with CeO_2.

(a) CeO_2- $LaO_{1.5}$

Lanthanum oxide (La_2O_3) adopts A-type modification. It was observed that CeO_2 can solubilize up to 30 mol% $LaO_{1.5}$ in its lattice retaining the fluorite-type structure. This is followed by a biphasic region consisting of an A-type La_2O_3 phase and an F-type solid solution of lanthana with ceria. The representative X-ray diffraction patterns showing all the phase fields in this system are depicted in Figure 3. The fluorite type peaks shift to lower angle on adding 40 mol% $LaO_{1.5}$ because the average cationic size increases on substitution of Ce^{4+} ions by La^{3+} ions. However, the peaks due to La_2O_3 do not show any shift implying that it is primarily pure La_2O_3 phase.

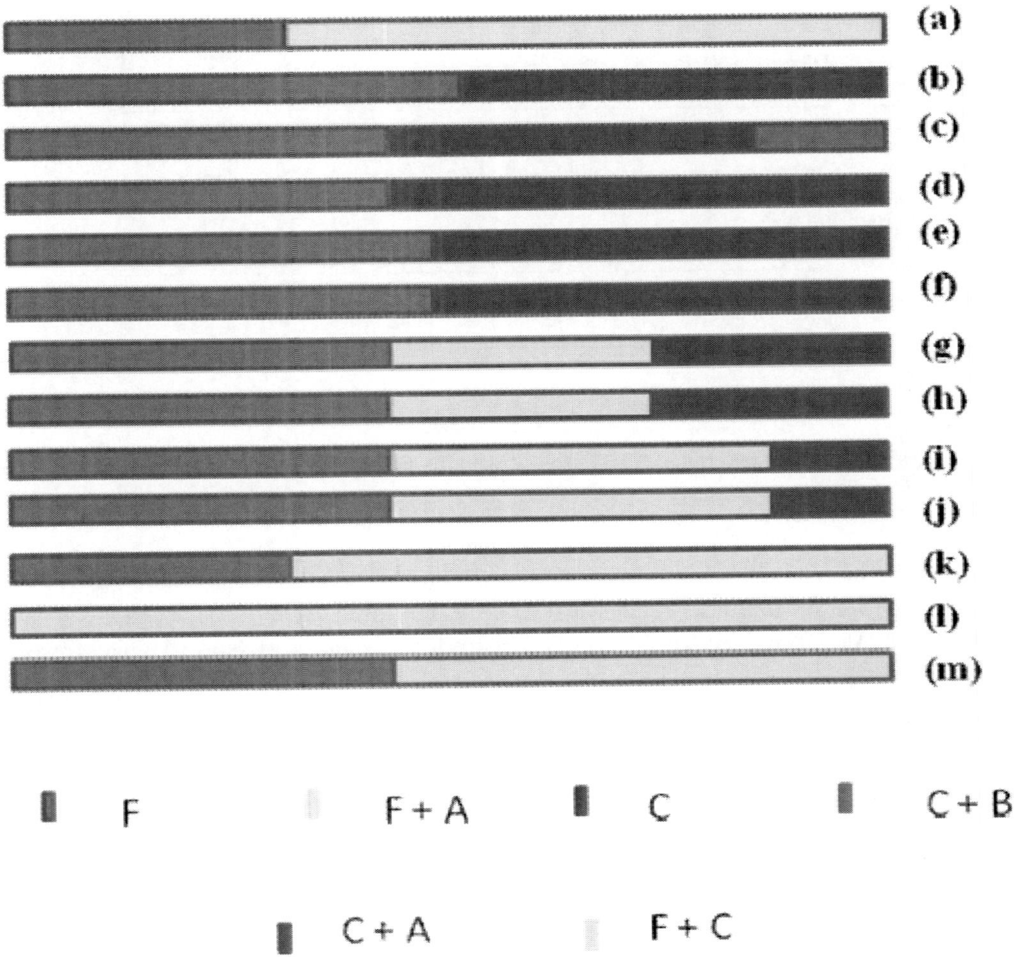

Figure 2. Bar diagram depicting phase relations in $CeO_2 - MO_{1.5}$ system where M is (a)La, (b)Nd, (c)Sm, (d)Eu, (e)Gd, (f)Dy, (g)Ho, (h)Er, (i)Tm, (j)Yb, (k)Lu, (l)Sc, (m)Y. Synthesis conditions: Three step heating (at 1200°C, 1300°C and 1400°C), heating/cooling rate: 2°C/min).

This implies that under the synthesis conditions employed (multi-step heating, final heating at 1400°C, slow cooling) in our study, there is no solubility of CeO_2 in $LaO_{1.5}$. There have been reports supporting no solubility of CeO_2 in La_2O_3 [42] and vice versa [43]. Though the latter case appears unlikely in view of the present study, it depends on the reaction conditions, primarily the temperature which dictates the solubility limits.

(b) $CeO_2 - NdO_{1.5}$ [44]

In this particular case, the cubic fluorite type structure of CeO_2 was retained till the composition $Ce_{0.5}Nd_{0.5}O_{1.75}$. At and beyond 52.5 mol% Nd^{3+} substitution, super-lattice peaks appear in the XRD patterns showing the presence of a C-type phase which persists up to 67.5 mol% Nd^{3+}, followed by a biphasic field of C-type and hexagonal (A) phases.

This biphasic nature continued till the end and pure Nd_2O_3 is hexagonal (A) phase.

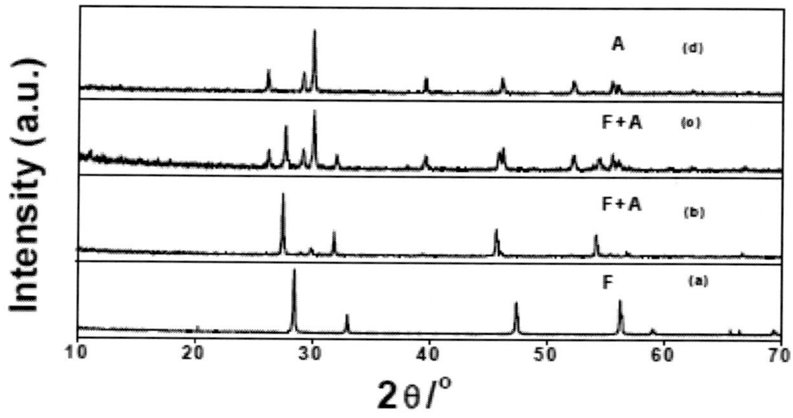

Figure 3. X-ray diffraction patterns in $Ce_{1-x}LaxO_{2-x/2}$ where x is (a) 0 (b) 0.4 (c) 0.8 (d) 1.0.

F: Fluorite-type. A: Hexagonal-type.

The lattice parameters were found to increase with increase in Nd^{3+} content which can be readily understood on the basis of the difference in the ionic sizes of Ce^{4+} and Nd^{3+}, being 0.87 and 0.98 A° [2], respectively, in the eight-fold coordination (CN=VIII). This result is somewhat different from that reported by Bevan et al. [41] who performed the same phase relation study by quenching of the sample prepared at much higher temperatures (>1600°C). The latter favors the wider homogeneity of disordered F-type phase, kinetically hindering its C-type ordering.

(c) $CeO_2 – SmO_{1.5}$ [45]

The phase relations in CeO_2-$SmO_{1.5}$ are characterized by an F-type (ceria-like) phase till 40 mol% of Sm^{3+} which is followed by the appearance of small superstructure peaks indicating the onset of ordering of oxygen vacancies to yield a C-type phase which persisted upto 80 mol% $SmO_{1.5}$. It was followed by a bi-phasic field consisting of a C-type phase and a monoclinic (B) phase. Pure Sm_2O_3 exists in monoclinic (B) phase at room temperature. The lattice parameters increase with increase in Sm^{3+} content in both F-type and C-type regions.

(d) $CeO_2 – EuO_{1.5}$ [45]

This particular system exhibited single-phasic behavior throughout the entire range according to the X-ray diffraction result, being F-type solid solution till 40 mol% Eu^{3+} and a C-type one beyond that. The lattice parameters were found to increase for the F-type solid solution which can be explained based on the relative ionic size considerations on incorporation of Eu^{3+} ion at the Ce^{4+} site (Eu^{3+} being larger than Ce^{4+} causes expansion of unit cell). On the other hand, lattice parameters remained almost constant for the C-type region. Some reports have shown that the lattice parameters of the C-type solid solution in CeO_2 – $EuO_{1.5}$ system slightly decrease [46].

(e) $CeO_2 – GdO_{1.5}$ [47]

This system was observed to be similar to the above CeO_2-$EuO_{1.5}$ system in the sense that they both exhibit a single-phasic behavior, consisting of an F-type solid solution till 45 mol% $GdO_{1.5}$ and a C-type phase beyond that. Herein, an interesting lattice parameter trend is observed where the lattice parameter of the F-type phase still increased on Gd^{3+} doping but it decreases in the C-type region. The reason for this lies in the creation of oxygen (anionic) vacancies and would be discussed later in this chapter. The structural analysis of this anion-rich C-type solid solution phase, i.e., $Gd_{1-x}Ce_xO_{1.5+x/2}$ (x = 0.20 and 0.40) was reported [48]. This is basically isostructural with C-type rare earth oxide itself (Figure 1(b)) (space group Ia-3, No. 206, Z = 32), containing excess oxygens required for charge balance on the built-in *16c* oxygen vacancy site. The structural analysis reveals it to have two different kinds of cation sites, *8b* and *24d*, and two different kinds of anion sites, *48e* and *16c,* the latter being originally just the above-mentioned empty oxygen-vacancy site in pure $LnO_{1.5}$.

(f) $CeO_2 – DyO_{1.5}$

Here again, an F-type region was observed till 30 mol% Dy^{3+} doping which was followed by a C-type region, the lattice parameter increased very slightly in the F-type region and then it continued to decrease in C-type region with increased Dy^{3+} substitution. Interplay of the average cation ion size(s) and the presence of oxygen vacancy is the controlling factor in determining such lattice parameter behavior.

(g) $CeO_2 – HoO_{1.5}$ and (h) $CeO_2 – ErO_{1.5}$ [49]

Some differences in the phase relation(s) creep in when rare-earth ions are Er^{3+} and Ho^{3+}. The single-phasic F-type phase field still persists till 40 mol% Er^{3+} and Ho^{3+} substitution but now it is followed by a biphasic phase field of F-type and C-type phases which co-exist till 70 mol% of them. This is then followed by a single-phasic C-type region. The lattice parameter trends are also different from that observed for previous series, as now it decreases for the F-type solid solution(s) with increase in the rare-earth content, in contrast to that observed for the above-mentioned lighter rare-earth ions. The lattice parameter for C-type solid solution continues to decrease with increase in RE^{3+} content. The lattice parameters, however, were constant for both the phases in the biphasic region, as expected.

(i) $CeO_2 – TmO_{1.5}$ and (j) $CeO_2 – YbO_{1.5}$ [50]

The phase relations observed with substitution of these two rare-earth ions in ceria constituted of an F-type phase field till 40 mol% rare earth ion followed by a biphasic region till 90 mol% of rare earth ion content. Beyond that, single-phasic C-type solid solution phase field was observed.

The lattice parameters of both F- and C-type phase fields were found to decrease with increasing rare earth concentration despite the fact that the ionic sizes of both Tm^{3+} and Yb^{3+} are still larger than that of Ce^{4+} (at the same oxygen-CN).

(k) CeO_2 –$LuO_{1.5}$ [51]

$LuO_{1.5}$ showed solubilility of 30 mol% in CeO_2 thus forming an oxygen (anion) deficient fluorite-type (F-type) solid solution which was followed the biphasic region (F + C). The refinement of lattice parameters showed that the C-type phase is pure Lu_2O_3, for no variation in lattice parameter of C-type phase was observed with change in CeO_2 (ceria) content. The absence or disappearance of C-type solid solution phase on the $REO_{1.5}$–rich side on going from RE^{3+} = Yb to Lu in Figure 2 indicates that Lu^{3+} is too small to accommodate Ce^{4+} in the same C-type solid solution lattice.

(l) CeO_2- $ScO_{1.5}$ [51]

This system was found to be biphasic throughout of pure F-type CeO_2 and pure C-type $ScO_{1.5}$. However, careful cell parameter refinement showed that there is a small but discernible decrease in the lattice parameter of F-type phase which indicated a small solubility of Sc_2O_3 into CeO_2.

Also, no variation was observed in the lattice parameter of C-type phase implying that there is no solubility of CeO_2 into scandia. The wider biphasic range and lesser solubility observed in CeO_2-$ScO_{1.5}$ system compared to above-mentioned other series can also be attributed to larger difference in ionic sizes of Ce^{4+} and Sc^{3+} as compared to that between Ce^{4+} and other RE^{3+} which affects their miscibility behavior. This system, as reported here, was synthesized by solid state route.

The synthesis route can have important bearing on phase relation. This is exemplified by the fact that, whereas by solid state route almost no mutual solubility was obtained, gel combustion route could incorporate even 50 mol% of $ScO_{1.5}$ into CeO_2 yielding a meta-stable fluorite (F)-type solid solution [52]. Furthermore, this solid solution (at ~50 mol% $ScO_{1.5}$) under controlled heat treatments yielded a perovskite-type phase.

(m) CeO_2 –$YO_{1.5}$ [53]

In case of $Ce_{1-x}Y_xO_{2-x/2}$, 45 mol% Y^{3+} could be dissolved in CeO_2 lattice retaining fluorite structure and beyond that a biphasic phase field consisting of F-type and C-type phases was observed in our study. Yet, many studies reports that C-type solid solution also forms here in the Y-rich region, as seen in the similar medium-size RE^{3+}-bearing d) CeO_2-$EuO_{1.5}$ to j) CeO_2-$YbO_{1.5}$ systems in Figure 2. The aliovalent Y^{3+} ion substitution in CeO_2 lattice causes gradual decrease in the lattice parameter.

3. Variation of Lattice Parameters in Doped Ceria Systems

An interesting interplay of various factors is revealed when the trends in lattice parameters for various rare earth cation systems are carefully inspected in F-type and C-type solid solution regions summarized above in Figures 2 and 3. First, consider the F-type region of all the rare-earth doped ceria systems; it is observed that the lattice parameters for the F-type phase increases with the substitution of the RE^{3+} ion into CeO_2 lattice till Dy^{3+}. Beyond that it decreases with increase in rare earth ion content. In fact, it was observed that the rate of increase of the lattice parameters decreased from La^{3+} to Dy^{3+} where the increase was really small. This can be explained if we consider that substitution of Ce^{4+} by RE^{3+}, leads to creation of oxygen vacancies. Since RE^{3+} are generally larger than Ce^{4+}, the cell dimensions should increase as a bigger ion substitutes Ce^{4+} ion. However, an oxygen vacancy is also introduced in the lattice for every two RE^{3+}s to maintain electroneutrality. One possibility is that these vacancies tend to cluster together and decrease the average ionic size. It is obvious that these two factors act in opposite directions, i.e., the average ion size increases which would result in expansion of the cell while the vacancies would lead to contraction. It is known that the ionic size decrease across the lanthanide series due to lanthanide contraction. Therefore, it could be inferred that for the lighter (and larger) rare-earth ions, the average ionic size dominates and we observe the net increase in lattice parameters, whereas for heavier (and smaller) rare-earths there is the predominance of the vacancy effect leading to decrease in ionic radii of the fluorite-type solid solutions. Apparently for Dy^{3+}, these two factors are almost equal in magnitude which is manifested in a very small increase in lattice parameter for the F-type solid solution. Similar behavior was observed for the lattice parameters for C-type solid solution. Herein, it would be easier to understand the trend if the converse ceria substitution in rare-earth oxides is considered. The substitution of tetravalent Ce^{4+} in place of trivalent RE^{3+} would be accompanied with an additional oxide ion into the lattice to maintain electroneutrality. On increasingly substituting the Ce^{4+} ion in place of RE^{3+} ion, the average cation size decreases but at the same time, there is extra repulsion induced in the lattice among the interstitial oxygen anions introduced along with Ce^{4+} substitution. The former will result in the lattice contraction while the latter will cause lattice expansion and it is reasonable to assume that the overall lattice parameter trend will result as a sum of of these two factors. It was mentioned previously that the cell parameters of C-type solid solution will decrease with substitution of Ce^{4+} in place of RE^{3+} when rare earth ion is Nd^{3+} and Sm^{3+}, remains almost constant in case of Eu^{3+} and increases for later heavier rare earth ions. This observation can be explained by keeping in view the opposing factors discussed above. For lighter, bigger rare earth cations, the size effect predominates and the lattice parameters were found to decrease with increasing Ce^{4+} substitution. For heavier and smaller rare-earth cations which approach Ce^{4+} in terms of size, the inter-anionic repulsion takes over and the cell shows a net expansion with increase in Ce^{4+} substitution.

The variation of C-type lattice parameters for $CeO_2 - RE_2O_3$ (RE: Gd, Eu, Sm) is shown in Figure 4. Nakamura et al. have done detailed and thorough studies relating the macroscopic properties, i.e., lattice parameters and the microscopic defect studies using Mössbauer spectroscopy on these defective oxide systems with the aim of elucidating the local structures in these systems and have come up with a model to describe the non-vegardian behavior of these defect fluorite oxides. [46]

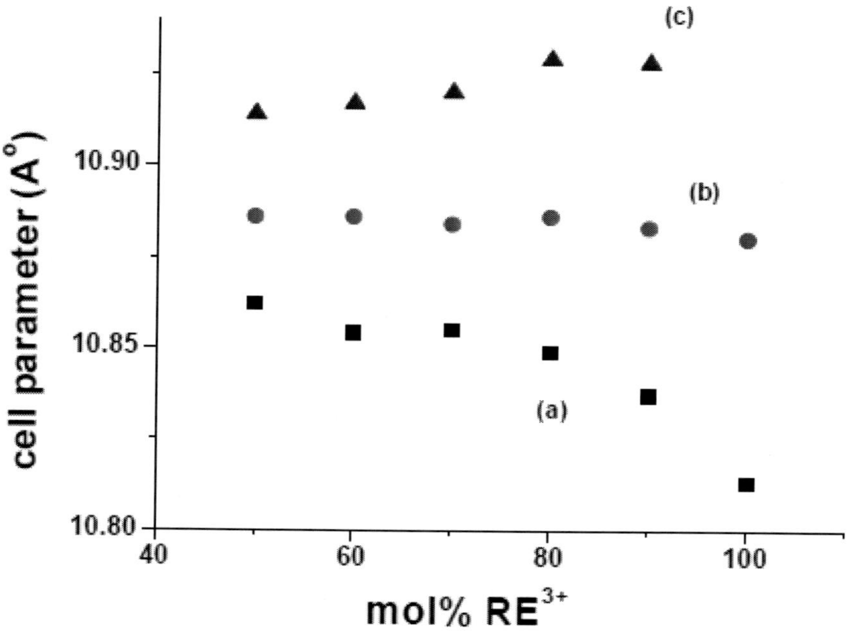

Figure 4. Variation of lattice parameter of C-type solid solution with increase in RE content in $Ce_{1-x}RE_xO_{2-x/2}$ where RE: (a) Gd, (b) Eu, and (c) Sm.

In case of rare-earth substituted ceria, another feature that crops up especially in case of gadolinia and dysprosia substituted one, is that due to similarity in ionic size (lattice parameter) of these ions (systems) to Ce^{4+} (CeO_2), the peak positions in the X-ray diffraction pattern of their F and C-type solid solutions are so close that it becomes difficult to detect the presence of bi-phasic region in these systems using lab source XRD machines. Moreover, owing to large coherence length of X-rays, it is not often easy to detect parasitic micro domains of altogether different new phases amidst a bulk phase using XRD technique [54].

Of course, it might be able to solve this problem by using high intensity X-ray sources like synchrotrons. These sources, however, are not readily available. Further, X-ray diffraction technique, in general, is insensitive to microscopic phase changes initiated by oxygen displacement, because X-ray scattering efficiency of oxygen is significantly lower compared with that of the high Z rare-earth cations.

Raman spectroscopy becomes useful in such cases and promises better results because of its sensitivity to both oxygen displacement due to large polarizability [55] and intermediate range order without long range periodicity [56,49]. In particular, the Raman spectrum of the perfect crystal lattice undergoes drastic changes in the presence of any disorder in the system and proves to be an excellent probe for local disorder. The C-type lattice differs from F-type lattice in terms of ordering of vacancies in anionic sublattice, as discussed above, and Raman spectroscopy is quite sensitive to oxygen-cation vibrations (and in turn anionic sub-lattice) as opposed to X-ray diffraction method (which is more sensitive to disorder in the cation sublattice). Hence, Raman spectroscopy is an excellent probe for studying local structure. There have been quite a few examples wherein Raman spectroscopic investigations have provided unequivocal information to distinguish between a defect-fluorite structure and the so-called pyrochlore type-ordered one for various compositions with cationic radius ratio lying at the pyrochlore / defect-fluorite borderline.

As above mentioned, CeO_2-Gd_2O_3 series is single-phasic in F-type (till 45 mol% Gd^{3+} substitution) and C-type regions (50 mol% Gd^{3+} and beyond) by X-ray diffraction studies. Raman spectroscopic studies were carried out on a complete range of ceria-gadolinia solid solutions ($Ce_{1-x}Gd_xO_{2-x/2}$) to determine the intricate microscopic changes that occur in the system [57]. Interestingly, here it is revealed that the system is characterized by the presence of F-type Raman modes until $x \leq 0.2$, the co-existence of Raman modes of both F-type and C-type in the range ($0.3 \leq x \leq 0.8$) and mostly C-type Raman mode for $x \geq 0.8$. Several broad modes indicating a high degree of disorder in the F-type lattice could also be observed. Hence, Raman spectroscopy reveals a very wide range of co-existence of F-type and C-type regions in this system rather than single-phasic F- or C- type as revealed by X-ray diffraction studies. The comparison of X-ray diffraction pattern and Raman spectrum for the nominal composition $Ce_{0.70}Gd_{0.30}O_{1.85}$ is displayed in Figure 5. The X-ray diffraction shows a clean F-type lattice for this composition whereas in the Raman spectrum there is a peak corresponding to C-type ordering. However, it is possible that there lie small micro-domains of one type of lattice in another and hence to call them distinct phases is still a question. Similarly, in case of $Ce_{1-x}Tm_xO_{2-x/2}$ system [50], the presence of C-type ordering was confirmed from Raman spectroscopic studies in the composition containing 40 mol% Tm^{3+} which was otherwise shown to be completely single-phasic F-type in X-ray studies. Here, the discrepancy is believed to have originated because of the smaller domain size (smaller than the coherence length of X-rays) of the C-type lattice that are inhomogeneously distributed within the F-type solid solution, which could not be directly detected from XRD measurements. Likewise, in case of $CeO_2 - HoO_{1.5}$ and $CeO_2 - ErO_{1.5}$ [49], it was deduced from XRD studies that the biphasic region commences from 50 mol% rare-earth ion containing composition. Raman studies, however, revealed that the biphasic region actually crept into the 40 mol% rare-earth ion containing composition also wherein, the Raman bands belonging to C-type region could be clearly observed in Raman spectra of the nominal composition $Ce_{0.60}RE_{0.40}O_{1.80}$ (RE: Ho, Er). The photoluminescence studies were also performed on Ho^{3+} containing compositions and the results were in concurrence with the Raman results. Hence, this reveals the utility of a simple lab-based technique like Raman spectroscopy in determining the local structure.

4. SYNTHESIS OF CERIA AND CERIA BASED SOLID SOLUTIONS AND TUNING OF POWDER PROPERTIES

All the applications of ceria and related systems e.g., in catalysis, glass polishing materials, electrolyte in solid oxide fuel cell materials etc. require ceria in each finest and purest particular form. The powder properties like particle size, surface area, sinterability, porosity etc. have tremendous impact on the suitability of the powder product for various applications. These powder properties can be tuned by selecting the appropriate synthetic technique as well as by suitably modifying the synthetic technique to obtain the product (with desired properties). Ceria-based systems have been conventionally fabricated by solid state synthesis [47] and soft chemical routes like combustion [58], precipitation [59], complex polymerization method [60], ultrasonication [61], microwave synthesis [62], hydrothermal synthesis [63] etc. Conventional solid state synthesis involves multi-step heating of reactants with intermediate grindings.

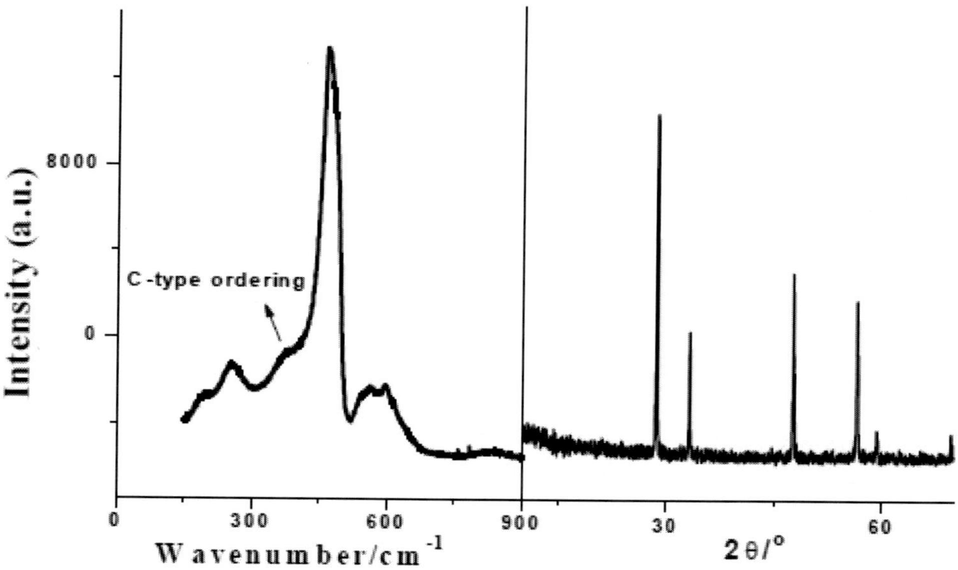

Figure 5. Raman spectrum (a) (left) and X-ray diffraction pattern (b) (right) for the nominal composition $Ce_{0.70}Gd_{0.30}O_{1.85}$.

However, with ceria there is an additional problem of reduction of Ce^{4+} to Ce^{3+} at elevated temperatures. It has been reported by Zhou and Rahman [64] convincingly using thermogravimetry data that CeO_2 does not show an appreciable oxygen loss up to 1400°C. In fact, these authors have shown that it is nano-crystalline ceria, which shows an appreciable weight loss at about 1300–1400°C. Even though the solid state route has advantages of the ready availability of starting oxide and low cost for a large scale powder production, it has some disadvantages such as the possibility of obtaining non-uniform sample with local stoichiometric variations arising from diffusion-limited insufficient homogeneization of the reaction mixture, which requires high temperature heating and long annealing times. In case of ceria, the controlled heating has to be performed at high temperatures to avoid reduction to trivalent state. Further, in view of the specific applications of ceria, the final material is required to have a definite shape, size and microstructure. Hence, often the material is desired to be in the nanocrystalline form so that its powder properties can be precisely tuned, which is possible only if the reaction temperature and time are considerably reduced. Thus, there arises a need for the homogeneity of the reacting constituents to be achieved at a molecular level. For this reason, the soft chemical routes have an edge over conventional solid state synthesis technique. In these processes, mixing is achieved at a molecular level in the liquid state, or in a molecular compound containing stoichiometric ratio of constituent atoms and an intimate blending of constituents is assured. Due to the proximity of the reactants, reaction at lower temperatures is possible.

In addition, the required final calcination / decomposition temperature are also generally lower, thereby resulting in very fine powders.

Among the available solution chemistry routes, the combustion technique is capable of producing nanopowders of oxide ceramics at lower calcinations temperatures in a surprisingly shorter time [65,66]. It is basically a redox reaction between an oxidant *viz.*, metal nitrate and a fuel such as citric acid, glycine etc.

The final product obtained via the combustion process is of the highest degree of phase purity with improved powder characteristics such as (like) narrow distribution of nanoparticle size, higher surface area and better sinterability [67-70]. The combustion technique involves an exothermic decomposition of a fuel-oxidant precursor which results in either the finely divided powder with required phase or semi-decomposed precursor having a considerable carbonaceous residue depending on the nature of fuel and fuel-to-oxidant ratio used in the process [71]. The size and properties of nanopowders can be controlled by controlling the fuel to oxidant ratio. This is because it determines the amount of gases evolved and the temperature reached during the combustion. It is the interplay of these two factors which determine the powder properties, e.g., a fuel-rich combustion will lead to higher temperatures and will cause partial local sintering thus leading to higher average particle sizes.

A considerable work has been done in our group on this aspect which is briefly discussed particularly in the context of synthesis of ceria and ceria-based system by gel combustion.

Ultrafine ceria nanopowders were synthesized by a glycine-nitrate combustion synthesis wherein the fuel to oxidant ratio was varied and the effects on powder properties were examined. Purohit et al. [72,73] synthesized ultra-fine ceria powders by three different routes; citrate–nitrate combustion, hydroxide and oxalate decomposition. The various processes gave ceria powders with different properties. It was observed that the nano-powders obtained with the citrate–nitrate combustion route have the smallest particle size and maximum surface area. In order to observe the effect of powder properties on sinterability, sintering experiment was performed under the identical conditions, i.e., 1250°C for 1 hr. SEM investigations revealed that the combustion derived ceria powder yielded the highest sintered density with ultra-fine grains. It may be noted that such a high sintered density was achieved without any further noteworthy grain growth from the green particles. This study demonstrated the superior powder properties of the nano-powders obtained by using combustion route. Another study by Mukherjee et al. [74] explored the effect of various fuels employed in combustion synthesis on powder properties of ceria using hitherto unexplored amino acid fuels such as aspartic acid, glutamic acid, arginine, tryptophan, phenyl alanine, valine, etc. The obtained powders were characterized by X-ray diffraction, surface area analysis, sinterability, dynamic light scattering, scanning electron microscopy (SEM) and small angle X-ray scattering (SAXS). The highest sintered densities were observed with the aspartic and glutamic acid. An interesting observation was the fact that there were larger variations in the agglomerate sizes as compared to the crystallite size and the biggest agglomerate size was obtained with the fuel giving smallest particle size [74]. Grover et al. studied the powder properties of technologically important $Ce_{0.8}Zr_{0.2}O_2$ solid solution using different fuel to oxidant ratios [75]. Higher amount of fuel leads to higher reaction temperature which enhances local sintering and thus relatively lower surface area and larger particle size. That is why fuel deficient ratio is preferredin case it is intended to obtain smaller particle size.

On the other hand, more fuel liberates more gaseous product which promotes higher surface area and hence smaller particle size. Also, there is a minimum fuel to oxidant ratio needed for the combustion to proceed. Usually an optimum fuel to oxidant ratio is needed to obtain the desired particle size and surface area. TEM images of $Ce_{0.8}Zr_{0.2}O_2$ composition synthesized with fuel deficient and stoichiometric fuel to oxidant (ratio) condition are shown in **Figure 6**, wherein the effect of fuel to oxidant ratio on particle size is evident.

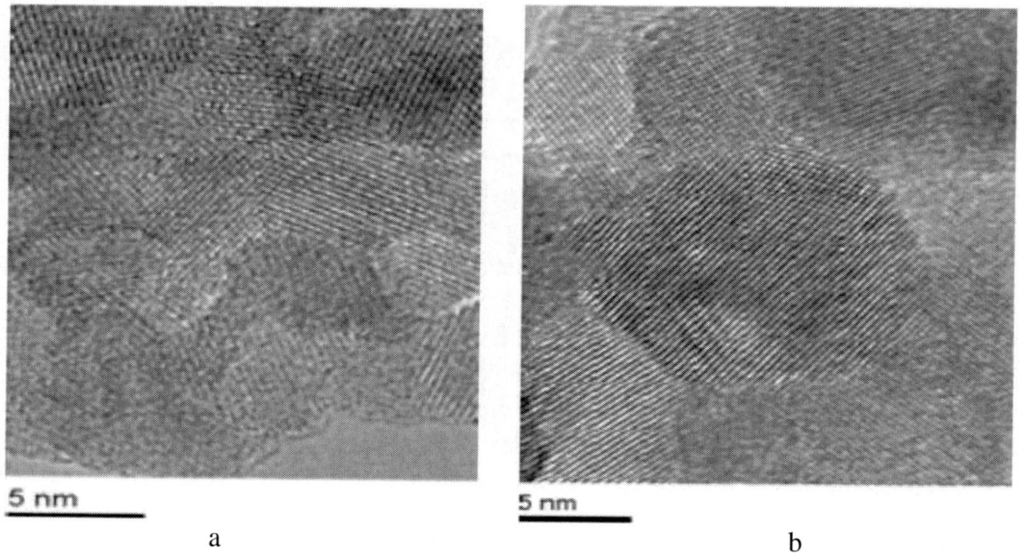

Figure 6. TEM micrographs of $Ce_{0.80}Zr_{0.20}O_{2.00}$ synthesized by gel combustion in (a) fuel deficient conditions, (b) stoichiometric amounts of fuel and oxidant.

Sometimes, slight modification of the existing synthetic route leads to superior powder properties. Mandal et al. [76] could improve the powder properties of Sm- and Gd-doped CeO_2 by addition of H_2O_2 during the co-precipitation synthesis of these materials.

This led to softer agglomerates, smaller crystallite size and higher surface area of the nanopowders. Better powder properties also resulted in the better catalytic properties of the products.

5. Ceria and Ceria–Based Materials As Ionic Conductors

Sustainability of the current methods of energy production and their environmental aspects are a major cause of concern today. One of the most environment-friendly means to produce energy, in the present-day world, is using fuel cells. Fuel cells are the electrochemical devices that generate power by conversion of chemical energy into electrical energy through electrochemical reactions. There are different types of fuel cells such as Alkaline Fuel Cell (AFC), Solid Oxide Fuel Cell (SOFC), Direct Methanol Fuel Cell (DMFC), Phosphoric Acid Fuel Cell (PAFC), Polymer Electrolyte Fuel Fell (PEFC) etc. The SOFC technology has certain advantages over others.

Foremost among them is that SOFC can use many common hydrocarbon fuels such as natural gas, diesel, gasoline and alcohol without the need to reform the fuels into pure hydrogen. In other fuel cells, such as the polymer electrolyte fuel cell, which uses pure hydrogen, the carbon dioxide is a poison.

Further, SOFC has a potentially lower cost as compared to proton exchange membrane and phosphoric acid fuel cells which use platinum as a catalyst. Some fuel cell types like those similar to battery acid use liquid electrolytes which would have corrosion of components, whereas since SOFCs use one piece solid state ceramic cells, they are easier to maintain due to the absence of such corrosion problem.

Potential life efficiency and extremely low emissions of CO in exhaust gases are other advantages of SOFC. Candidate electrolyte should possess minimum activation enthalpy (E_a) for the oxide ion motion and should have good ionic conductivity. Zirconia-based electrolytes, such as yttria-stabilised-zirconia (ZrO_2) (YSZ), are the most popular materials employed as the electrolyte in SOFCs nowadays because of reasonable good ionic conductivity, stability in both oxidizing and reducing environments, and compatibility with the electrode materials [77,78]. However, at temperatures below 1000°C the ionic conductivity of YSZ is not satisfactory. For reasons like long-term stability with low cost, decreasing the operation temperature down to 500–700°C range (termed as Intermediate Temperature SOFC) has been a worldwide interest for the researchers [77]. This is where, ceria-based electrolytes such as gadolinia-doped ceria (GDC), samaria-doped ceria (SDC) and yttria-doped ceria (YDC) have an important role to play. Ceria and doped ceria have been regarded as a lucrative class of electrolyte materials [79], as they exhibit high ionic conductivity coupled with the low activation energy in 400-800°C temperature range. In these doped fluorite-structured electrolytes, the current is carried by oxygen ions that are transported via oxygen vacancies, present to compensate for the lower charge of the dopant cations. The ionic conduction in ceria-based electrolytes is hence, closely related to the energetics of oxygen-vacancy formation and migration properties.

There has been a tremendous research in the past on the relationship between various factors, such as grain boundary, local structure, microdomain, grain size, composition, dopants, impurity, and processing and ionic conductivity of the system [77,78,80-82].

Various concepts were evoked to explain the ionic conductivity behavior with ionic size of the dopant and its concentration. It has been argued by various researchers [83-86] that to obtain high ionic conductivity, there has to be minimum size mismatch between dopant and the host lattice ion.

Also, the ionic conductivity values reach a maximum for a certain concentration of dopant [87]. The following is the brief discussion about the dependence of ionic conductivity on dopant size and concentration. Even though charge carriers are generated by aliovalent doping, it was shown that such vacancies were not free but bound to dopants in the form of defect clusters [81,88]. It has been shown, based on various conductivity and structural studies, that these defect clusters would have considerable influence on ionic conductivity in yttria doped ceria [89,90]. Thus, it is evident that dopant incorporation has a major influence on the transport behavior of oxygen conductors. The interactions between dopants and oxygen vacancies decreases the mobility of the vacancies. When larger defect clusters arise from dopant–vacancy interactions, they can act as deeper traps for oxygen vacancies and correspondingly influence the total ionic conductivity [91]. In many studies it has been verified that such clusters may order up and may further evolve into microdomains with short-range order, considerably affecting bulk conductivity in rare-earth doped ceria [92-94]. It is usually supposed that the energetics of the dopant–vacancy complex is dependent on two factors: The Coulombic term which refers to the electrostatic attraction between defects and dopants and the elastic interactions which depends on the relative sizes of the dopants and the host cations [91,95,96]. These two factors would in turn be controlled by the valence and ionic radius of the dopant cations relative to those of the host tetravalent cerium ions (Ce^{4+}).

There have been many studies exploring the relationship between different dopant radii and ionic conductivity [86, 97-99].

For a given dopant valence and concentration, the oxygen ionic conductivity of fluorite oxides has been found to vary with the dopant size [86,98,100]. If there is considerable size difference between the host and the dopant ions (bigger or smaller), a stress is accumulated in the host lattice due to size mismatch which blocks the path of migrating oxygen ions and consequently increase the ion-migration enthalpy [101]. This will eventually lead to a decrease of the ionic conductivity in doped fluorite oxides. Hence, it has been postulated that maximum ionic conductivity can be achieved with substituent dopants that cause little expansion or contraction of the host fluorite lattice [102], for the greater similarity between cationic sizes (host and dopant) will most plausibly discourage structural ordering . The size effect has also been explained based on the fact that at lower dopant concentrations, vacancy tends to lie within the first co-ordination shell of the smaller cation and the ordering is driven by substitutional disorder on cationic sublattice (cation-vacancy ordering) [103,104]. For higher dopant concentrations, vacancy-vacancy interactions are predominant and there is a tendancy for vacancy ordering along <111> direction [105]. However, it is also essential to take into account the effect of valence state on conductivity, since, as mentioned earlier, this factor will also affect defect cluster formation. This effect is however not so extensively researched and there is still debate on the nature of dopant–vacancy interaction and their relation to form distinct defect clusters. A new approach to the analysis of defect chemistry and its impact on ionic conductivity has been recently attempted by Nakamura [106].

The details about the conductivity studies on ceria doped with different rare earth ions and their comparison can be found in literature [87, 107-108].

The effects of multiple doping on the conductivity of ceria-based electrolytes have also been reported. Ralph et al. [109] and [110] have studied the effect of oxide additions on the total and grain boundary conductivity ($\sigma(\text{ion})_{gb}$) for Gd-doped ceria with compositions of $Ce_{1-x-y}Gd_xM_yO_2$ (M=Ca, Fe, Pr; x=0.09, 0.10, 0.18, 0.19, 0.20; y=0.01, 0.02). Small amounts of M-oxide additions have been found to significantly enhance the grain boundary conductivity compared to that of Gd-doped ceria without additions. Maricle et al. [111] have studied similar electrolyte systems with the composition of $Ce_{0.8}Gd_{0.19}RE_{0.01}O_{2-x}$ (RE=Pr, Sm) and demonstrated that these RE oxide additions also extended the electrolyte domain boundary to the lower oxygen partial pressure range (limit) by two orders of magnitude. The effect of addition of refractory oxides like alumina to ceria-based electrolytes has also been explored. Zhang et al. [112] examined the effect of alumina addition (0–10%) to 20% Gd doped ceria and found overall conductivity to be lesser in this range. This is explained due to a slight increase in grain interior conductivity but a rapid decrease in grain boundary conductivity observed with increasing alumina content. This is largely attributed to the solid state reaction between alumina and gadolinia, which in turn decreased the actual amount of gadolinia dissolved in ceria. Riess et al. [113] have clearly demonstrated that the ionic conductivity of substituted ceria strongly depends on the synthesis routes of the ceramics.

Also, electrolyte material should possess highest sintered density so as to allow smooth transfer of the conducting oxygen ions through the grain boundary. High sintered density is obtainable generally by high temperature heating. However, in case of ceria solid solutions, heating at high temperatures may also induce the problem of change of valence state of cerium ($Ce^{4+} \rightarrow Ce^{3+}$). One novel way of reducing the sintering temperatures is to use nanopowders of the candidate materials, for they can be sintered at relatively lower temperatures due to their high surface-to-volume ratio. A number of soft-chemical routes are employed to prepare nanopowders of solid oxide electrolyte materials.

The combustion technique is capable of producing nano-crystalline powders of oxide ceramics with highest degree of phase purity coupled with improved powder characteristics [114,115]. Bedekar et al. synthesized eight compositions in the $Ce_{1-x}Nd_xO_{2-x/2}$ system with varying Nd^{3+} concentrations ($0.0 \leq x \leq 0.50$) by the gel-combustion method and studied their ionic conductivity [116]. Ultra fine powders in the form of hard agglomerates were obtained which had profound effect on the various powder properties including sinterability. The conductivity studies were however performed only on systems containing 10 and 20 mol% Nd^{3+}, since at high Nd^{3+} concentration range the conductivity decreases with increasing concentration [80,117]. This has been attributed to higher defect concentration at higher substitution level, which leads to ordering due to dopant-vacancy interaction and vacancy-vacancy interaction [105], detrimental to ionic conductivity. The conductivity studies on neodymium (Nd) doped ceria systems showed that at lower temperatures (below 450°C) the conductivity was controlled by the grain boundary which has high activation energy (~ 1 eV). Properties of grain boundaries in the electrolyte such as impurity segregation and space charge play an important role in the ionic conductivity, especially at low temperatures. Certain additives can be used to minimize the blocking effect of grain boundaries to enhance the ionic conductivity. At higher temperatures (450-700°C) the ionic conductivity was contributed mainly by the grain interiors with lower activation energy (~0.7eV). The conductivity studies were also performed in hydrogen and oxygen atmosphere. Both the compositions were found to be stable in reducing atmosphere, thus implying negligible contribution of electronic conduction to the total electrical conductivity in the Nd doped ceria system. This aspect is very important since electronic conductivity is detrimental for electrolytes in SOFC applications as it leads to short-circuiting. It has also been reported and confirmed in many cases that doped ceria electrolytes exhibit electrical conductivity in air primarily due to oxide ion conduction and the effect of electronic conduction is almost negligible [118].

Andersson et al. [95] have performed a quantum-mechanical first-principles study of the influence of different trivalent dopants on dopant - oxygen vacancy interaction behavior and their results reveal remarkable correspondence between vacancy properties at the atomistic level and the macroscopic ionic conductivity. Their approach offers a simple way to select superior combination of dopants, and according to them the ideal dopant should have the atomic number between 61 (Pm) and 62 (Sm), and the combinations of Nd/Sm and Pr/Gd should show enhanced ionic conductivity, as compared with that for each single-kind of element. The important factors here are the migration barrier for bulk vacancy diffusion and vacancy–dopant interaction, the latter being represented by association energy of vacancy–dopant cluster(s). The molecular dynamics simulations, also, on ceria-yttria system were recently reported by Burbano et al. [105]

6. Ceria and Ceria Based Oxides As Catalysts

Ceria and ceria-based systems have been extensively investigated as catalysts particularly in the treatment of emission gases from internal combustion engines. Various potential applications include removal of soot from diesel engine emission and organics from wastewater and also as an additive for combustion catalysts for redox reactions.

The catalytic ability of ceria is due to its very novel property of an elevated oxygen transfer capacity associated with the redox (+3 ↔+4) property (ability to shift between reduced and oxidized states). In other words, the applicability of ceria as exhaust emission catalystis is due to its ability to undergo easy, fast, and reversible reduction to substoichiometric phases. Namely, an important function of ceria in three way catalyst is to take up and release oxygen in immediate response to the oxygen potential of the feed gas stream, being continuously and rapidly reduced and oxidized in cycles with its oxygen composition alternating between CeO_2 and CeO_{2-x}. This function is called oxygen storage/ release capacity and more generally it is referred to as *redox behavior* which accounts for all the characteristics of the reduction and oxidation processes of CeO_2. Structurally, this can be attributed to the stable rugged fluorite-type structure adopted by ceria because even after being stripped of oxygen ions and creation of lots of vacancies during reduction, the fluorite lattice of ceria is intact and that is primarily the reason for rapid filling of these vacancies upon oxygen uptake retaining the structure. Nevertheless, pure ceria is associated with various drawbacks like low thermal resistance and poor low-temperature activity. There have been numerous attempts to overcome these drawbacks by exploring various options like doping ceria with transition metal elements and rare earth ions and also utilizing noble metals in conjunction with ceria/doped ceria to modify the kinetics and enhance the redox properties at much lower temperatures.

Fluorite-structured ceria doped with zirconia are a new generation of three-way catalysts (TWC) for the treatment of noxious pollutants from car exhaust [119]. The prime features which contribute to the success of ceria–zirconia system are the associated higher thermal resistance, a higher reduction efficiency of the redox couple Ce(IV) –Ce(III) and the associated better oxygen storage and release capacity. These features result in higher efficiency of pollutant removal especially at low temperature during start up, where the majority of emissions are produced. In the metastable tetragonal phase, its t" phase (of zirconia) forms in $Ce_xZr_{1-x}O_2$ (for $x \leq 0.8$) [120]. In this particular phase the oxygen atoms are displaced from their ideal positions (of cubic lattice) in an ordered fashion, thus introducing distortion in oxide lattice at the same time maintaining the value c/a (cell parameters) at 1 despite this distortion. It has been reported that cubic lattice have better storage properties as compared to tetragonal ones [121]. This material is unique in that it has a high storage property typical of cubic lattice while at the same time it possesses distortion and structural defects necessary for oxide ion diffusion. It is this combination that bestows it with these enhanced catalytic properties. There is a huge amount of literature available for exploring the redox properties of ceria in relation to oxygen storage capacity [119, 122-129]. Likewise CeO_2-SiO_2 based fluorite type solid solutions have also been observed to possess [130-132] good oxygen storage capacities. This has also been attributed to textural and thermal stabilization imparted on ceria by introduction of silica. Mandal et al. [76] explored the physicochemical and the redox behavior of Sm^{3+} and Gd^{3+} - doped ceria and compared the behavior of these doped nanopowders synthesized in the presence and absence of H_2O_2. The presence of H_2O_2 was found to have a profound effect on powder properties such as surface area and particle size of these doped ceria samples and results in smaller crystallite size, softer agglomerates, and larger surface area attributable to the presence of peroxo-linkages formed in samples prepared in the presence of H_2O_2. Temperature programmed reduction studies showed the enhancement in the reduction properties of Sm-doped ceria and more reproducible behavior on performing multiple TPR cycles compared with pure ceria itself.

It was also shown that the systems prepared in the presence of H_2O_2 can lower the conversion temperature of CO to CO_2, proving them to be better catalysts [76].

Recently, some research works have also been initiated towards exploring doped ceria as photocatalysts for water splitting. The f-block metal oxides usually combined with other metal oxides act as photocatalysts. Pure CeO_2 powder was observed to show a consistent activity toward O_2 production in aqueous solutions containing Fe^{3+} and Ce^{4+} as electron acceptors [133]. It was observed that CeO_2 doped with Sr^{2+} was an active photocatalyst for overall water splitting when loaded with RuO_2 as a promoter [134]. Even zeolites supported with Ce(III) showed high photocatalytic activity for water splitting [134]. Photo-irradiation of Ce^{3+} species generated electrons that were captured effectively by water molecules for the nonstoichiometric H_2 and O_2 evolution.

7. Magnetic and Luminescent Properties of Ceria Based Systems

In the last 3-4 years, there has been a spur in the research exploring room temperature ferromagnetism in semiconducting or insulating oxides. In this respect, several groups have explored the potential of doped ceria in developing materials for dilute magnetic semiconductors. Dilute magnetic semiconductors have been projected to have applications in spintronic materials. CeO_2 is paramagnetic at ambient temperature and used as a key material in high K capacitors and as buffer layer in silicon on insulator devices [135–138].

Significantly, since CeO_2 is a face-centred cubic system possessing a fluorite crystal structure with closely matched lattice parameters to silicon, it has potential to facilitate the integration of next generation spintronic devices with current silicon based microelectronic devices. So, the existence of magnetism in this material would create a new direction for potential spintronic devices. As a consequence, there has been an upsurge in the studies exploring ferromagnetism in doped-ceria compounds. In fact there have been reports on magnetic properties of ceria at nano-size level. Fernandes et al. [139] have reported that the room temperature ferromagnetism in undoped ceria thin films is driven by both cerium and oxygen vacancies.

Dimri et al. [140] studied the magnetic properties of rare earth (RE) doped ceria (RE: Sm, Gd, Tb, Er and Dy). Room temperature ferromagnetism (FM) was observed in calcined powders as well as in sintered samples of Nd and Sm doped CeO_2, whereas other rare-earth dopants (Gd, Tb, Er and Dy) in CeO_2 exhibit paramagnetic behavior.

The origin of magnetism in these samples can be related to oxygen vacancies in the fluorite crystal structure. Jayakumar et al. [141] studied nanocrystalline $Ce_{1-x}Fe_xO_{2-x}$ particles with different Fe concentrations (x = 0.0, 0.05, 0.10, 0.15, and 0.20) synthesized by gel-combustion method. X-ray diffraction data revealed the formation of an impurity free $Ce_{1-x}Fe_xO_{2-x}$ products up to x = 0.15.

This observation is further confirmed from the detailed studies conducted on 10 atom% Fe doped CeO_2 using High-resolution transmission electron microscopy (HRTEM) imaging, Selected-Area Electron Diffraction (SAED) and Raman spectroscopy. DC magnetization studies as a function of field strength and temperature indicate that they are ferromagnetic with Curie temperature (T_c) well above room temperature.

As a new class of luminescent materials, rare earth-doped nanoparticles have technological applications in lighting, display, scintillator devices, and fluorescent probes as well as light sources for cancer treatment. It has been a strategy to find a suitable host lattice for a phosphor material and to dope it with visible light emitters (activators) to obtain solid state phosphors.

CeO_2, with no 4f electron has been explored as a promising photoluminescence host material because of strong light absorption through the charge transfer from O^{2-} to Ce^{4+}. If this energy can be transferred to the doped activators in ceria, characteristic emissions from the latter would be observed and a correct combination of the dopants can lead to a white light source. Rare earth ions are the activators of choice because of sharp f-f transitions. Moreover, the comparable ionic radius of cerium ions and the rare earth ions favors the latter's extensive solubility in ceria matrix.

There has been a decent number of studies exploring ceria as the host lattice for phosphor. Yoshida et al. [142] studied hydrothermally synthesized CeO_2: Sm^{3+} nanopowders, wherein they showed that luminescence properties are greatly affected by their shapes. The different photoluminescent properties were attributed to the oxygen defects. Li et al. [143] investigated Gd^{3+}-doped CeO_2 and showed that the luminescence properties were enhanced due to oxygen vacancies. For the past few years, the upconversion luminescence of infrared light to visible light of different materials has been investigated extensively for various applications like all-solid compact laser devices operating in the violet–blue–green region and upconversion phosphors etc.

In this process, two (or more) low-energy photons from excitation source are converted into one photon with higher energy. Ceria is a promising host material for upconversion because of its excellent thermal stability and low phonon cutoff energy (465cm^{-1}) [144-146] which would decrease the possibility of nonradiative transitions and in turn result in a high quantum yield of upconverion process. Among all the options available, Er^{3+} ion is the most popular as well as one of the most efficient ions for upconversion [147-149]. Er^{3+}-doped CeO_2 has been investigated as upconversion phosphor which showed quantum yield of 3% and the upconversion efficiency increased with increasing temperature. Doping a suitable host lattice by different activators which emit complementary colors is an important way to design a white light emitting phosphor. Dutta et al. [150] studied doped ceria nanophosphors with the aim of synthesizing white light phosphors wherein they could alter the content of Dy^{3+}, Tb^{3+} and Eu^{3+} to control the light emission.

8. Nuclear Applications

It would be appropriate to mention that fluorite-type structured material forms the backbone of nuclear industry. The basic structural type adopted by fuels in majority of present day reactor is fluorite. The fluorite structure is chosen because of the fact that there are many unoccupied interstices in the fluorite structure. During the progress of fission reaction to produce energy in the reactors, many fission products are produced. The fluorite structure has the ability to accommodate fission products in these voids, thereby reducing problematic fuel swelling.

Other than possessing favorable characteristics as nuclear materials like high melting point, chemical stability etc., fluorite structure also displays tremendous radiation stability which can be attributed to its capability to accommodate large number of defects without undergoing amorphisation. Because of these characteristics it is the structure of choice for nuclear materials. However, most of the nuclear materials are associated with radio-toxicity and hence cannot be handled easily. As mentioned earlier, cerium is widely used as an inactive surrogate for plutonium. Tetravalent cerium has a very similar ionic radius to tetravalent plutonium (for a given coordination number) making it a good surrogate from a crystallographic viewpoint. Cerium oxide has thus also attracted a lot of attention in the past as a good surrogate material for studying uranium and plutonium oxides and mixed oxide (MOX) fuels. Namely, CeO_2 has very similar physico-chemical properties such as high melting temperature (UO_2: ~2870°C, CeO_2: ~2600°C, PuO_2: ~2662 °C), and thermal diffusivity etc. to those of these very important nuclear materials [151-155]: Hence, ceria provides a viable non radiotoxic option for studying PuO_2 and UO_2 based systems like MOX fuel.

Tremendous work has been done on studying the irradiation stability of undoped and doped ceria from point of view of studying the behavior of ceria-based matrices under the radiation environment in reactors as well as its application in waste management. Oxides with fluorite structure form a large class of compounds with a high radiation tolerance, which is again related to their unique ability to accommodate a variety of defects and to form nonstoichiometric compounds with large homogeneity range [156]. Stabilized zirconia which has the fluorite structure similar to ceria is the most favored host lattice for nuclear materials in the advanced concepts like inert matrix fuel (IMF) for plutonium utilization and actinide transmutation. Numerous studies have been performed on stabilized zirconia, zirconia-ceria (to simulate zirconia – plutonia/urania), zirconia and ceria doped with rare earth oxides like Er_2O_3 (to simulate burnable poisons) etc. They have been explored for their chemical stability, mechanical stability, leach resistance, thermal conductivity etc. [157-160]. X-ray diffraction, electron microscopy (TEM and SEM), Raman spectroscopy etc. have been employed to study the nature of post-irradiation defects introduced in those systems. All the studies have supported the relatively high radiation tolerance of these fluorite-based matrices [161-163]. It has, in fact, been speculated that that compounds with superstructures related to fluorite, such as bixbyite and pyrochlore, will transform to disordered defect-fluorite structures upon irradiation. Then, provided compound chemistry is favorable to the accommodation of lattice disorder, these compounds will persist as crystalline fluorites to very high radiation doses [164].

CONCLUSION

Ceria and ceria-based materials have attracted great interest of the material science community worldwide for the tremendous application potential. The outcomes of such research has had in the past, and continues to have large impact on the associated technologies, be it the enhancement of ionic conductivities of the currently-available electrolytes for intermediate-temperature solid oxide fuel cell (IT-SOFC) technology or improvement of the catalytic efficiency for removing noxious pollutants, and so on. All these ultimately originate from the very versatile fluorite structure adopted by ceria and ceria-based solid solutions.

As is described here, ceria with fluorite structure is highly tolerant to dissolution of various lower-valent metal oxides which have the capability to greatly enhance the functionalities of ceria. The facts that the enhancement in catalytic or the condution properties could be attained while retaining the fluorite-structure make these ceria-based solid solutions particularly attractive. It is the amenability of fluorite structure to oxygen ion movement and the capability of Ce^{4+} to switchover between +3 and +4 oxidation states backed up by the inherent stability of fluorite structure that has bestowed ceria and related systems with these marvelous properties. In this chapter we tried to review and summarize the past and ongoing research and development on ceria and ceria-based systems in terms of their basic fluorite structure in various (nano-particle and sintered bulk) forms and its bearing on various chemical (electrochemical, catalytic etc.) properties and the practical applications based on them.

ACKNOWLEDGMENTS

Dr. B. P. Mandal, Dr. S. V. Chavan, Dr. Vinila Bedekar and Mr. Rakesh Shukla, are thanked for discussion on various ceria-based systems.

REFERENCES

[1] M. O'Keeffee and B.G. Hyde, *Crystal Structures I. Patterns and Symmetry*, Mineralogical Society of America, Washington, US, 1996.
[2] Z.C. Kang and L. Eyring, *J. Alloys Compd.* (1997) 249, 206-212.
[3] Z.C. Kang and L. Eyring, *J. Alloys Compd.* (1998) 275, 30-36.
[4] G. Adachi and N. Imanaka, *Chem. Rev.* (1998) 98, 1479-1514.
[5] S.V. Krivovichev, *Solid State Sciences* (1999) 1, 221-231.
[6] G. Adachi and N. Imanaka, *Chem. Rev.* (1998) 98, 1479-1514.
[7] R. Rao and B.G. Mishra, *Bulletin of the Catalysis Society of India* (2003) 2, 122-134.
[8] G.R. Rao, *Bull. Mater. Sci.* (1999) 22, 89-94.
[9] G.R. Rao, J. Kašpar, S. Meriani, R.D. Monte and M. Graziani, *Catal. Lett.* (1994) 24, 107-112.
[10] P. Fornasiero, R. Di Monte, G.R. Rao, J. Kašpar, S. Meriani, A. Trovarelli and M. Graziani, *J. Catal.* (1995) 151, 168-177.
[11] S. Zec, S. Boskovic, M. Hrovat, M. Kosec, *J. European Ceram. Soc.* (2007) 27, 523–526.
[12] G.R. Rao, P. Fornasiero, R. Di Monte, J. Kašpar, G. Vlaic, G. Balducci, S. Meriani, G. Gubitosa, A. Cremona and M. Graziani, *J. Catal.* (1996) 162, 1-9.
[13] P. Fornasiero, G.R. Rao, J. Kašpar, F.L. Erario and M. Graziani, *J. Catal.* (1998) 175, 269-279.
[14] V. Perrichon, A. Laachir, G. Bergeret, R. Frety and L. Tournayan, *J. Chem. Soc. Faraday Trans.* (1994) 90, 773-781.

[15] J. Kašpar, M. Graziani and P. Fornasiero, "*Handbook on the Physics and Chemistry of Rare Earths*", K.A. Gschneidner, Jr. and L. Eyring, eds, Elsevier Science B.V., Amsterdam (2000) 29, 159-267.
[16] A. Troveralli, *Catal. Rev. Sci. Eng.* (1996) 38, 439-520.
[17] W. Liu, and M. Flytzani-Stephanopoulos, *Chem. Eng. J.* (1996) 64, 283-294.
[18] Y. Li, Q. Fu and M. Flytzani-Stephanopoulos, *Appl. Catal. B: Environ.* (2000) 27, 179-191.
[19] A. Trovarelli, C. de Leitenburg, M. Boaro and G. Dolcetti, *Catal. Today* (1999) 50, 353 -367.
[20] X. Wang and R.J. Gorte, *Appl. Catal. A: Gen.* (2003) 247, 157-162.
[21] J.A. Montoya, E. Romero-Pascual, C. Gimon, P. Del Angel and A. Monzón, *Catal. Today* (2000) 63, 71-85.
[22] N. Hickey, P. Fornasiero, J. Kapar, M. Graziani, G. Martra, S. Coluccia, S. Biella, L. Prati and M. Rossi, *J. Catal.* (2002) 209, 271-274.
[23] S. Imamura, Y. Okumura, T. Nishio, K. Utani and Y. Matsumura, *Ind. Eng. Chem. Res.* (1998) 37, 1136-1139.
[24] G. Neri, A. Pistone, C. Milone and S. Galvagno, *Appl. Catal. B: Environ.* (2002) 38, 321-329.
[25] M.A. Henderson, C.L. Perkins, M.H. Engelhard, S. Thevuthasan and C.H.F. Peden, *Surf. Sci.* (2003) 526, 1-18.
[26] Z. Xu, Z. Qi and A. Kaufman, *J. Power Sources* (2003) 115, 40-43.
[27] S. Zha, J. Cheng, Q. Fu and G. Meng, *Mater. Chem. Phys.* (2003) 77, 594-597.
[28] S. Meriani, *Mat. Sci. Eng. A* (1989) 109, 121-130.
[29] J.W. Holmes and B.H. Pilsner, "Thermal spray", *Proceedings of the National Thermal Spray Conference*, ASM (1988) 259-270.
[30] G. Kim, *Ind. Eng. Chem. Prod. Res. Dev.* (1982) 21, 267-274.
[31] K. Eguchi, T. Setoguchi, T. Inoue and H. Arai, *Solid State lonics* (1992) 52, 165-172.
[32] M.P. Woods, P. Gawade, B. Tan, U.S. Ozkan, *Appl. Catal. B: Environ.* (2010) 97, 28-35.
[33] G.J. Kašpar, P. Fornasiero and N. Hickey, *Catal. Today* (2003) 77, 419-449.
[34] D.R. Ou, T. Mori, F. Ye, J. Zou, G. Auchterlonie and J. Drennan, *Phys. Rev. B* (2008) 77, 024108.
[35] F. Esch, S. Fabris, L. Zhou, T. Montini, C. Africh, P. Fornasiero, G. Comelli and R. Rosei, *Science* (2005) 309, 752-755.
[36] X.W. Liu, K.B. Zhou, L. Wang, B.Y. Wang and Y.D. Li, *J. Am. Chem. Soc.* (2009) 131, 3140-3141.
[37] S.P. Li, J.Q. Lu, P. Fang and M.F. Luo, *J. Power Sources* (2009) 193, 93-98.
[38] T. Arima, S. Yamasaki, S. Torikai, K. Idemitsu, Y. Inagaki and C. Degueldre, *J. Alloys Compd.* (2005) 398, 296-303.
[39] J. Porta and M. Asou, *Prog. Nucl. Ener.* (2001) 38, 347-350.
[40] R.D. Shannon, *Acta Cryst. A* (1976) 32, 751-767.
[41] D.J.M. Bevan, E. Summerville, In: K.A. Gschneider, L. Eyring (Eds.), *Handbook on the Physics and Chemistry on Rare Earths* (1979) 3, 401-524.
[42] M. Hrovat, J. Holc, S. Bernik and D. Makovec, *Materials Research Bulletin* (1998) 33, 1175-1183.

[43] E.R. Andrievskaya, O.A. Kornienko, A.V. Sameljuk and A. Sayir, *J. European Ceram. Soc.* (2011) 31, 1277-1283.
[44] S.V. Chavan, M.D. Mathews and A.K. Tyagi, *Mater. Res. Bull.* (2005) 40, 1558-1568.
[45] B.P. Mandal, V. Grover and A.K. Tyagi, *Mat. Sci. Eng. A.* (2006) 430, 120-124.
[46] A. Nakamura, N. Masaki, H. Otobe, Y. Hinatsu, J. Wang and M. Takeda, *Pure Appl. Chem.* (2007) 79, 1691–1729.
[47] V. Grover and A.K. Tyagi, *Mater. Res. Bull.* (2004) 39, 859-866.
[48] V. Grover, S.N. Achary and A.K. Tyagi, *J. Appl. Cryst.* (2003) 36, 1082-1084.
[49] B.P. Mandal, M. Roy, V. Grover and A.K. Tyagi, *J. Appl. Phys.* (2008) 103, 033506.
[50] B.P. Mandal, V. Grover, M. Roy and A.K. Tyagi, *J. Am. Ceram. Soc.* (2007) 90, 2961-2965.
[51] V. Grover, A. Banerji, P. Sengupta and A.K. Tyagi, *J. Solid State Chem.* (2008) 181, 1930-1935.
[52] R. Shukla, A. Arya and A.K. Tyagi, *Inorg. Chem.* (2010) 49, 1152-1157.
[53] S.V. Chavan, M.D. Mathews and A.K. Tyagi, *J. Am. Ceram. Soc.* (2004) 87, 1977-1980.
[54] A. Nakajima, A. Yoshihara and M. Ishigame, *Phys. Rev. B.* (1994) 50, 13297.
[55] C. Kittel, *Introduction to solid state physics*, 4th Ed. Wiley, New York (1971) p. 460.
[56] M. Yashima, H. Arashi, M. Kakihana and M. Yoshimura, *J. Am. Ceram. Soc.* (1994) 77, 1067-1071.
[57] A. Banerji, V. Grover, V. Sathe, S.K. Deb and A.K. Tyagi, *Solid State Comm.* (2009) 149, 1689-1692.
[58] M.M. A. Sekar, S.S. Manoharan and K.C. Patil, *J. Mater. Sci. Lett.* (1990) 9, 1205-1209.
[59] Pei-Lin and I.-W. Chen., *J. Am. Ceram. Soc.* (1993) 76, 1577-1583.
[60] V.A. Sadykov, T.G. Kuznetsova, G.M. Alikina, Y.V. Frolova, A.I. Lukashevich, Y.V. Potapova, V.S. Muzykantov, V.A. Rogov, V.V. Kriventsov, D.I. Kochubei, E.M. Moroza, D.I. Zyuzin, V.I. Zaikovskii, V.N. Kolomiichuk, E.A. Paukshtis, E.B. Burgina, V.V. Zyryanov, N.F. Uvarov, S. Neophytides and E. Kemnitz, *Catal. Today* (2004) 93, 45-53.
[61] D. Zhang, H. Fu, L. Shi, C. Pan, Q. Li, Y. Chu and W. Yu, *Inorg. Chem.* (2007) 46, 2446-2451.
[62] D.Y. Chung and E.H. Lee, *J. Alloys and Compds.* (2004) 374, 69-73.
[63] C.S. Wright, R.I. Walton, D. Thompsett, J. Fisher and S.E. Ashbrook, *Adv. Mater.* (2007) 19, 4500-4504.
[64] Y. Zhou and M.N. Rahaman, *Acta Mater.* (1997) 45, 3635-3639; Z. Rahman, *J. Mater. Res.* (1993) 8, 1680-1686.
[65] J.J. Kingsley, K. Suresh and K.C. Patil, *J. Mater. Sci.* (1990) 25, 1305-1312.
[66] V.M. Ferreira, F. Azough, J.L. Baptista and R. Freer, *Ferroelectrics* (1992) 133, 127-132.
[67] S.S. Manoharan and K.C. Patil, *J. Am. Ceram. Soc.* (1992) 75, 1012-1015.
[68] J.J. Kingsley and L.R. Pederson, *Mater. Lett.* (1993) 18, 89-96.
[69] Y. Zhang and G.C. Stangle, *J. Mater. Res.* (1994) 9, 1997-2004.
[70] S. Bhaduri, S.B. Bhaduri and E. Zhou, *J. Mater. Res.* (1998) 13, 156-165.
[71] S.V. Chavan and A.K. Tyagi, *J. Mater. Res.* (2004) 19, 3181-3188.

[72] R.D. Purohit, B.P. Sharma, K.T. Pillai and A.K. Tyagi, *Mater. Res. Bull.* (2001) 36, 2711-2721.
[73] R.D. Purohit, S. Saha and A.K. Tyagi, *Ceramics International* (2006) 32, 143-146.
[74] S.T. Mukherjee, V. Bedekar, A. Patra, P.U. Sastry and A.K. Tyagi, *J. Alloys Compd.* (2008) 466, 493-497.
[75] V. Grover, S.V. Chavan, P.U. Sastry and A.K. Tyagi, *J. Alloys Compds.* (2008) 457, 498-505.
[76] B.P. Mandal, V. Grover, M.R. Pai and A.K. Tyagi, *J. Mater. Res.* (2009) 24, 2845-2854.
[77] N.Q. Minh, *Science,Technology of Ceramic Fuel Cells*, Elsevier, Amsterdam, 1995.
[78] J.B. Goodenough, *Ann. Rev. Mater. Res.* (2003) 33, 91-128.
[79] A. Tsoga, A. Naoumidis, W. Jungen and D. Stover, *J. Eur. Ceram. Soc.* (1999) 19, 907-912.
[80] H. Inaba and H. Tagawa, *Solid State Ionics* (1996) 83, 1-16.
[81] M. Mogensen, N.M. Sammes and G.A. Tompsett, *Solid State Ionics* (2000) 129, 63-94.
[82] B.C.H. Steele, *Solid State Ionics* (2000) 129, 95-110.
[83] J.A. Kilner, In *Solid State Chemistry*, R. Metselaar, H.J.M. Heijingers, J. Schoonman eds., Elsevier Science Ltd., Amsterdam, 1982.
[84] D.J. Kim, *J. Am. Ceram. Soc.* (1989) 72, 1415-1421.
[85] J.A. Kilner and R.J. Brook, *Solid State Ionics* (1982) 6, 237-252.
[86] C.R.A. Catlow, *Solid State Ionics* (1984) 12, 67-73.
[87] R.V. Kumar, H. Iwahara, Eds. K.A. Gschneidner, L. Eyring, *Handbook on the Physics and Chemistry of Rare-earths*, 28 (2000)pp. 131-185.
[88] K. Huang, M. Feng, J.B. Goodenough, *J. Am. Ceram. Soc.* (1998) 81, 357-362.
[89] P. Li, I. Chen, J.E. Hahn, T. Tien, *J. Am. Ceram. Soc.* (1991) 74, 958-967.
[90] S.B. Adler, J.W. Smith, *J. Chem. Soc. Faraday Trans.* (1993) 89, 3123-3128.
[91] J.A. Kilner, *Chem. Lett.* (2008) 37, 1012-1015.
[92] Z.P. Li, T. Mori, F. Ye, D.R. Ou, J. Zou and J. Drennan, *Microsc. Microanal.* (2011) 17, 49-53.
[93] Z.P. Li, T. Mori, G. Auchterlonie, J. Zou and J. Drennan, *Appl. Phys. Lett.* (2011) 98, 093104.
[94] Z.P. Li, T. Mori, G. Auchterlonie, J. Zou and J. Drennan, *Mater. Res. Bull.* (2012) 47, 763-767.
[95] D.A. Andersson, S. I. Simak, N.V. Skorodumova, I.A. Abrikosov, and B. Johansson, *Proc Natl. Acad. Sci., US.* (2006) 103, 3518-3521.
[96] K. Muthukkumaran, R. Bokalawela, T. Mathews, S. Selladurai, *J. Mater. Sci.* (2007) 42, 7461-7466.
[97] D.Y. Wang, D.S. Park, J. Griffith, A.S. Nowick, *Solid State Ionics* (1981) 2, 95- 105.
[98] J. Kimpton, T.H. Randle and J. Drennan, *Solid State Ionics* (2002) 149, 89-98.
[99] M. Hartmanova, E.E. Lomonova, F. Kubel, J. Schneider, V. Bursikova, M. Jergel, V. Navratil and F. Kundracik, *J. Mater. Sci.* (2009) 44, 234-243.
[100] M. Yashima and T. Takizawa, *J. Phys. Chem. C* (2010) 114, 2385-2392.
[101] R.J. Stafford, S.J. Rothman and J.L. Routbort, *Solid State Ionics* (1989) 37, 67-72.
[102] S.P.S. Badwal and F.T. Ciacchi, *Ionics* (2000) 6, 1-21.
[103] S.T. Norberg, S. Hull, I. Ahmed, S.G. Eriksson, D. Marrocchelli, P.A. Madden, P. Li, and J.T.S. Irvine, *Chem Mater.* (2011) 23, 1356-1364.

[104] D. Marrocchelli, P.A. Madden, S.T. Norberg and S. Hull, *Chem. Mater.* (2011) 23, 1365–1373.
[105] B.M. Burbano, S.T. Norberg, S. Hull, S.G. Eriksson, D. Marrocchelli, P.A. Madden, and G.W. Watson, *Chem. Mater.* (2012) 24, 222–229.
[106] A. Nakamura, *Solid State Ionics* (2010) 181, 1543–1564; A. Nakamura, *Solid State Ionics* (2010) 181, 1631–1653.
[107] J. Faber, G. Geoffroy, A. Roux, A. Sylvestre and P. Abelard, *Appl. Phys. A* (1989) 49, 225-232.
[108] H. Yahiro, Y. Eguchi, K. Eguchi and H. Arai, *J. Appl. Electrochem.* (1988) 18, 527-531.
[109] J.M. Ralph and J.A. Kilner, In *Proc. of SOFC-V.,* U. Stimming, S.C. Singhal, H. Tagawa, W. Lehnert (Eds.), *The Electrochem. Soc.* (1997) p. 1021.
[110] J.M. Ralph, J.A. Kilner and B.C.H. Steele, *MRS Symp. Proc.* (2000) 575, 309-314.
[111] D.L. Maricle, T.E. Swarr and S. Karavolis, *Solid State Ionics* (1992) 52, 173-182.
[112] T. Zhang, Z. Zeng, H. Huang, P. Hing and J.A. Kilner, *Mater. Lett.* (2002) 57, 124-129.
[113] I. Riess, D. Braunshtein and D.S. Tannhauser, *J. Am. Ceram. Soc.* (1981) 64, 479-485.
[114] J.J. Kingsley, K. Suresh and K.C. Patil, *J. Mater. Sci.* (1990) 25, 1305-1312.
[115] S. Bhaduri, S.B. Bhaduri and E. Zhou, *J. Mater Res.* (1998) 13, 156-165.
[116] V.V. Bedekar, S. Patra, A. Dutta, R.N. Basu and A.K. Tyagi, *Int. J. Nanotechnol.* (2010) 7, 1178-1187.
[117] J. Hormes, M. Pantelouris, G.B. Balazs and B. Rambabu, *Solid State Ionics* (2000) 136, 945-954.
[118] Z. Zhan, T.L. Wen, H. Tu and Z.Y. Lu, *J. Electrochem. Soc.* (2001) 148, 427-432.
[119] J. Kaspar, P. Fornasiero, M. Graziani, *Catal. Today* (1999) 50, 285-298.
[120] A. Trovarelli, F. Zamar, J. Llorca, C. de Leitenburg, G. Dolcetti, J. Kiss, *J. Catal.* (1997) 169, 490-502.
[121] B.K. Cho, *J. Catal.* (1991) 131, 74-87.
[122] A. Trovarelli, *Commun. Inorg. Chem.* (1999) 20, 263-284.
[123] P. Fornasiero, R. Di Monte, G. Ranga Rao, J. Kaspar, S. Meriani, A Trovarelli and M. Graziani, *J. Catal.* (1995) 151, 168-177.
[124] C.E. Hori, H. Permana, K.Y.S. Ng, A. Brenner, K. More, K.M. Rahmoeller, D. Belton, *Appl. Catal., B* (1998) 16, 105-117.
[125] H. Vidal, J. Kaspar, M. Pijolat, G. Colon, S. Bernal, A. Cordon, V. Perrichon and F. Fally, *Appl. Catal. B: Environ.* (2000) 27, 49-63.
[126] Y. Madier, C. Descorme, A.M. Le Govic, D. Duprez, *J. Phys. Chem. B* (1999) 103, 10999-11006.
[127] H.-W. Jen, G.W. Graham, W. Chun, R.W. McCabe, J.-P. Cuif, S.E. Deutsch and O. Touret, *Catal. Today* (1999) 50, 309-328.
[128] J. R. Gonzales-Velasco, M.A. Gutierrez-Ortiz, J.-L. Marc, J.A. Botas, M. Pilar Gonzales Marcos and G. Blanchard, *Appl. Catal. B: Environ.* (2000) 25, 19-29.
[129] M. Ozawa and C.-K. Loong, *Catal. Today* (1999) 50, 329-342.
[130] L. Kepinski and M. Wolcyrz, *J. Solid State Chem.* (1997) 131, 121-130.
[131]M. Haneda, T. Mizushima, N. Kakuta and A. Ueno, *Bull. Chem. Soc. Jpn.* (1994) 67, 2617.
[132] R. Craciun, *Solid State Ionics* (1998) 110, 83-89.

[133] G.R. Bamwenda, T. Uesigi, Y. Abe, K. Sayama and H. Arakawa, *Appl. Catal. A: Gen.* (2001) 205, 117-128.
[134] H. Kadowaki, N. Saito, H. Nishiyama and Y. Inoue, *Chem. Lett.* (2007) 36, 440-441.
[135] A. Thurber, K.M. Reddy, V. Shutthanandan, M.H. Engelhard, C. Wang, J. Hays and A. Punnoose, *Phys.Rev. B* (2007) 76, 165206.
[136] T. Inoue, T. Ohsuna, L. Luo, X.D. Wu, C.J. Maggiore, Y. Yamamoto, Y. Sakurai and J. H. Chang, *Appl. Phys. Lett.* (1991) 59, 3604-3606.
[137] Y. Nishikawa, T. Yamaguchi, M. Yoshiki, H. Satake and N. Fukushima, *Appl. Phys. Lett.* (2002) 81, 4386-4388.
[138] J. Kang, X. Liu, G. Lian, Z. Zhang, G. Xiong, X. Guan, R. Han and Y. Wang, *Microelectron. Eng.* (2001) 56, 191-194.
[139] V. Fernandes, P. Schio, A. J.A. de Oliveira, W.A. Ortiz, P. Fichtner, L. Amaral, I.L. Graff, J. Varalda, N. Mattoso, W.H. Schreinerand and D.H. Mosca, *J. Phys.: Condens. Matter* (2010) 22, 216004.
[140] M. Chandra Dimri, H. Khanduri, H. Kooskora, J. Subbi, I. Heinmaa, A. Mere, J. Krustok and R. Stern, *Phys. Status Solidi A* (2012) 209, 353-358.
[141] O.D. Jayakumar, A. Vinu, K.V. Guduru, T. Sakuntala and A.K. Tyagi, *J. Nanosci. Nanotechnol.* (2010) 10, 2299-2303.
[142] A.Y. Yoshida, S. Fujihara, *Eur. J. Inorg. Chem.* (2011) 10, 1577–1583.
[143] G.Li, D. Qu, L. Arurault and Y. Tong, *J. Phys. Chem. C* (2009) 113, 1235–1241.
[144] C. Ho, J.C. Yu, T. Kwong, A.C. Mak and S. Lai, *Chem. Mater.* (2005) 17, 4514-4522.
[145] S. Wang, W. Wang, J. Zuo and Y. Qian, *Mater. Chem. Phys.* (2001) 68, 246-248.
[146] W.H. Weber, K.C. Hass and J.R. McBride, *Phys. Rev. B* (1993) 48, 178-185.
[147] D. Matsura, *Appl. Phys Lett.* (2002) 81, 4526-4528.
[148] H. Lin, G. Meredith, S. Jiang, X. Peng, T. Luo, N. Peyghambarian and E. Pun, *J. Appl. Phys.* (2003) 93, 186-191.
[149] H. Guo, N. Dong, M. Yin, W. Zhang, L. Lou and S. Xia, *J. Phys. Chem. B* (2004) 108, 19205-19209.
[150] D.P. Dutta, N. Manoj, A.K. Tyagi, *J. Luminescence* (2011) 131, 1807–1812.
[151] T. Sonoda, M. Kinoshita, Y. Chimi, N. Ishikawa, M. Sataka, A. Iwase, *Nuclear Instruments and Methods in Physics Research B* (2006) 250, 254–258.
[152] R.J.M. Konings, K. Bakker, J.G. Boshoven, R. Conrad and H. Hein, *J. Nucl. Mater.* (1998) 254, 135–142.
[153] K. Yasunaga, K. Yasuda, S. Matsumura and T. Sonoda, *Nuclear Instruments and Methods in Physics Research B* (2006) 250, 114–118.
[154] K. Yasunaga, K. Yasuda, S. Matsumura and T. Sonoda, *Nuclear Instruments and Methods in Physics Research B* (2008) 266, 2877–2881.
[155] T. Sonoda, M. Kinoshita, N. Ishikawa, M. Sataka, Y. Chimi, N. Okubo, A. Iwase and K. Yasunaga, *Nuclear Instruments and Methods in Physics Research B* (2008) 266, 2882–2886.
[156] G. Baldinozzi, L. Desgranges, D. Simeone, D. Gosset and L. Luneville, *MRS Proceedings* (2012) 1383.
[157] P.G . Medvedev, M.J. Lambregts and M.K. Meyer, *J. Nucl. Mater.* (2006) 349, 167-177.
[158] C. Ronchi, J.P. Ottaviani, C. Degueldre and R. Calabrese, *J. Nucl. Mater.* (2003) 320, 54-65.

[159] C. Degueldre, T. Arima and Y.W. Lee, *J. Nucl. Mater.* (2003) 319, 6-14.

[160] N.K. Kulkarni, K. Krishnan, U.M. Kasar, S.K. Rakshit, S.K. Sali and S.K. Aggarwal, *J. Nucl.Mater.* (2009) 384, 81-86.

[161] K. E. Sickafus, H. Matzke, T. Hartmann, K. Yasuda, J. A. Valdez, P. Chodak III, M. Nastasi and R. A. Verrall, *J. Nucl. Mater.* (1999) 274, 66-77.

[162] N. Yu, K.E. Sickafus, P. Kodali and M. Nastasi, *J. Nucl. Mater.* (1997) 244 266-272.

[163] K. Yasuda, M. Nastasi, K.E. Sickafus, C.J. Maggiore and N. Yu, *Nucl. Instr. and Meth. B* (1998) 136–138, 499-504.

[164] K.E. Sickafus, J.A. Valdez, J.R. Williams, R.W. Grimes and H.T. Hawkins, *Nucl. Instr. and Meth. Phys. Res. B* (2002) 191, 549–558.

In: New Research Trends of Fluorite-Based Oxide Materials ISBN: 978-1-63117-350-9
Editors: Akio Nakamura and Junichiro Mizusaki

Chapter 3

CRYSTAL STRUCTURE, STRUCTURAL DISORDER AND OXIDE-ION DIFFUSIONAL PATHWAY OF THE FLUORITE-TYPE OXIDES AND FLUORITE-RELATED PHASES

Masatomo Yashima*
Department of Chemistry and Materials Science,
Tokyo Institute of Technology, Tokyo, Japan

ABSTRACT

Many fluorite-type structured oxides and fluorite-related materials such as zirconia, ceria and bismuth oxide exhibit excellent electrical and mechanical properties. The material properties are strongly dependent on the crystal structure. In this chapter, I describe and review the crystal structure, structural disorder and oxide-ion diffusional pathway of fluorite-type structured oxides. The combined procedure of the Rietveld method, maximum-entropy method (MEM) and MEM-based pattern fitting (MPF) using the high-temperature neutron powder diffraction data is powerful to deduce the accurate nuclear-density distribution, which enables the visualization of positional disorder and oxide-ion diffusional pathway. The diffusional pathway along the <100> directions is observed in the oxide-ion conductors with the fluorite-type structure such as $Bi_{1.4}Yb_{0.6}O_3$, $Ce_{0.927}Y_{0.073}O_{1.962}$ and $Y_{0.785}Ta_{0.215}O_{1.715}$.

1. INTRODUCTION

The fluorite is the mineral name of calcium fluoride CaF_2. The fluorite-type structure is simple: In the fluorite-structured AX_2 compound, the A and X atoms are located at 4*a* (0, 0,0) and 8*c* (1/4, 1/4, 1/4) sites, respectively, in the cubic $Fm\bar{3}m$ space group (Figure. 1). The A

* E-mail address: yashima@cms.titech.ac.jp, Department of Chemistry and Materials Science, Tokyo Institute of Technology, 2-12-1-W4-17, O-okayama, Meguro-ku, Tokyo, 152-8551, Japan.

atom is located at the corner and at the face-centered position. Namely, the A atoms form the fcc lattice, and the X atoms form the simple cubic lattice. The coordination numbers of A and X atoms are 8 and 4, respectively. The fluorite-type AX_2 structure can be drawn with AX_8 cubes (Figure. 2(a)). This AX_2 structure can be expressed by XA_4 tetrahedra (Figure. 2(b)).

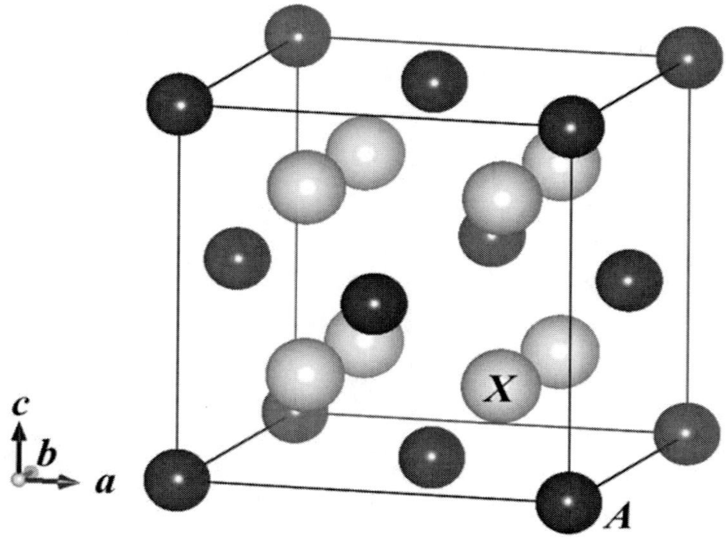

Figure 1. Crystal structure of a fluorite-type structured compound AX_2. The figure was drawn with a computer program VESTA [1]. A cation and X anion are located at the $4a$ (0, 0, 0) and $8c$ (1/4, 1/4, 1/4) sites, respectively.

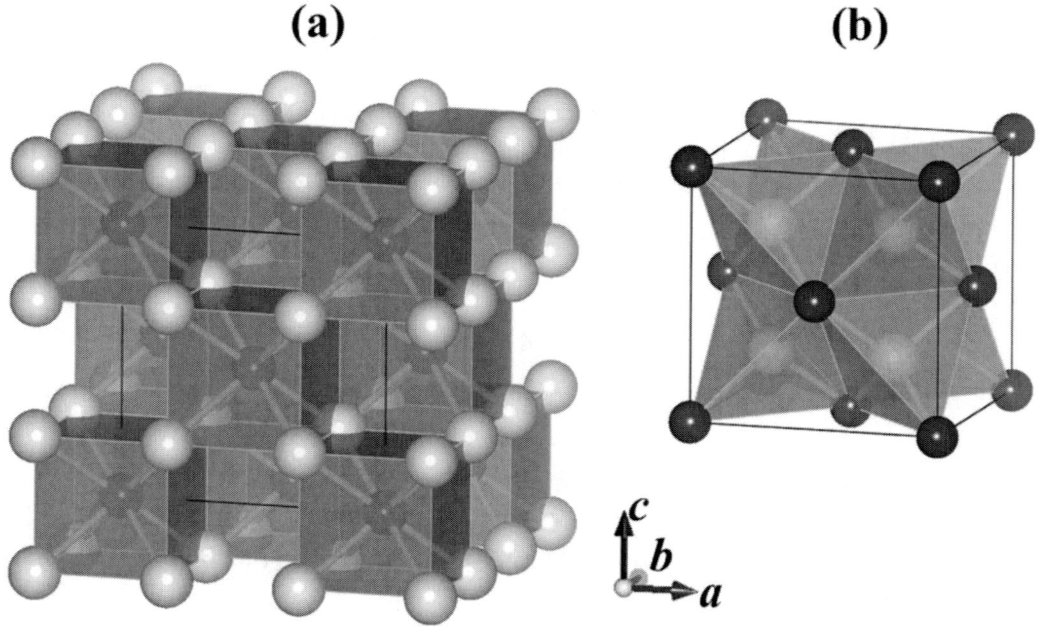

Figure 2. Coordination polyhedra of a fluorite-type structured compound AX_2. (a) AX_8 cubes. (b) XA_4 tetrahedra. The figure was drawn with a computer program VESTA [1].

In the fluorite-type AO_2 oxide, the A and O atoms are located at $4a$ (0, 0, 0) and $8c$ (1/4, 1/4, 1/4) sites, respectively, in the cubic $Fm\bar{3}m$ space group (Figure. 1). The fluorite-type AO_2 oxide can be described by AO_8 cubes or by OA_4 tetrahedra (Figure. 2). The fluorite-type $AO_{2-\delta}$ oxide with oxygen deficiencies is called as defect fluorite-type oxide. Here the δ denotes the concentration of oxygen deficiency. It is well known that the oxide ions move through the oxygen defects in the oxides $AO_{2-\delta}$. The positional disorder of oxide ions in the fluorite-type oxides is essential for the oxide-ion diffusion and conduction, which will be explained in details later for various fluorite-type structured oxides and fluorite-related materials. The coordination number (CN) of 8 for A cation requires the atomic species A with relatively large ionic radius. The ratio of ionic radii $r(A,8)/r(O,4)$ expected from the Pauling's 1st rule is ranged from 0.782 to 1.0, where the $r(A,8)$ and $r(O,4)$ are ionic radii of A cation with the coordination number of 8 (CN = 8) and of O^{2-} anion with CN = 4.

$$0.782 \leq r(A,8)/r(O,4) \leq 1.0 \quad (1)$$

In this review, I utilize the ionic radii after Shannon [2]. The radius of oxide ion is $r(O,4)$ = 1.38 Å. Thus, the equation (1) becomes

$$1.079 \leq r(A,8) \leq 1.38(Å) \quad (2).$$

Table 1. Ionic species A, its ionic radius, crystal structure and coordination number CN(A) in some fluorite-type and distorted fluorite-type AO_2 oxides

Ionic species A, Compound	Ionic radius (Å)	Crystal structure	Coordination number of A cation CN(A)
Hf^{4+}, HfO_2	$r(Hf^{4+},7) = 0.76$	Monoclinic baddeleyite-type structure at room temperature	7
	$r(Hf^{4+},8) = 0.83$	Tetragonal distorted fluorite-type structure and cubic fluorite-type structure at high temperatures	8
Zr^{4+}, ZrO_2	$r(Zr^{4+},7) = 0.78$	Monoclinic baddeleyite-type structure at room temperature	7
	$r(Zr^{4+},8) = 0.84$	Tetragonal distorted fluorite-type structure and cubic fluorite-type structure at high temperatures	8
Ce^{4+}, CeO_2	$r(Ce^{4+},8) = 0.97$	Cubic fluorite-type structure	8
U^{4+}, UO_2	$r(U^{4+},8) = 1.00$	Cubic fluorite-type structure	8
Th^{4+}, ThO_2	$r(Th^{4+},8) = 1.05$	Cubic fluorite-type structure	8

As shown in Table 1, however, this is only a guideline. The AO_2 oxides for relatively larger-sized A^{4+} cation (A = Ce, U and Th) have the cubic fluorite-type structure. The AO_2 oxides for relatively smaller-sized A^{4+} cation (A =Hf and Zr) have a largely-distorted monoclinic baddeleyite-type structure at room temperature with the coordination number of A^{4+} cation of 7. At high temperatures the AO_2 oxides (A = Hf and Zr) have the less-distorted tetragonal and cubic fluorite-type structures.

The unit-cell parameter is an important factor for material properties and design. The unit-cell parameter a of cubic fluorite-type oxides can be expressed by ion-packing model and ionic radii [3,4] as;

$$a = 4\cdot\left(r(A,8) + r(\mathrm{O},4)\right)/\sqrt{3} \quad (3).$$

Detailed analysis of unit-cell parameter of the fluorite-type structured oxides was done by Yashima et al. [3,4] and by Nakamura [5].

2. Crystal Structure and Structural Disorder of Bismuth Oxide Bi_2O_3

Bismuth oxide (Bi_2O_3)-based materials are fast oxide-ion conductors, which enables the use as materials for applications to solid-oxide fuel cells (SOFCs) and oxygen sensors [6-13]. The bismuth oxide δ-Bi_2O_3 has a cubic defect fluorite-type structure where the Bi cation and oxide ions are located at the 4*a* and 8*c* sites in Figure. 1, respectively, and in addition one-fourth of the oxygen 8c site is vacant (i.e., CN(Bi) = 6). Due to the presence of such high concentration of oxygen vacancy in the 'disordered' oxygen sublattice, the δ-Bi_2O_3 and its solid solutions are materials with the highest oxide-ion conductivities in the known compounds with fluorite-type and other structures. The crystal structure of the δ-Bi_2O_3 has therefore been studied for more than 75 years by many researchers [7-9, 14-17]. Gattow [15] analyzed the crystal structure of δ-Bi_2O_3 using the simple defect fluorite model where the Bi and O atoms are put at 4*a* (0,0,0) and 8*c* (1/4,1/4,1/4) sites, respectively (Figure. 3(a)). Harwig [16] refined the structure of δ-Bi_2O_3 with a split-atom model where the Bi and O atoms are located at 4*a* (0,0,0) and 32*f* (*x*, *x*, *x*) sites, respectively (Figure. 3(b)). Battle et al. [17] analyzed the crystal structure of δ-Bi_2O_3 with the other split-atom model where the Bi atom is put at 4*a* (0,0,0) position and oxygen atom is located at 8*c* (1/4,1/4,1/4) and 32*f* (*x*, *x*, *x*) sites (Figure. 3(c)). Yashima and Ishimura [9] refined the crystal structure of δ-Bi_2O_3 using Rietveld method [18-20] and neutron powder diffraction data measured at 778°C. The sample was heated by a home-made furnace [21] during the neutron diffraction measurements with HERMES diffractometer [22]. The weighted reliability factors in the analyses with the Gattow, Harwig and Battle models were 4.75%, 3.81% and 3.70%, respectively [9]. Maximum-entropy method (MEM) [23-27] was used to obtain the accurate nuclear-density distribution of δ-Bi_2O_3. It is well known that the MEM can produce a nuclear-density distribution map from the neutron diffraction data. In the MEM analysis, any kind of complicated nuclear-density distribution is allowed as long as it satisfies the symmetry requirements. The R factor based on the Bragg intensities, R_B, was improved from 4.56% in Rietveld analysis to 1.82% in MEM-based pattern fitting (MPF) [9]. In the MEM analysis, the R factor based on the structure factors, R_F, was also improved from 1.54% in Rietveld analysis to 0.89% of MPF analysis. The MEM map provided much information on the complicated disorder of oxide ions (Figure 3d), compared with the previous models (Figures. 3(a)-3(c)). The previous simple models consisting of atom spheres are no longer appropriate to describe the positional disorder of oxide ions. The oxide ions in δ-Bi_2O_3 have a complicated positional disorder and spread over a wide area, compared with Bi cations [9]. Bulges in the <111> directions indicate a possible diffusion mechanism where the oxide ions move through a face of the surrounding Bi tetrahedron to the octahedral cavity and then re-entering a neighboring anion site. Bulges in the [001] directions suggest another diffusion path where the oxide ions move along the *c* axis. The complicated disorder of oxide ions would be responsible for the fast oxide-ion conduction [9].

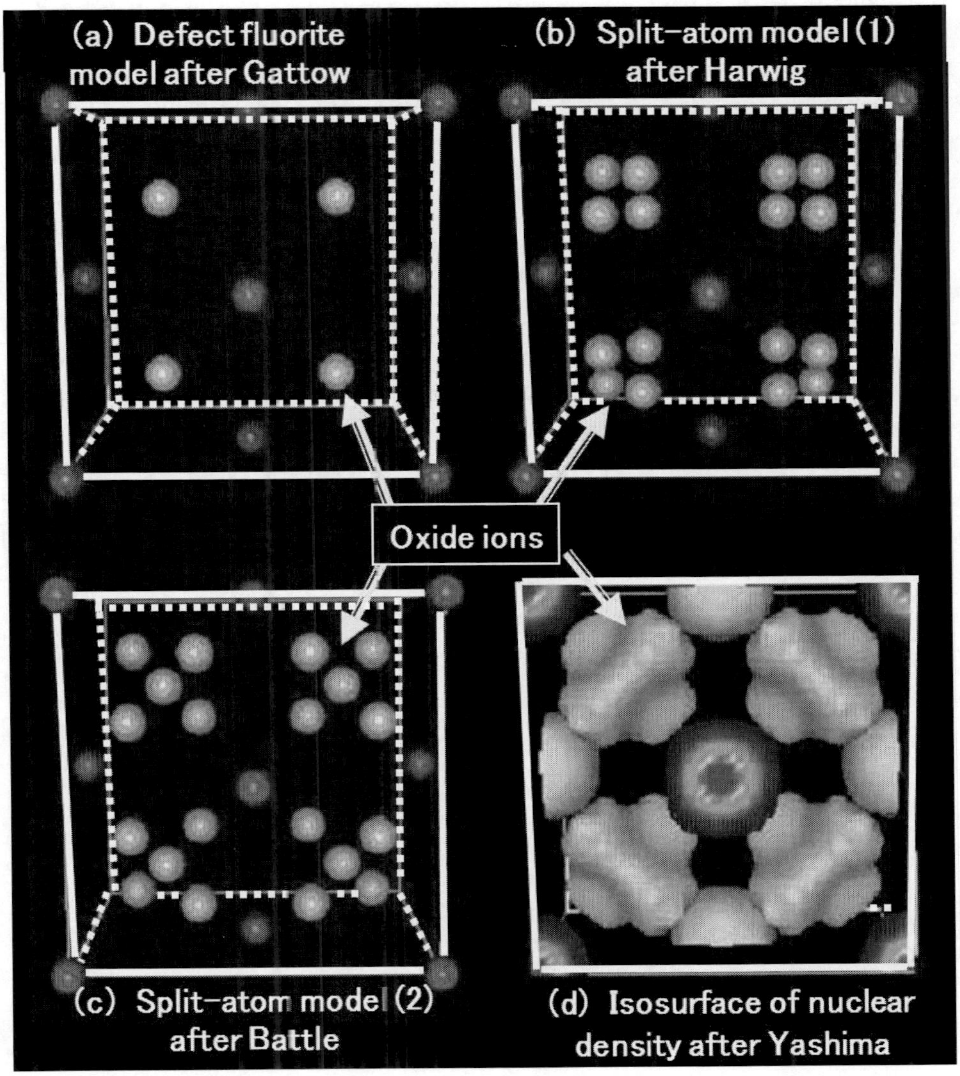

Figure 3.(a-c) Structural models and (d) isosurface of nuclear density of δ-Bi_2O_3 [9]. (a) Gattow model [15] based on an ideal defect fluorite-type structure: Average distribution of six oxygen atoms at the 8*c* site of $Fm\overline{3}m$ (*x*,*x*,*x*; *x*=1/4). (b) Harwig model [16]: Average distribution of six oxygen atoms at the 32*f* site of $Fm\overline{3}m$ (*x*,*x*,*x*;*x*=1/4+δ) in the unit cell. (c) Battle model [17]: Six oxygen atoms are distributed at the 8*c* and 32*f* sites. (d) Nuclear-density distribution of δ-Bi_2O_3 at 778°C with isosurface at 0.3 fmÅ^{-3} obtained by the combination technique of Rietveld refinement and the maximum-entropy-method-based pattern fitting [9]. To simplify, only the region of $0 \le y \le 1/2$ is drawn. Red and yellow spheres denote the bismuth and oxygen atom sites, respectively. The figures were drawn with a computer program VENUS [24]. Copyright 2003 Elsevier.

3. CRYSTAL STRUCTURE AND OXIDE-ION DIFFUSIONAL PATHWAY OF A Bi_2O_3-BASED MATERIAL $Bi_{1.4}Yb_{0.6}O_3$

The bismuth oxide δ-Bi_2O_3 and its solid solutions δ-$Bi_{2(1-x)}M_{2x}O_3$ with the cubic fluorite-type structure are some of the highest oxide-ion conductors, where M is a trivalent dopant such as Yb and Y. Their crystal structures have been the subjects of many investigations for more than 75 years [7-17, 25-29]. Accurate nuclear-density distribution of $Bi_{1.4}Yb_{0.6}O_3$ compound has been studied at 384°C and at 738°C by the maximum-entropy method (MEM) and MEM-based pattern fitting combined with the Rietveld method using neutron powder diffraction data [27]. The results reveal that the oxide ions have a complicated disorder spreading over a wide area. The oxide ions shift to the <111> directions from the ideal fluorite 8*c* position, and then move to the nearest neighbor 8*c* site along the <100> directions (Figures.4(b) and 5). The MEM nuclear-density distribution map provided much information on the complicated disorder of oxide ions (Figure. 4(b)), compared with the simple spherical atom model (Figure. 4(a)). The latter consisting of atom spheres is not appropriate to describe the positional disorder of oxide ions in δ-$Bi_{1.4}Yb_{0.6}O_3$. The oxide ions have a complicated disorder and spread over a wide area, compared with cations in δ-$Bi_{1.4}Yb_{0.6}O_3$. The disorder is more significant at higher temperature (738°C) compared with that at lower temperature (384°C), which is consistent with higher oxide-ion conductivity and higher atomic displacement parameters at higher temperatures [27]. The oxide ions shift to the opposite direction from the cations Bi and Yb probably due to the repulsion force between the cation and anion. The shift of oxide ions to the <111> directions observed in δ-$Bi_{1.4}Yb_{0.6}O_3$ by Yashima and Ishimura [27] is consistent with those in the δ-Bi_2O_3 and δ-$Bi_{2(1-x)}M_{2x}O_3$ (M=Y and Er) in the literature [8,9,17,28,29].

The most striking feature is the diffusion path of oxide ions along the <100> directions on the {100} planes (Figures. 4(b) and 5) [27]. An oxide ion migrates three-dimensionally to the nearest-neighbor sites along the [100], [010] or [001] directions (Figure. 5).

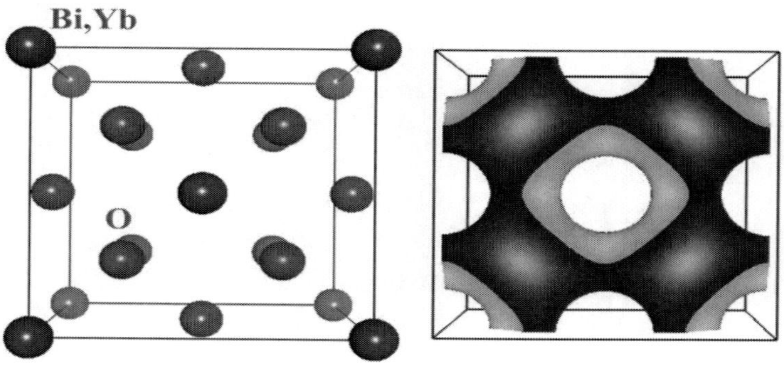

(a) (b)

Figure 4. (a) Structural model and (b) the isosurface of nuclear density at 0.1 fm Å^{-3} for δ-$Bi_{1.4}Yb_{0.6}O_3$ [27]. (a) A defect fluorite-type structure: Average distribution of four cations and six oxide ions at the 4*a* and 8*c* sites of $Fm\overline{3}m$, respectively, in a fluorite unit cell. To describe the distribution of the oxide ions on the *ab* plane at around z = 3/4, only the region of 0.70 ≤ z ≤ 0.80 is drawn in the figure (b) [27].

It is interesting that the diffusion path of oxide ions exists along the <100> directions, although the oxide ions shift to the <111> directions near the stable position. In the previous studies [8,17,28,29], it was difficult to deduce the diffusion path. On the contrary, Yashima and Ishimura [27] have succeeded in the visualization of the diffusion path. The disorder and the diffusion pathway along the <100> direction of oxide ions would be responsible for the fast oxide-ion conduction in δ-$Bi_{1.4}Yb_{0.6}O_3$.

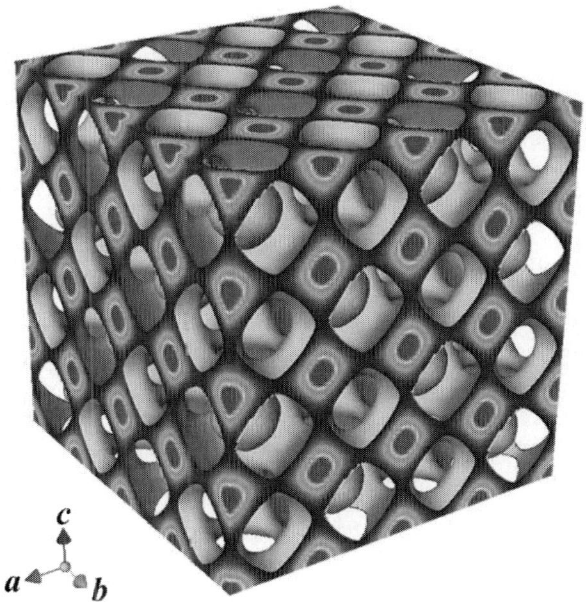

Figure 5. Isosurface of the nuclear density at 0.1 fmÅ^{-3} in nuclear-density distribution on the {400} planes of δ-$Bi_{1.4}Yb_{0.6}O_3$ measured at 738°C [27]. Diffusion paths of oxide ions are observed along the <100> directions on the {400} planes. The oxide ions usually exist at the red portion around the 8*c* site, and migrate to the nearest-neighbor 8*c* site through the blue diffusion pathway [27].

4. Crystal Structure and Disorder of Ceria CeO_2

Cerium dioxide (ceria, CeO_2)-based materials are of considerable interest for potential use as electrolytes in solid oxide fuel cells (SOFCs) [13,30-34]. These materials have a higher oxide-ion conductivity than that of stabilized zirconia and a lower cost compared with lanthanum gallate-based perovskite-type phases [13,30-34]. Other promising applications of ceria-based materials include automotive catalysts, SOFCs anode materials, solid-electrolyte oxygen pumps, and mixed-conducting membranes for oxygen separation and partial oxidation of hydrocarbons [33]. The crystal structure of the $CeO_{2-\delta}$ at high temperatures has been investigated by Faber et al. [35-37]. However, the previous studies are not sufficient to clarify the structural disorder of oxide ions in CeO_2.

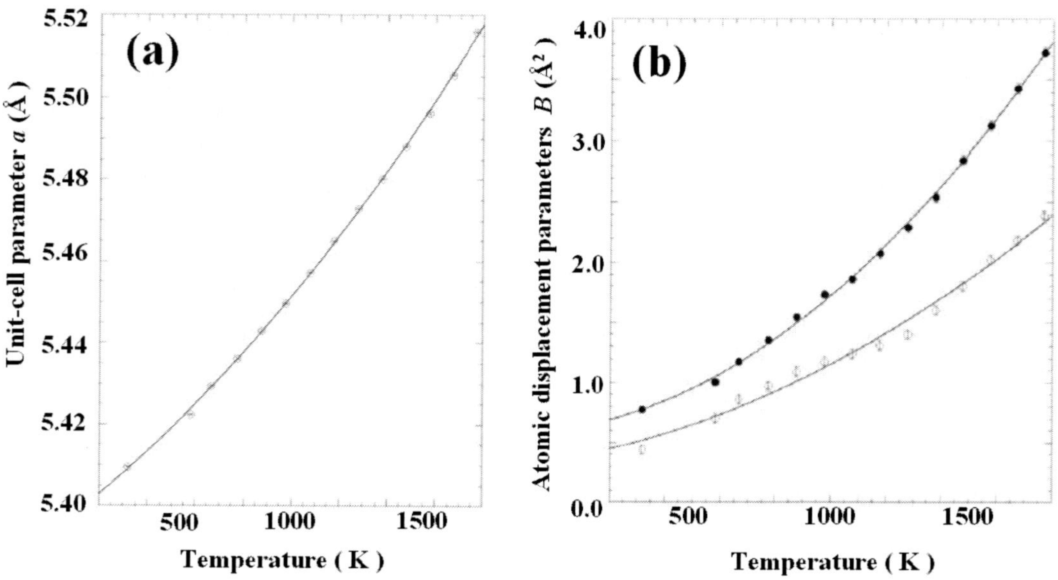

Figure 6. Temperature dependence of (a) unit-cell and (b) atomic displacement parameters of ceria, CeO_2 [34]; B(O) (●), B(Ce) (○). Copyright 2003 Elsevier.

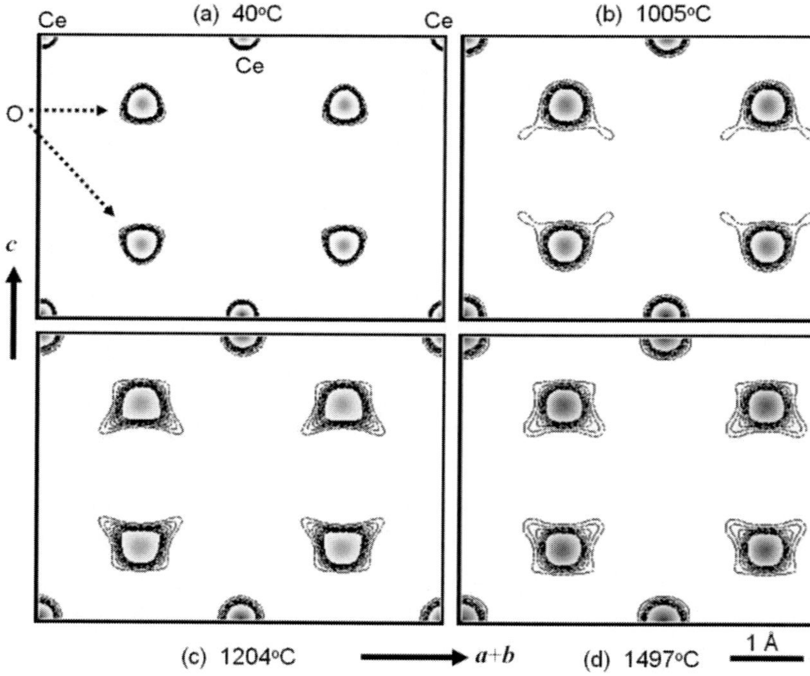

Figure 7. Nuclear-density distributions on the (110) plane of CeO_2 (a) at 40°C, (b) at 1005°C, (c) at 1204°C and (d) 1497°C with contours in the range from 1.0 to 5.0 fmÅ^{-3} (0.5 fmÅ^{-3} step) [39]. Copyright 2006 Elsevier.

Therefore, we have performed the neutron powder diffraction measurements from room temperature to 1770K and analyzed the crystal structure and MEM nuclear-density distributions of ceria (CeO_2, space group: $Fm\overline{3}m$) [38,39]. The unit-cell parameter increases continuously with temperature (Figure. 6(a)) [39]. The atomic displacement parameter of oxygen atom, *B*(O), is higher than that of cerium, *B*(Ce), in the whole temperature range from room temperature to 1770K (Figure. 6(b)), which suggests a higher mobility of oxide ions compared with the Ce cations [39]. Both *B*(O) and *B*(Ce) increase with an increase of temperature, which indicates the increase of thermal vibration with temperature [39].

Nuclear-density distribution of ceria (CeO_2, space group: $Fm\bar{3}m$) has been investigated between 40 and 1497°C by the MEM and MEM-based pattern fitting (MPF) combined with the Rietveld method using neutron powder diffraction data [38,39]. The results of the MEM analysis reveal that the oxide ions have a positional disorder spreading over a wide area and shift to the <111> directions from the ideal position 8*c*(1/4,1/4,1/4) of the fluorite-type structure (Figure. 7). The positional disorder of the oxide ions is more significant at higher temperatures. The spatial distribution of oxide ions in this almost stoichiometric $CeO_{2-\delta}$ ($\delta \sim 0$) in Figures. 7(b), (c) and (d) is much smaller than those of grossly-defective Bi_2O_3-based systems (Figures. 3, 4 and 5) even at higher temperatures.

5. Crystal Structure and Oxide-Ion Diffusional Pathway of a CeO_2-Based Material $Ce_{0.927}Y_{0.073}O_{1.962}$

The development of the ceria-based materials requires the knowledge of the disorder and diffusion path of mobile ions in the ceria-based materials at high temperatures where they work efficiently [40]. Berber et al. [41] studied the nuclear-density distribution in the $Ce_{1-x}Y_xO_{2-x/2}$ samples at high temperatures (x=0 to 0.18). However, the previous work was insufficient to clarify the diffusional pathway of oxide ions in $Ce_{1-x}Y_xO_{2-x/2}$. Here I describe the crystal structure, disorder and diffusion path of a ceria-based defect-fluorite-type oxide-ion conductor, yttria-doped ceria solid solution, $Ce_{0.927}Y_{0.073}O_{1.962}$ [40]. We have studied this chemical composition, $Ce_{0.927}Y_{0.073}O_{1.962}$, because it exhibits the maximum electrical conductivity in the ceria-yttria solid solutions [42].

The ceria-yttria solid solution, $Ce_{0.927}Y_{0.073}O_{1.962}$, has a cubic defect fluorite-type structure with the oxygen vacancies on the anion sublattice, as shown by the refined structure in Figure. 8(a) [40]. The MEM map provided much information on the positional disorder of oxide ions (Figures. 8(b) and 9(b)), compared with the structural model obtained in the Rietveld analysis (Figure. 8(a)). The conventional simple model consisting of atom spheres is no longer appropriate to describe the positional disorder of oxide ions at high temperatures. The spatial distribution of oxide ions at 1434°C (Figure. 9(b)) was by far larger than that at 23°C (Figure. 9(a)), corresponding to the much higher atomic displacement parameters [40]. There were two types of bulges in the MEM distribution map. One bulge leads to the diffusion path along the <100> directions. The other exists along the <110> directions. These directions of oxide-ion displacements are opposite to those of the Ce and Y cations. The bulges are attributable to the repulsion between the cations and oxide ions. Such anisotropic feature in the nuclear-density distributions has been observed in pure ceria itself (Figure. 7 [39]). The shift of oxide ions in the <111> directions has been reported also in the above-discussed bismuth oxide-based compounds [8,9,17,25-29] and β' phase in the Mg-Zr-O-N system [43]. These compounds have the cubic defect fluorite-type structure as well as ceria-based materials. Thus, this feature would be common in the fluorite-type structured materials.

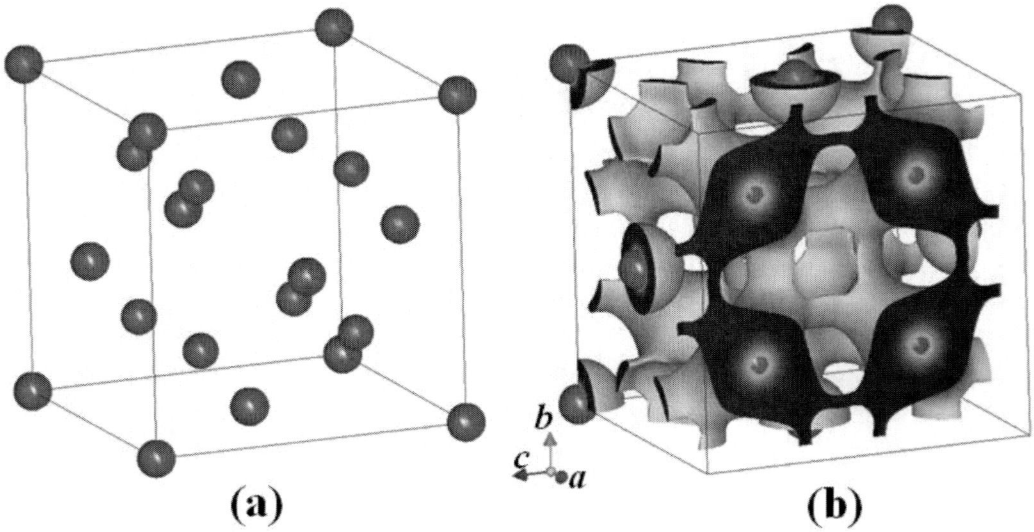

Figure 8.(a) Refined crystal structure and (b) isosurface of nuclear density at 0.003 fmÅ^{-3} with nuclear-density distribution on the *bc* plane of $Ce_{0.927}Y_{0.073}O_{1.962}$ obtained by the combination technique of Rietveld refinement and the maximum-entropy method of neutron powder diffraction data measured at 1434°C [40]. Red and black spheres stand for the (O, Vo) anion and (Ce,Y) cation, respectively, where the Vo is the oxygen vacancy. Copyright 2007 Ceramic Society of Japan.

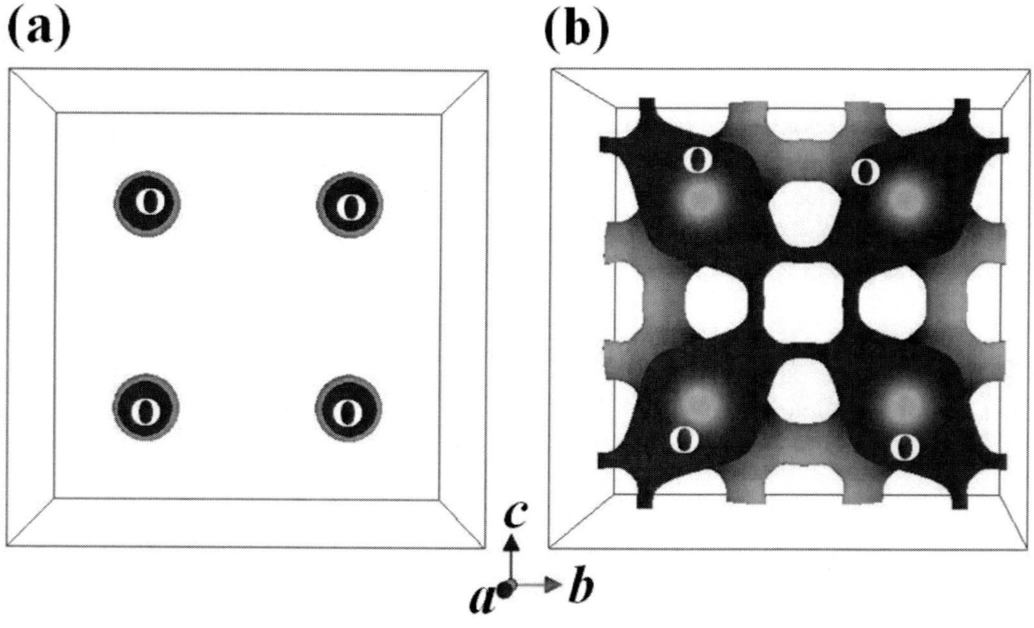

Figure 9. Isosurface of the nuclear density at 0.003 fmÅ^{-3} with nuclear-density distribution on the *bc* plane of $Ce_{0.927}Y_{0.073}O_{1.962}$ obtained by the combination technique of Rietveld refinement and the maximum-entropy method of neutron powder diffraction data measured at (a) 23°C and (b) 1434°C [40]. Copyright 2007 Royal Society of Chemistry.

6. Atomic Displacement Parameters and Structural Disorder of Oxide Ions in the $Ce_xZr_{1-x}O_2$ Solid Solutions: Possible Factors of High Catalytic Activity of Ceria-Zirconia Catalysts

Ceria-zirconia ($Ce_xZr_{1-x}O_2$) catalysts are widely used for the cleaning of exhaust gases from automobiles [44,45]. The development of improved catalysts requires a better understanding of crystal structure and oxide-ion diffusion in ceria-zirconia materials [46-49]. The crystal structures of $Ce_xZr_{1-x}O_2$ solid solutions have been investigated by some techniques such as laboratory-based and synchrotron X-ray diffraction [49-53], Raman scattering [53], EXAFS [54] and neutron diffraction [46,47,49]. Nevertheless, the structural disorder in $Ce_xZr_{1-x}O_2$ solid solutions remains poorly understood.

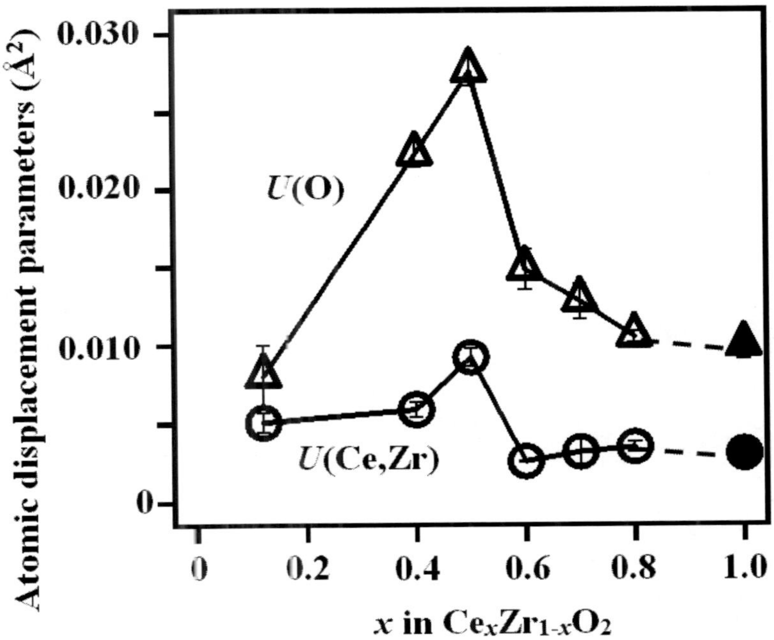

Figure10. Variation of the isotropic atomic displacement parameters of Ce and Zr atoms U(Ce,Zr) and of O atoms U(O) in $Ce_xZr_{1-x}O_2$ (x=0.12-1.0) with the composition x at 299 K [48]. Open circles and triangles denote U(Ce,Zr) and U(O) for tetragonal $Ce_xZr_{1-x}O_2$ solid solutions, respectively. Filled circle and triangle denote U(Ce) and U(O) for cubic CeO_2, respectively. The solid line connects the data points of tetragonal $Ce_xZr_{1-x}O_2$. The dashed line connects the data points of tetragonal $Ce_{0.8}Zr_{0.2}O_2$ and cubic CeO_2. Copyright 2009 American Institute of Physics.

We have investigated the structural disorder of $Ce_xZr_{1-x}O_2$ solid solutions by synchrotron X-ray powder diffraction, Rietveld method and MEM [48]. The isotropic atomic displacement parameter of the oxygen atoms U(O) in $Ce_xZr_{1-x}O_2$ and CeO_2 is higher than that of Ce and Zr atoms U(Ce,Zr) over the whole $0.12 \leqslant x \leqslant 1.0$ range, which suggests higher mobility of oxide ions compared with that of cations. The U(O) in $Ce_xZr_{1-x}O_2$ increases with an increase of CeO_2 content x in the compositional range of $0.12 \leqslant x \leqslant 0.5$, while it turns to decrease with x in the $0.5 \leqslant x \leqslant 1.0$ range (Figure. 10). Thus, the $Ce_{0.5}Zr_{0.5}O_2$ composition has the highest U(O) value in the $Ce_xZr_{1-x}O_2$ solid solutions (x = 0.12, 0.4, 0.5, 0.6, 0.7, 0.8 and 1.0), most plausibly suggesting the highest bulk diffusivity of the oxide ions in $Ce_{0.5}Zr_{0.5}O_2$, compared with those of other compositions.

Figures.11(a), 11(b) and 11(c) show the electron-density contour maps at 299 K on the (100) and (110) planes of tetragonal $Ce_xZr_{1-x}O_2$ (x=0.12 and 0.5) and cubic CeO_2, respectively. These figures provide lots of information on the positional disorder of oxide ions, which are mobile at high temperatures. The electron-density distribution of CeO_2 is consistent with the literature data [38,39,47,55]. The results reveal that the oxide ions in $Ce_{0.5}Zr_{0.5}O_2$ are spread over a significantly wider area compared with other $Ce_xZr_{1-x}O_2$ (x=0.12 and 1.0), which suggests the high bulk diffusivity of the oxide ions in $Ce_{0.5}Zr_{0.5}O_2$. Yashima and Wakita [48] demonstrated that at room temperature the atomic displacement parameter of oxygen atoms in $Ce_{0.5}Zr_{0.5}O_2$ (U(O) = 0.0276(9) $Å^2$) is higher than that in pure CeO_2 (U(O) = 0.0092(6) $Å^2$) (Figure. 10), which is consistent with the results at 1832K [47]. Boaro et al. [56] reported that the oxygen diffusivity of $Ce_{0.5}Zr_{0.5}O_2$ is higher than that of CeO_2, corresponding well to the higher (in fact the highest) U(O) in $Ce_{0.5}Zr_{0.5}O_2$ [48]. Our work [48] has also indicated that the atomic displacement parameter of the oxygen atoms U(O) in $Ce_xZr_{1-x}O_2$ increased with CeO_2 content x in the compositional range of $0.12 \leq x \leqslant 0.5$, and the U(O) in $Ce_xZr_{1-x}O_2$ decreased with x in the $0.5 \leqslant x \leqslant 1.0$ range (Figure. 10). Thus, the U(O) in $Ce_{0.5}Zr_{0.5}O_2$ has a maximum value in the $0.12 \leqslant x \leqslant 1.0$ range as well as the oxygen diffusivity.

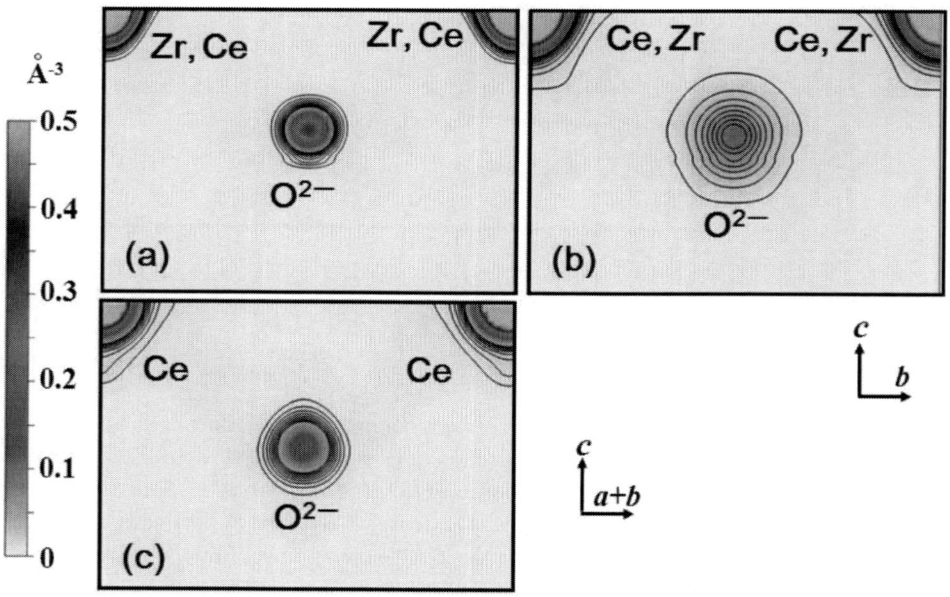

Figure 11. Parts of the electron-density distributions on the (100) planes of (a) $Ce_{0.12}Zr_{0.88}O_2$ and (b) $Ce_{0.5}Zr_{0.5}O_2$ at 299 K with black contours in the range from 4.0 to 20.0 $Å^{-3}$ (2.0 $Å^{-3}$ step) ($1/2 \leqslant z \leqslant 1$) [48]. (c) A part of the electron-density distribution on the (110) plane of CeO_2 at 299 K with black contours in the range from 4.0 to 20.0 $Å^{-3}$ (2.0 $Å^{-3}$ step) ($1/2 \leqslant z \leqslant 1$) [48]. Copyright 2009 American Institute of Physics.

The electron-density distributions of $Ce_xZr_{1-x}O_2$ (x=0.12, 0.5 and 1) at room temperature in Figure. 11 have shown that the spatial distribution of oxide ions in $Ce_{0.5}Zr_{0.5}O_2$ is larger than those of the others. These results altogether indicate that the oxygen diffusivity is higher when the U(O) and spatial distribution of oxide ions are larger. Figure. 12 shows the relationship between the U(O) and oxygen diffusivity. The oxygen diffusivity increases with increasing U(O). Thus, the atomic displacement parameter of oxide ion U(O) is a good indicator of the oxygen diffusivity. High U(O) value indicates the high dynamic thermal vibration and/or the high static positional disorder, which enhance the movement of oxide ions across the unit cell and hence the associated catalytic activity of $Ce_{0.5}Zr_{0.5}O_2$. The high U(O) is regarded as a structural origin of high oxide-ion diffusivity and catalytic activity of $Ce_{0.5}Zr_{0.5}O_2$ [48].

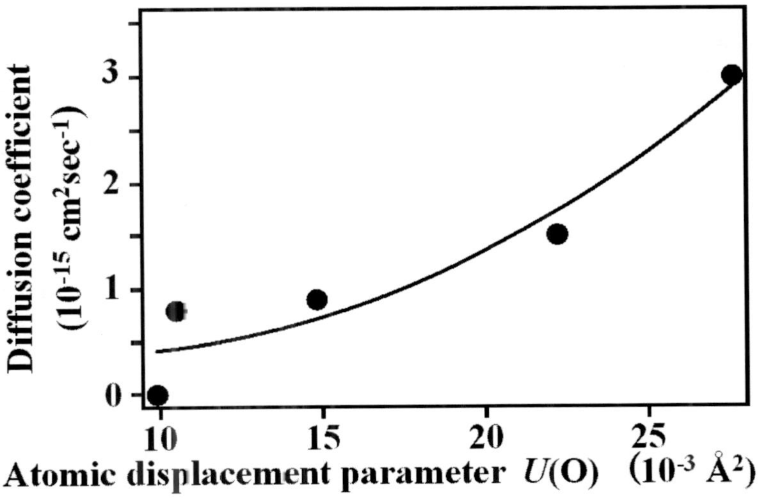

Figure 12. Relationship between the isotropic atomic displacement parameter of O atoms U(O) and diffusion coefficient of O atoms in $Ce_xZr_{1-x}O_2$ solid solutions [48]. The solid line was obtained by a least-squares fit (quadratic polynomial).

7. Crystal Structure, Disorder and Diffusion Path of the Oxide-Ion Conductors $Y_{1-x}Ta_xO_{1.5+x}$ (x = 0.215 and 0.30)

Rare earth tantalates $R_{1-x}Ta_xO_{1.5+x}$ (R: rare earth, Y and Sc elements) have different crystal structures depending on the R species, chemical composition x, temperature, and preparation method, varying from a cubic defect-fluorite-type structure to an orthorhombic fluorite-related structure [57-56]. The oxygen lattice diffusion coefficient and its activation energy in the fluorite-type $Y_{0.8}Ta_{0.2}O_{1.7}$ are close to those of a stabilized zirconia [67]. The $Y_{0.8}Ta_{0.2}O_{1.7}$ composition exhibits the highest oxide-ion conductivity in the $Y_{1-x}Ta_xO_{1.5+x}$ materials ($0 \le x \le 0.25$) [68,69]. However, the structural origin of the high oxygen diffusion coefficient and conductivity of $Y_{0.8}Ta_{0.2}O_{1.7}$ has not been understood satisfactorily. To solve this problem, we have investigated the crystal structure and MEM nuclear-density distribution of $Y_{0.8}Ta_{0.2}O_{1.7}$ and $Y_{0.7}Ta_{0.3}O_{1.8}$ [66].

Figure. 13(a) shows the crystal structure of the *Cmmm* $Y_{0.7}Ta_{0.3}O_{1.8}$ depicted with the $TaO_{6.6}$, YO_8 and $(Y_{0.9}Ta_{0.1})O_{7.2}$ polyhedra, using the refined crystallographic parameters obtained by the neutron diffraction and Rietveld method [66]. Relationship of the lattice vectors between the *Cmmm* and pseudo-fluorite-type structures can be described as $\mathbf{a}_{\mathrm{ortho}} = 2\cdot\mathbf{c}_{\mathrm{F}}$, $\mathbf{b}_{\mathrm{ortho}} = \mathbf{a}_{\mathrm{F}} - \mathbf{b}_{\mathrm{F}}$, $\boldsymbol{c}_{\mathrm{ortho}} = (\boldsymbol{a}_{\mathrm{F}} + \boldsymbol{b}_{\mathrm{F}})/2$, where the $\boldsymbol{a}_{\mathrm{ortho}}$ and c_{F} are the lattice vectors of the *Cmmm* and pseudo-fluorite cells, respectively.

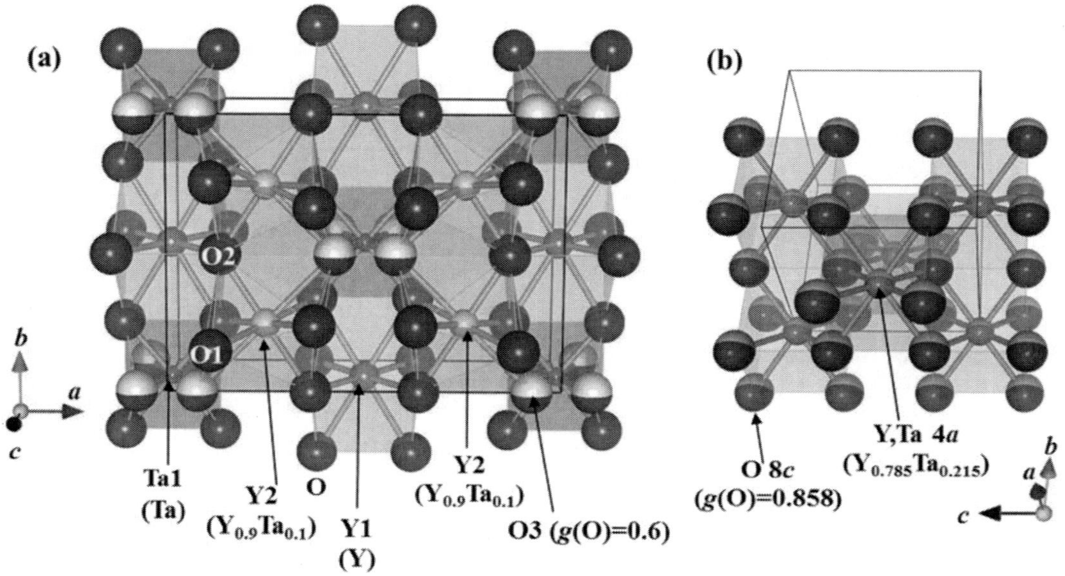

Figure 13. (a) Refined crystal structure of the orthorhombic *Cmmm* $Y_{0.7}Ta_{0.3}O_{1.8}$ depicted using the $TaO_{6.6}$, YO_8 and $(Y_{0.9}Ta_{0.1})O_{7.2}$ polyhedra (299 K) [66]. There exist occupational ordering: Occupancy factor of Y atom at *X* site is $g_Y(X)$; $g_Y(\mathrm{Ta1}) = 0$, $g_Y(\mathrm{Y1}) = 1$, $g_Y(\mathrm{Y2}) = 0.9$. There is also anion occupational ordering: $g_O(\mathrm{O1}) = g_O(\mathrm{O2}) = 1$ and $g_O(\mathrm{O3}) = 0.6$. (b) Crystal structural model of the cubic defect fluorite-type $Y_{0.785}Ta_{0.215}O_{1.715}$ phase (Space group: $Fm\overline{3}m$) depicted with the $(Y,Ta)O_{6.86}$ cubes. Average distribution of 6.86 oxygen atoms at the 8*c* site (1/4, 1/4, 1/4) of $Fm\overline{3}m$ (Occupancy of oxygen atom: $g(O) = 0.858$). Solid line denotes the unit cell of the crystal structure [66]. Copyright 2007 American Chemical Society.

Most of Ta atoms are localized at the Ta1 2*d* site (Wyckoff notation in the *Cmmm* space group, fractional coordinate: (0, 0, 1/2)). The Y atoms exist at the Y1 2*c* (0, 1/2, 1/2) and Y2 4*e* (1/4, 1/4, 0) sites, indicating the occupational order of Y and Ta atoms (Figure. 13(a)). The oxygen vacancy exists at the O3 4*g* site (0.756(10), 0, 0), which is the nearest neighbor site of Ta atom, making the $TaO_{6.6}$ and $(Y_{0.9}Ta_{0.1})O_{7.2}$ polyhedra. On the contrary, the O1 8*q* (0.1239(5), 0.2060(5), 1/2) and O2 4g (0.1368(6), 1/2, 0), which are the nearest neighbor sites of the Y1 site, are fully occupied by oxygen atoms, making the YO_8 polyhedron. These results indicate the occupational order of oxygen atoms and their vacancies among the O1, O2 and O3 sites. The occupational ordering would be responsible for the lower oxygen diffusivity in *Cmmm* $Y_{0.7}Ta_{0.3}O_{1.8}$.

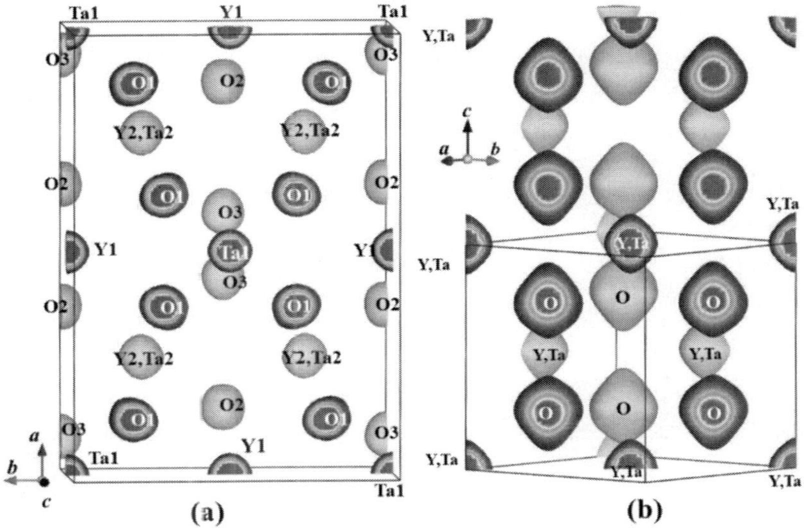

Figure 14. Nuclear-density distribution of (a) the orthorhombic *Cmmm* $Y_{0.7}Ta_{0.3}O_{1.8}$ and (b) cubic fluorite-type $Y_{0.785}Ta_{0.215}O_{1.715}$ at 299 K with isosurface at 0.2 fmÅ^{-3} obtained by the combination technique of Rietveld refinement, MEM and MPF analyses [66]. Copyright 2007 American Chemical Society.

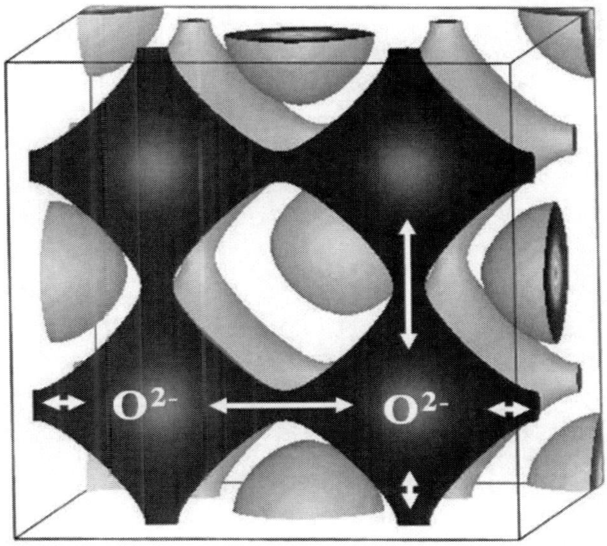

Figure 15. Nuclear-density distribution of the cubic fluorite-type $Y_{0.785}Ta_{0.215}O_{1.715}$ at 808 K [66].

Neutron-diffraction profile of the $Y_{0.785}Ta_{0.215}O_{1.715}$ material confirms the presence of a single defect fluorite-type structured phase at 299 and 808 K. The oxygen atoms in $Y_{0.785}Ta_{0.215}O_{1.715}$ exist at the 8*c* site of the $Fm\overline{3}m$ symmetry. Y and Ta atoms are located at the special position 4*a* of (0, 0, 0) where the occupancy factors of Y and Ta were 0.785 and 0.215, respectively, indicating almost the complete occupational disorder of Y and Ta atoms.

MEM nuclear-density analysis demonstrates that the oxygen spatial distribution in $Fm\overline{3}m$ $Y_{0.785}Ta_{0.215}O_{1.715}$ (Figure. 14(b)) is larger than that in *Cmmm* $Y_{0.7}Ta_{0.3}O_{1.8}$ (Figure. 14(a)).

At a higher temperature of 808 K, the spatial distribution of oxide ions is larger and the diffusion paths of oxide ions are observed along the <100> directions in $Fm\bar{3}m$ $Y_{0.785}Ta_{0.215}O_{1.715}$ (Figure. 15). High diffusion coefficient and conductivity of oxide ions in the cubic $Y_{0.8}Ta_{0.2}O_{1.7}$ are attributable to the occupational and positional disorder.

CONCLUSION

Many fluorite-type structured oxides and fluorite-related materials such as zirconia, ceria and bismuth oxide exhibit excellent electrical and mechanical properties. The material properties are strongly dependent on the crystal structure. In the present review I have described the crystal structure, structural disorder and oxide-ion diffusional pathway of some representative fluorite-type structured oxides and fluorite-related materials. The combined procedure of the Rietveld method, maximum-entropy method (MEM) and MEM-based pattern fitting (MPF) is powerful to deduce the accurate nuclear-density distribution, which enables the visualization of positional disorder and oxide-ion diffusional pathway. As shown in Figures. 16(a)-(d) [25], the diffusion paths along the <100> directions are observed in various ionic conductors with the fluorite-type structure such as $Ce_{0.927}Y_{0.073}O_{1.962}$ [40] (a), δ-$Bi_{1.4}Yb_{0.6}O_3$ [27] (b), $Y_{0.785}Ta_{0.215}O_{1.715}$ [66] (c), and α-CuI (d) in Figure. 16 [70]. Recently we have demonstrated that the oxide-ion diffusional pathway of nano-sized tetragonal ceria-zirconia $Ce_{0.5}Zr_{0.5}O_2$ powders is also curved and along the <100>$_F$ directions where the subscript F denotes the pseudo fluorite lattice [71].

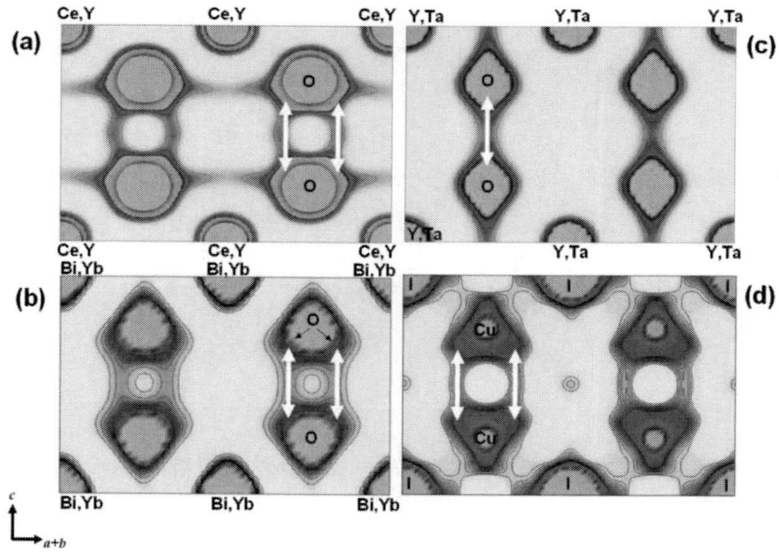

Figure 16.(a–c) Nuclear-density distribution on the (110) plane of the fluorite-type (a) $Ce_{0.927}Y_{0.073}O_{1.962}$ at 1434°C [40], (b) δ-$Bi_{1.4}Yb_{0.6}O_3$ at 738°C [27] and (c) $Y_{0.785}Ta_{0.215}O_{1.715}$ at 535°C [66]. (d) Electron-density distribution on the (110) plane of the fluorite-type α-CuI at 487 °C [70]. Copyright 2008 Elsevier.

ACKNOWLEDGMENTS

A part of this work was financially supported by the Ministry of Education, Culture, Sports, Science and Technology of Japan, through Grant-in-Aid for Scientific Research (A) and Challenging Exploratory Research (Nos. 24240617, 23655190) from the Ministry of Education, Culture, Sports, Science and Technology of Japan. This work was carried out under the Joint-use Research Program for Neutron Scattering, Institute for Solid State Physics (ISSP), the University of Tokyo, at the Research Reactor JRR-3, JAEA (Proposal Nos. 8766, 7768, 6723, 6721, 5681, 5682, 4688, 4689, 3653). The synchrotron data were collected at the BL-4B2 beam line of Photon Factory under the projects (Nos. 2008G084, 2006G264, 2005G158) and at the BL02B2 beam line of SPring-8 (Project Nos. 2011B1995 and 2011A1442).

REFERENCES

[1] K. Momma and F. Izumi, *J. Appl. Crystallogr.* (2008) 41, 658.
[2] R.D. Shannon, *Acta Crystallogr.* (1976) 32, 751.
[3] M. Yashima, N. Ishizawa and M. Yoshimura, *J. Am. Ceram. Soc.* (1992) 75, 1541.
[4] M. Yashima, N. Ishizawa and M. Yoshimura, *J. Am. Ceram. Soc.* (1992) 75, 1550.
[5] A. Nakamura, *Solid State Ionics* (2010) 181, 1514.
[6] T. Takahashi and H. Iwahara, *J. Appl. Electrochem.* (1972) 3, 65.
[7] P. Shuk, H.-D. Wiemhöfer, U. Guth, W. Göpel and M. Greenblatt, *Solid State Ionics* (1996) 89, 179.
[8] N.M. Sammes, G.A. Tompsett, H. Näfe and F. Aldinger, *J. Eur. Ceram. Soc.*19 (1999), 1801.
[9] M. Yashima and D. Ishimura, *Chem. Phys. Lett.* (2003) 378, 395.
[10] R. Punn, A.M. Feteira, D.C. Sinclair and C. Greaves, *J. Am. Chem. Soc.* (2006) 128, 15386.
[11] M. Drache, P. Roussel and J-.P. Wignacourt, *Chem. Rev.* (2007) 107, 80.
[12] D.W. Jung, K.L. Eric and D. Wachsman, *Acta Mater.* (2010) 58, 355.
[13] E.C. Wachsman and K.T. Lee, *Science* (2011) 334, 935.
[14] L.G. Sillen, *Ark. Kemi.Mineral. Geol.* (1937)12A, 1.
[15] G. Gattow and H. Schroder, *Z. Anorg. Allg. Chem.* (1962) 318, 176.
[16] H.A. Harwig, *Z. Anorg. Allg. Chem.* (1978) 444, 151.
[17] P.D. Battle, C.R.A. Catlow, J. Drennan and A.D. Murray, *J. Phys. C: Solid State Phys.* (1983)16, L561.
[18] H.M. Rietveld, *J. Appl. Crystallogr.* (1968) 2, 65.
[19] F. Izumi and T. Ikeda, *Mater. Sci. Forum* (2000) 321-324, 198.
[20] F. Izumi and K. Momma, *Solid State Phenom.* (2007)130, 15.
[21] M. Yashima, *J. Am. Ceram. Soc.* (2002) 85, 2925.
[22] K. Ohoyama, T. Kanouchi, K. Nemoto, M. Ohashi, T. Kajitani and Y. Yamaguchi, *Jpn. J. Appl. Phys.* (1998) *Part 1*37, 3319.
[23] M. Sakata, T. Uno, M. Takata and C.H. Howard, *J. Appl. Cryst.* (1993) 26, 159.
[24] F. Izumi and R.A. Dilanian, *Recent Res. Development Physics* (2002) 3, 699.

[25] M. Yashima, *Solid State Ionics*, 179, 797 (2008).
[26] M. Yashima, *J. Ceram. Soc. Jpn.* (2009) 117, 1055.
[27] M. Yashima and D. Ishimura, *Appl. Phys. Lett.* (2005) 87, 221909.
[28] P.D. Battle, C.R.A. Catlow, J.W. Heap, and L.M. Moroney, *J. Solid State Chem.* (1986) 63, 8.
[29] P.D. Battle, C.R.A. Catlow, J.W. Heap, and L.M. Moroney, *J. Solid State Chem.* (1987) 67, 42.
[30] M. Mogensen, N.M. Sammes and G.A. Tompsett, *Solid State Ionics* (2000) 129, 63.
[31] H. Inaba and H. Tagawa, *Solid State Ionics* (1996) 83, 1.
[32] T. Kudo and Y. Obayashi, *J. Electrochem. Soc.* (1976) 123, 415.
[33] M. Goedickemeier and L.J. Gauckler, *J. Electrochem. Soc.* (1998) 145, 414.
[34] M. Yashima, D. Ishimura, Y. Yamaguchi, K. Ohoyama and K. Kawachi, *Chem. Phys. Lett.* (2003) 372, 784.
[35] J. Faber Jr., M.A. Seitz and M.H. Mueller, *J. Phys. Chem. Solids* (1976) 37, 903.
[36] J. Faber Jr., M.A. Seitz and M.H. Mueller, *J. Phys. Chem. Solids* (1976) 37, 909.
[37] J. Faber Jr., *Physica* B (1988) 150, 241.
[38] M. Yashima and S. Kobayashi, *Appl. Phys. Lett.* (2004) 84, 526.
[39] M. Yashima, S. Kobayashi and T. Yasui, *Solid State Ionics* (2006) 177, 211.
[40] M. Yashima, S. Kobayashi and T. Yasui, *Faraday Discussions* (2007) 134, 369; M. Yashima, *Bull. Ceram. Soc. Jpn.* (2007) 42, 932 (in Japanese).
[41] K. Berber, U. Martin, Z, Mursic, J. Schneider, H. Boysen and F. Frey, *Mater. Sci. Forum* (1991) 79-82, 685.
[42] D.Y. Wang, D.S. Park, J. Griffth and A.S. Nowick, *Solid State Ionics* (1981) 2, 95.
[43] M. Lerch, H. Boysen and P.G. Radaelli, *J. Phys. Chem. Solids* (1997) 58, 1557.
[44] M. Ozawa, M. Kimura, and A. Isogai, *J. Alloys Compd.* (1993) 193, 73; A. Trovarelli, *Catal. Rev. Sci. Eng.* (1996) 38, 439.
[45] M. Yashima, Chapter 1, pp.1-45 in *Catalysis by Ceria and Related Materials 2nd Edition*, ed. A. Trovarelli and P. Fornasiero, Catalytic Science Series, Vol. 12, Imperial College Press, London (2013).
[46] T. Wakita and M. Yashima, *Acta Crystallogr.* (2007) B63, 384.
[47] T. Wakita and M. Yashima, *Appl. Phys. Lett.* (2008) 92, 101921.
[48] M. Yashima and T. Wakita, *Appl. Phys. Lett.* (2009) 94, 171902.
[49] M. Yashima, S. Sasaki, Y. Yamaguchi, M. Kakihana, M. Yoshimura and T. Mori, *Appl. Phys. Lett.* (1998) 72, 182.
[50] M. Yashima, K. Morimoto, N. Ishizawa, and M. Yoshimura, *J. Am. Ceram. Soc.* (1993) 76, 1745.
[51] M. Yashima, K. Morimoto, N. Ishizawa, and M. Yoshimura, *J. Am. Ceram. Soc.* (1993) 76, 2865.
[52] M. Yashima, H. Takashina, M. Kakihana, and M. Yoshimura, *J. Am. Ceram. Soc.* (1994) 77, 1869.
[53] M. Yashima, H. Arashi, M. Kakihana, and M. Yoshimura, *J. Am. Ceram. Soc.* 77, 1067 (1994).
[54] S. Lemaux, A. Bensaddik, A.M.J. van der Eerden, J.H. Bitter, and D.C. Koningsberger, *J. Phys. Chem.* (2001) B 105, 4810.
[55] M. Yashima, K. Oh-uchi, M. Tanaka and T. Ida, *J. Am. Ceram. Soc.* (2006) 89, 1395.
[56] M. Boaro, C. Leitenburg, G. Dolcetti, and A. Trovarelli, *J. Catal.* (2000) 193, 338.

[57] H.P. Rooksby and E.A.D. White, *J. Am. Ceram. Soc.* (1964) 47, 94.
[58] J.G. Allpress and H.J. Rossell, *J. Solid State Chem.* (1979) 27, 105.
[59] H.J. Rossell, *J. Solid State Chem.* (1979) 27, 115.
[60] H.J. Rossell, *J. Solid State Chem.* (1979) 27, 287.
[61] V.P. Shirotinkin,A.A. Evdokimov, V.K. Trunov, *Russ J. Inorg. Chem.* (1982) 27, 931.
[62] Y. Yokogawa and M. Yoshimura, *J. Am. Ceram. Soc.* (1991) 74, 2077.
[63] T. Tanaka, N. Ishizawa, M. Yoshimura, F. Marumo and H. Oyanagi, *J. Solid State Chem.* (1995) 114, 79.
[64] Y. Yokogawa and M. Yoshimura, *J. Am. Ceram. Soc.* (1997) 80, 1965.
[65] O. Yokota, M. Yashima, N. Yamamoto and M. Yoshimura, *J. Am. Ceram. Soc.* (1997) 80, 2429.
[66] M. Yashima and T. Tsuji, *Chem. Mater.* (2007) 19, 3539.
[67] Y. Ikuma, Y. Tsubaki, Y. Nakao, Y. Yokogawa and M. Yoshimura, *Solid State Ionics* (1990) 40/41, 258.
[68] S. Kim, M. Yashima, M. Kakihana and M. Yoshimura, *J. Alloys and Compounds* (1993) 192, 72.
[69] A. Petric and P. Huang, *J. Mater. Chem.* (1995) 5, 607.
[70] M. Yashima, Q. Xu, A. Yoshiasa and S. Wada, *J. Mater. Chem.* (2006) 16, 4393.
[71] M. Yashima, T. Sekikawa, D. Sato, H. Nakano and K. Omoto, *Cryst. Growth Des.* (2013) 13, 829.

In: New Research Trends of Fluorite-Based Oxide Materials ISBN: 978-1-63117-350-9
Editors: Akio Nakamura and Junichiro Mizusaki © 2015 Nova Science Publishers, Inc.

Chapter 4

NEW DEFECT-CRYSTAL-CHEMISTRY APPROACH TO FLUORITE-BASED MO_2-$LNO_{1.5}$ SOLID SOLUTIONS (M^{4+} = CE AND TH; LN^{3+} = LANTHANIDE)

Akio Nakamura*
Advanced Science Research Center,
Japan Atomic Energy Agency, Ibaraki, Japan

ABSTRACT

This chapter gives a comprehensive review of a new defect-crystal-chemistry approach to coupled non-Vegardianity and non-random defect structure of entitled parent fluorite-type oxide solid solutions as two sides of 'distortion-dilation' phenomenon of C-type $LnO_{1.5}$ [1-5]. The model provides a direct link to their controversial defect structure, enabling their possible real local-structure based systematic defect-property description: Besides compatibility with ion-packing model, these include presently (1) extraction of overall non-random oxygen coordination number of both cations from lattice-parameter data analysis, (2) derivation of almost all the mutually non-randomly coordinated cation ↔ anion concentrations and (3) a new interpretation for ionic-conductivity maximum behavior in their low $LnO_{1.5}$ range. All these are discussed with their reported crystal- and defect-structural, ionic-conductivity and defect-thermodynamic data, touching briefly also to its some next-step extension/elaboration directions in the future.

1. INTRODUCTION

Fluorite (F)-type MO_2 (M^{4+} = Hf, Zr, Ce, Th, U, Np and Pu *etc*) (Fig. 1a) and its oxygen-vacancy (Vo) ordered superstructure of C-type lanthanide (Ln) and light-Ln sesquioxide $LnO_{1.5}$ (Ln^{3+} = La-Lu, Y, In and Sc *etc*) (Fig. 1b) with doubled lattice parameter of $a_0(C) = 2a_0(F)$ form various grossly oxygen-deficit defect-fluorite (DF)-type solid solutions (ss) with

* E-mail address: nakamura.akio@jaea.go.jp, Advanced Science Research Center, Japan Atomic Energy Agency, 2-4 Shirakata, Shirane, Tokai, Naka, Ibaraki 319-1195, JAPAN

the genral formula; (1-y)MO_2 + $y$$LnO_{1.5}$ = $M_{1-y}Ln_yO_{2-y/2}(Vo_{y/2})$, being often abbreviated as M-Ln in the following. These so-called defect-fluorite (DF) oxides have been the object of enormous research efforts for over these several decades due to their well-known technological importance. Though the application of their actinide (M^{4+} = An = Th, U and Pu *etc*) counterparts is mostly limited to nuclear area, that of cubic-'stabilized' zirconia (SZ; M^{4+}= Zr) from the distorted monoclinic (M) ZrO_2 is far more widespread, such as radioactive nuclear-waste forms and inert-matrix fuels (IMF) in nuclear area and solid electrolytes for oxygen sensors/pumps and solid oxide fuel cells (SOFCs), refractory/structural ceramics, catalysts, thermal-barrier coatings (TBC), dielectric substrates and synthetic teeth / jewels *etc* in electrochemical and ceramic *etc* areas. Its Hf analog, stabilized hafnia (SH), is also of interest as neutron shielding / absorption devices. Shortly touching its one electrochemical use, yttria-stabilized zirconia (YSZ; Zr-Y) fabricated at the lower M/DF phase boundary is the representative solid electrolyte due to its superior mechanical property, high oxide-ion conductivity (σ(ion)) and long-term phase and structural stability. The YSZ based home SOFC system has launched in Japan in 2008. The development of higher-σ(ion) Sc-SZ and/or Ce-Ln based SOFC for lower-temperature (< 800°C) operation is a today's hot issue.

The knowledge of defect structure and thermodynamic properties is indispensable for better understanding and designing their key properties for various applications such as σ(ion), radiation tolerance, catalytic activity and phase and structural stability *etc.* As many of them, *e.g.,* σ(ion) are plausibly more sensitive to the system's short-range local structure in the vicinity of Vo rather than its long-range average structure, recently besides various X-ray, neutron and electron diffraction (XRD, ND and ED) [6-13], calorimetric [14-19] and computational [20-29] *etc* studies, a variety of spectroscopic (NMR [30-36], Mőssbauer [37-47], EXAFS [48-55], Raman [56-60] etc) studies have been widely made. Yet, much has remained elusive up to now as to the exact local-structure nature of these systems and its correlations with various key physicochemical properties. For instance, we still do not understand well why they almost commonly exhibit σ(ion) maximum (σ(ion)(max)) behavior in so low $LnO_{1.5}$ range (y < 0.10-0.20), despite numerous modeling attempts [61-72]. In recent years such modeling works are increasingly shifting more to molecular-dynamics (MD) and Monte-Carlo (MC) simulations type (either classical or ab-initio) due to the rapid progress of silicon (Si)-based computer technology. If Fermi-surface topology and band structure, i.e. Fermiology is the face of such (metallic and semiconducting) electronic conductors, we do not seem to know yet such detailed face of these, generally insulating, highly-defective DF oxides, to understand their various defect properties.

One could more concretely specify the problem here referring to the basic crystal structure of the two end members: While M^{4+} in F MO_2 (Fig. 1a) has a regular oxygen cube (CN = Ⅷ), Ln^{3+} in C $LnO_{1.5}$ (Fig. 1b) has two kinds of largely-distorted defective oxygen octahedrons (CN = Ⅵ) with two O^{2-}s regularly missing either face- or body-diagonally in 3 : 1 ratio, i.e., 24d or 8b site with each one <110> or <111> Vo-Vo pair, respectively. Where both 24d Ln^{3+} and 48e O^{2-} are largely shifted from each ideal-F position [73]. Such shift only of 24d Ln^{3+} is schematically depicted there by arrow. So, in $M_{1-y}Ln_yO_{2-y/2}$, the MO_2-rich side disordered DF-type phase seen in their phase diagram in Fig. 2, which can be viewed as an oxygen-excess disordered version of C-type ordered solid solution $Ln(M)O_{1.5+x}$ seen in the $LnO_{1.5}$-rich side, would have three kinds of O^{2-} 8-, 7- and 6-fold coordinated cation sites, whose concentrations, {M^{4+}(Ⅷ,Ⅶ,Ⅵ)} and {Ln^{3+}(Ⅷ,Ⅶ,Ⅵ)}, metal-O^{2-} bond lengths, atomic positional shifts and local distortions etc are largely changing with *y* in some mutually closely correlated and plausibly strongly non-random manner. It might be still a simplified picture not valid for the latter ordered C-type solid solution phase (Fig. 2), for it does not distinguish between the 24d and 8b sites, whose inclusion makes the situation much more complex. It might not be valid even in the former, for it is also often reported to involve extensive C-type short-range order [1-3,5,6]. Another complication exists in SZ(SH) with intermediate pyrochlore (P)-type ordering; ideal P $Ln_2Zr(Hf)_2O_7$ at y=0.50 formed for the larger (early) Lns (=La-Gd) (Fig. 3) [74] has long-range ordered unique cation-anion arrangement, in which the smaller M^{4+}=Zr(Hf) and the larger Ln^{3+} are CN = Ⅵ and Ⅷ, respectively, totally switched from those in F MO_2 and C $LnO_{1.5}$ (Figs. 1a-b). Such P-type extreme $Zr(Hf)^{4+}$-2Vo association is either entirely opposite to a naive electrostatic dopant (Ln')-defect ($Vo^{\cdot\cdot}$) associate scheme (in Kröger-Vink notation [75]) often supposed for parent F-type DF oxides of many large M^{4+} = Ce and An^{4+} = Th, U, Np and Pu *etc* adopting themselves cubic F structure (Fig. 1a). Also note that Dy-SZ and Er-SH with the smaller Ln^{3+}s than Gd in Fig. 2 are not P- but δ-type ordered, the ideal δ $Zr(Hf)_3Ln_4O_{12}$ (= $[Zr(Hf)^{4+}]_{VI}$ $[Zr(Hf)^{4+}{}_2Ln^{3+}{}_4]_{VII}$ O_{12}) at y = 4/7 being much less $Zr(Hf)^{4+}$-V_O associative than P; $CN(Zr(Hf)^{4+})$ = 6.67 < $CN(Ln^{3+})$ = 7 [76]. Such P or δ-type ordering over the extended ~0.33 ≤ y ≤ ~070 range is the most likely origin of a broad lattice-paramter a_0(ss) hump observed over there with these SZ(SH)s in Fig. 2 [1-5,47,77]. These facts imply that dopant Ln^{3+} - host M^{4+} ionic-radii mismatch is a dominant factor determining the basic a_0(ss), crystal- and local-structural, σ(ion) and thermodynamic etc properties of these DF oxides.

Since such full structural data can hardly be obtained with any diffraction or spectroscopic method, one should inevitably rely on some modeling tool here to integrate the vast amount of structural, σ(ion) and thermodynamic etc data into a coherent picture of their whole defect properties. We seem to have no reliable tool even today to do this in the sense that no previous model can provide reliable key concentration data of {M4+(Ⅷ,Ⅶ,Ⅵ)} and {Ln3+ (Ⅷ,Ⅶ,Ⅵ)} for any DF oxide as the most important basic shape (piece) of its face.

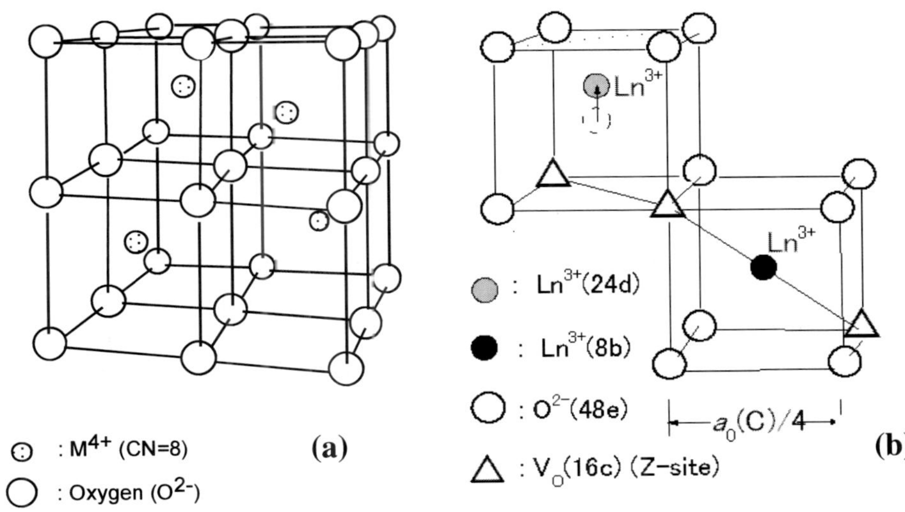

Figure 1. (a) Fluorite (F) structure of MO_2. (b) a part of C-type structure of $LnO_{1.5}$.

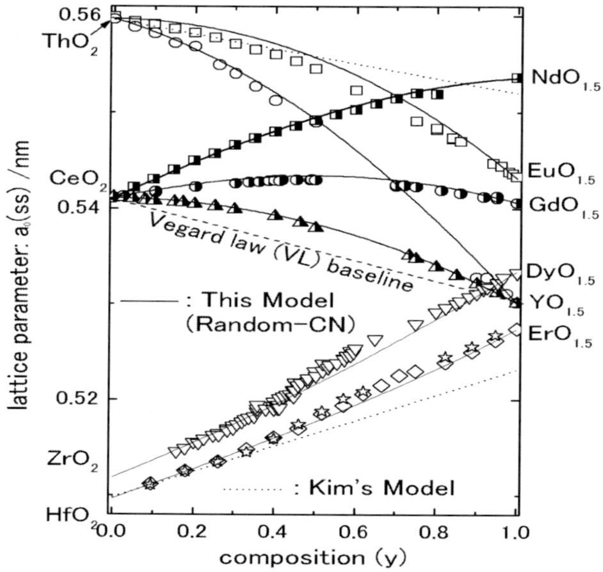

Figure 2. Phase-diagram lattice-parameter a_0(ss) - y plot of some DF MO_2-$LnO_{1.5}$ solid solutions (ss).

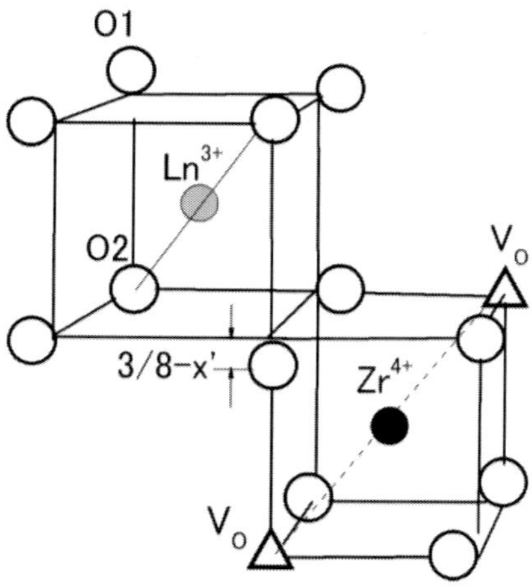

Figure 3. A part of pyrochlore (P) structure of $Ln_2Zr(Hf)_2O_7$ at $y = 0.50$.

As an alternative macroscopic approach to this problem, we have formerly proposed a random-CN lattice-parameter (a_0(ss)) model [47,77]: When viewed over the entire y range in Fig. 2, these DF M-Ln solid solutions (ss) show 'so-far almost completely overlooked' marked non-Vegardianity shifting from positive (>0) to negative (<0) with decreasing $r_C(M^{4+})$(Ⅷ)(nm) = 0.105 (Th) → 0.097 (Ce) → 0.084 (Zr) → 0.083 (Hf) [78], and this concise a_0(ss) model incorporating 'distortion dilation' of C $LnO_{1.5}$ relative to F MO_2 in Fig. 4; f_F (= a_0(F)) < f_C (= a_0(C)/2) (f_F and f_C are each a_0 functional), can describe fairly well such non-linear a_0(ss) data especially for parent-F type DF M^{4+} = Ce and Th systems using Shannon's $r_C(M^{4+}, Ln^{3+})$ data (Fig. 5), surmounting the limit of many y-linear Kim's [79] and ion-packing a_0(ss) models [80-83]. This fact suggests that the M^{4+} = Ce and Th are indeed the F MO_2 - C $LnO_{1.5}$ binary systems basically describable with this F-C binary a_0(ss) model.

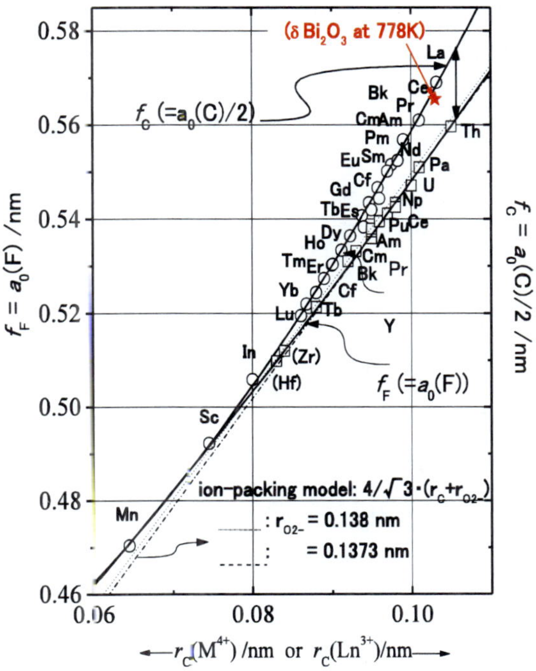

Figure 4. a_0(F) vs. r_C(M^{4+})(VIII) plot for F MO_2 and a_0(C)/2 vs. r_C(Ln^{3+})(VI) plots for C $LnO_{1.5}$.

Therefore, as the next step its elaboration to the more-general non-random oxygen-CN model has been attempted aiming at its possible establishment as a refined quantitative a_0(ss) model for parent F-type DF oxides that can extract their key non-random CN(Ln^{3+}, M^{4+}) data [1]. This is entirely different from fitting the a_0(ss)(random) curve to each a_0(ss)(exp) data fragmentally for some M-Lns by tuning Shannon's r_C(M^{4+}, Ln^{3+}) data as done in Fig. 2 [77] for the a_0(ss)(exp) ≠ a_0(ss)(random) of these presumably variously non-random systems. In [1] this refined a_0(ss) model, furnished with systematized Shannon's r_C(M^{4+}, Ln^{3+}) expressions (Fig. 5), has successfully attained this aim for total twelve M-Lns (M^{4+} = Ce and Th; Ln^{3+} = Y, Gd, Eu, Sm, Nd and La), providing a novel possible unified picture of non-Vegardianity and non-random defect structure in these DF oxides. The model, being compatible with ion-packing model, also provides a handy method to derive their metal-oxygen bond-length data. Furthermore, it has opened a new direct route to their detailed non-random local-structure and σ(ion) analysis [2]:

Full use of these CN(M^{4+}, Ln^{3+}) data has enabled the derivation of almost all the mutually non-random cation↔anion concentration data, {M^{4+}(VIII,VII,VI)}, {Ln^{3+}(VIII,VII,VI)}, and the counter Ln^{3+} n(=0-4)-fold coordinated {O^{2-}(n)} and {Vo(n)}, and the last two have further combined to provide a new plausible σ(ion)(max) description in their low y range.

This chapter gives a comprehensive review of this whole story of defect-crystal-chemistry (DCC) of parent F-type DF oxides basically following [1,2] but including also some more recent analysis results [3-5]. This also alludes briefly to its one major next challenge toward a new defect thermodynamic description of DF oxides. However, this does not cover the a_0(ss)-model extension to more complex P(δ)-type SZ(SH)s with a broad a_0(ss) hump (Fig. 2) [5], which needs long additional pages for its description, only briefly touching it in the last part. In this sense this is a good-spot interim report of our ongoing efforts to establish the DCC model as an unique integrated approach to these DF oxides, either parent-F or stabilized-P(δ) type, focusing mainly on the former.

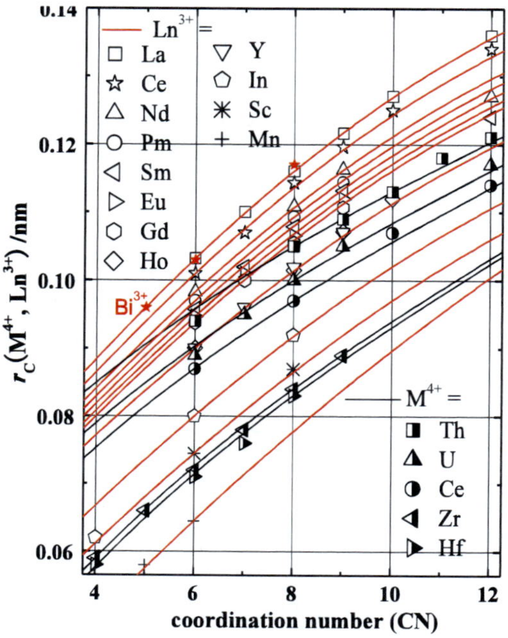

Figure 5. Reported (symbol) and systematized (line) Shannon's $r_C(M^{4+}, Ln^{3+})$ data as a function of CN.

2. Model Description and Lattice-Parameter Data Analysis

2.1. Generalized Vegard-Law Lattice-Parameter Representation

The new DCC non-random-CN a_0(ss) model for $M_{1-y}Ln_yO_{2-y/2}$ [1], as well as its previous random-CN version [77], has a generalized Vegard-law (VL) a_0(ss) formula of;

$$a_0(ss) = (1\text{-}y) \cdot f_F(r_C(ss)) + y \cdot f_C(r_C(ss)) \text{ at } r_C(ss) = (1\text{-}y) \cdot r_C(M^{4+}) + y \cdot r_C(Ln^{3+}) \quad (1\text{-}2).$$

using lattic–parameter functionals of f_F *and* f_C in Fig. 4. Namely, the intercept of the vertical straight line at the system's average cation radius r_C(ss) (drawn at around 0.105 nm in Fig. 4) with the f_F or the f_C curve provides the system's a_0, if this were at CN = 8 or 6 of the hypothetic F $M'O_2$ or C $Ln'O_{1.5}$ at its r_C(ss) (Eq. (2)), respectively. As the actual system is at CN = 8-2y as a (1-y) $M'O_2$ + y·$Ln'O_{1.5}$ hybrid mixture, its a_0(ss) is given by their arithmetic y average (Eq. (1)) rather than that of the two actual end members (MO_2 and $LnO_{1.5}$) as with the conventional VL;

$$a_0(VL) = (1\text{-}y) \cdot f_F(r_C(M^{4+})(\text{VIII})) + y \cdot f_C((r_C(Ln^{3+})(\text{VI})) \text{ (linear in } y) \quad (3),$$

drawn in Fig. 2 for Ce-Y. r_C(ss) in Eq. (2) can be calculated for the entire $0 \leq y \leq 1.0$ ($8 \geq CN \geq 6$)

range using Shannon's $r_C(M^{4+},Ln^{3+})$ data (Fig. 5), e.g. for the 'random'solid solution with;

$$CN(av) = CN(M^{4+}) = CN(Ln^{3+}) = CN(random) = 8 - 2\,y \quad (4)$$

Converting r_C(ss) to y provides the usual a_0(ss)(random) - y curves being drawn in Fig. 2. 'Non-random' here only means if Vo associates more to Ln^{3+} or M^{4+}, whether in ordered-C or disordered-DF type state. f_F and f_C in Eq. (1) and $r_C(M^{4+}, Ln^{3+})$ in Eq. (2) are therefore two sets of pair equations necessary for the a_0(ss) analysis. All these in Figs. 4-5 are gien by [1];

$$f_F(r_C) = a_0(F) = 0.3571+1.5016\cdot r_C+4.076\cdot r_C^2 \text{ (nm) (for } r_C = r_C(M^{4+})(\text{VIII})) \quad (5a),$$

$$f_C(r_C) = a_0(C)/2 = 0.40693+0.037411\cdot r_C+14.7973\cdot r_C^2 \text{ (nm)(for } r_C = r_C(Ln^{3+})(\text{VI})) \quad (5b),$$

$$r_C(M^{4+}) = r_C(\text{VIII})+\{4.46\cdot 10^{-4}(CN-8)-1.54610^{-5}\cdot(CN-8)^2+1.2\cdot 10^{-6}(CN-8)^3\}/r_C(\text{VIII}) \quad (6a),$$

$$r_C(Ln^{3+}) = r_C(\text{VI})\cdot\{1+6.7\cdot 10^{-2}\cdot\exp[\{r_C(La^{3+})(\text{VI})/r_C(Ln^{3+})(\text{VI})-1\}^{1.5}]\cdot(CN-6)$$

$$-\,2.3\cdot 10^{-3}\cdot(CN-6)^2-2\cdot 10^{-5}\cdot(CN-6)^3\}\text{(nm)} \quad (6b).$$

Eqs. (6a-b) in Fig. 5, 'systematized' Shannon's $r_C(M^{4+}, Ln^{3+})$ expressions, are introduced in [1] to make a systematic a_0(ss) analysis of total twelve M-Lns (M^{4+}=Th and Ce; Ln^{3+}=Y-La) and extract their non-random $CN(Ln^{3+}, M^{4+})$ data.

Such $CN(Ln^{3+}, M^{4+})$ pair should always fulfill the basic overall average CN condition that;

$$(1 - y)\cdot CN(M^{4+}) + y\cdot CN(Ln^{3+}) = CN(av) = CN(random) = (8 - 2\,y) \quad (7).$$

Their actual functional forms used to fit the a_0(ss) model curve to each a_0(ss) data are basically the following type with a suffix-A linear and a suffix-B non-linear term, which fulfill Eq. (7);

$$N(Ln^{3+})=CN(Ln^{3+})_A+CN(Ln^{3+})_B=\{(8-n)-(2-n)\cdot y\}+(1-y)\cdot A/[b+\{b'-\exp(y_m-y)\}^2] \quad (8a)$$

$$CN(M^{4+})=CN(M^{4+})_A+CN(M^{4+})_B=\{8-(2-n)\cdot y\}-y\cdot A/[b+\{b'-\exp(y_m-y)\}^2] \quad (8b).$$

As our prime interest is in solute (Ln^{3+}) - defect (V_O) interaction in the DF $y \leq\sim 0.50$ range, $CN(Ln^{3+})$ is given first. These n, A, b, b' and y_m values in Eqs. (8a-b) are chosen so as to satisfy $6 \leq CN(M^{4+}, Ln^{3+}) \leq 8$ at any $0 \leq y \leq 1.0$. This constraint, together with the obvious fact revealed in Fig. 5 that Shannon's $r_C(Ln^{3+}, M^{4+})$ data for CN≤5 are absent for M^{4+}=Th and Ce and Ln^{3+}=La-Y means that only the highest three CN = VIII, VII and VI sites are present for use by both cations in these DF oxides. This 'restricted (non-) randomness' is a key concept used in the DCC model. It, being fully consistent with some recent ^{89}Y MAS-NMR coordination-structure data on these systems [30-34], is utilized to show the compatibility of 'CN-continuous' Eqs. (6a-b) with their discrete-lattice model version [1] and further to make a new local-structure and σ(icn) analysis of these systems [2], as described later.

Eq. (7) means that only one CN(Ln^{3+}) can in principle describe all the macroscopic-a_0(ss) - microscopic non-randomness behavior of these systems in a fully-coupled unified manner. This unified non-Vegardianity - non-randomness picture of parent F-type DF oxides is first validated below aginast their actual a_0(ss)(exp) behavior.

2.2. Latter-Parameter Data Analysis of the M^{4+} = Th and Ce Systems

2.2.1. Contrastive Coupled Non-Vegardian and Non-Random Behavior

To know the systematic non-Vegardianity trends among total twelve (six each) M-Lns (M^{4+}= Th and Ce; Ln^{3+}= Y-La) over the wide range of $r_C(Ln^{3+})$(VI)(nm) = 0.090 (Y) → 0.0938 (Gd) → 0.0947 (Eu) → 0.0958 (Sm) → 0.0983 (Nd) → 0.1032 (La), each 'apparent' non-Vegardianity data were extracted by subtracting the linear a_0(VL) (Eq. (3)) from the raw a_0(ss)(exp) data**;**

$$\text{Apparent non-Vegardianity: } \Delta a_0(\text{ss})(\text{exp}) = a_0(\text{ss})(\text{exp}) - a_0(\text{VL}) \quad (9a),$$

and their detailed analysis made. Figs. 6a-b summarize in pair each Δa_0(ss) data in color symbol and each reference (ref.) random and best non-random Δa_0(ss) model curves for M^{4+} = Th and Ce, respectively. Fig. 7a-b summarize in pair the corresponding best non-random CN(M^{4+}, Ln^{3+}) curves used to draw the non-random Δa_0(ss) curves in Figs. 6a-b. Fig. 8a-b summarize each 'generalized' non-Vegardianity defined by;

$$\text{Generalized non-Vegardianity: } \delta(\Delta a_0(\text{ss})(\text{exp})) = \Delta a_0(\text{ss})(\text{exp}) - \Delta a_0(\text{ss})(\text{random}) \quad (9b),$$

together with each best $\delta(\Delta a_0$(ss)(non-random)) model curve, i.e., the random → non-random change of each Δa_0(ss)-model curve indicated with dot arrow in Figs. 6a-b. These three sets of Figures contain baisically all the analysis results and information of markedly non-Vegardian a_0(ss) data of both systems and their coupled non-random behavior clarified in [1], also including some more-recent analysis results. These data clearly illuminate the contrastive coupled non-Vegardian - non-random behavior of both systems in a systematic manner almost for the first time in their long history of thermodynamic and defect-property study.

In Figs. 6a-b the present F-C binary a_0(ss) model can indeed provide a first quantitative description of simply positively bowing-type non-Vegardian Δa_0(ss) data of both systems by incorporating each-specific non-random CN(Ln^{3+}, M^{4+}) structure shown in Figs. 7a-b. The M^{4+} = Th (Fig. 6a) are overall much more positively Δa_0(ss) than the M^{4+} = Ce (Fig. 6b), due to the location of the largest host Th^{4+} where C-type distortion dilation is the largest in Fig. 4. This however needs more careful consideration, for both exhibit just the opposite trend in Figs. 6a-b that Th-La(Y) and Ce-La(Y) with the most(least) distortion-dilated $LaO_{1.5}$($YO_{1.5}$) in Fig. 4 are in fact the least(most) Δa_0(ss), both Δa_0(ss) steadily decreasing with Ln^{3+} = Y → La. As shown in [1], it is the random → non-random CN(Ln^{3+}) change, δCN(Ln^{3+}) = CN(Ln^{3+}) - (8-2y) in Figs. 7a-b, and the generalized non-Vegardian $\delta(\Delta a_0$(ss)) in Figs. 8a-b that have a direct 1:1 correspondence via the random → non-random r_C(ss) change, δr_C(ss);

$$\delta(\Delta a_0(\text{ss})) \sim A \cdot \delta r_C(\text{ss}) = A \cdot y \cdot (\alpha_{Ln} - \alpha_M) \cdot \delta CN(Ln^{3+}) = A \cdot y \cdot \Delta\alpha_{Ln\text{-}M} \cdot \delta CN(Ln^{3+}) \quad (10).$$

Here A is a constant and $\Delta\alpha_{Ln-M}(= (\alpha_{Ln} - \alpha_M))$ the slope difference of $r_C(Ln^{3+}, M^{4+})$ - CN plot in Fig. 5. As drawn in dashed line in Figs. 8a-b, A = 2.65 (2.55) for M^{4+} = Th (Ce) gives each approximate $\delta(\Delta a_0(ss))$ - y curve. So, one can assert that $\delta CN(Ln^{3+})$ in Figs. 7a-b first induces $\delta r_C(ss)$ (Eq. (2)) in Fig. 4 and then produces $\delta(\Delta a_0(ss))$ in Figs. 8a-b, clarifying the onset of two modes of generalized non-Vegardianity - non-randomness coupling bordered by the ideally random generalized-Vegardian system with both $CN(Ln^{3+}) = 0$ and $\delta(\Delta a_0(ss)) = 0$;

case (i) Ln^{3+}-V_O associative $\delta CN(Ln^{3+}) < 0$ $(CN(Ln^{3+}) < (8-2y) < CN(M^{4+}))$ for $\delta(\Delta a_0(ss)) < 0$ (11a)

case (ii) M^{4+}-V_O associative $\delta CN(Ln^{3+}) > 0$ $(CN(Ln^{3+}) > (8-2y) > CN(M^{4+}))$ for $\delta(\Delta a_0(ss)) > 0$ (11b).

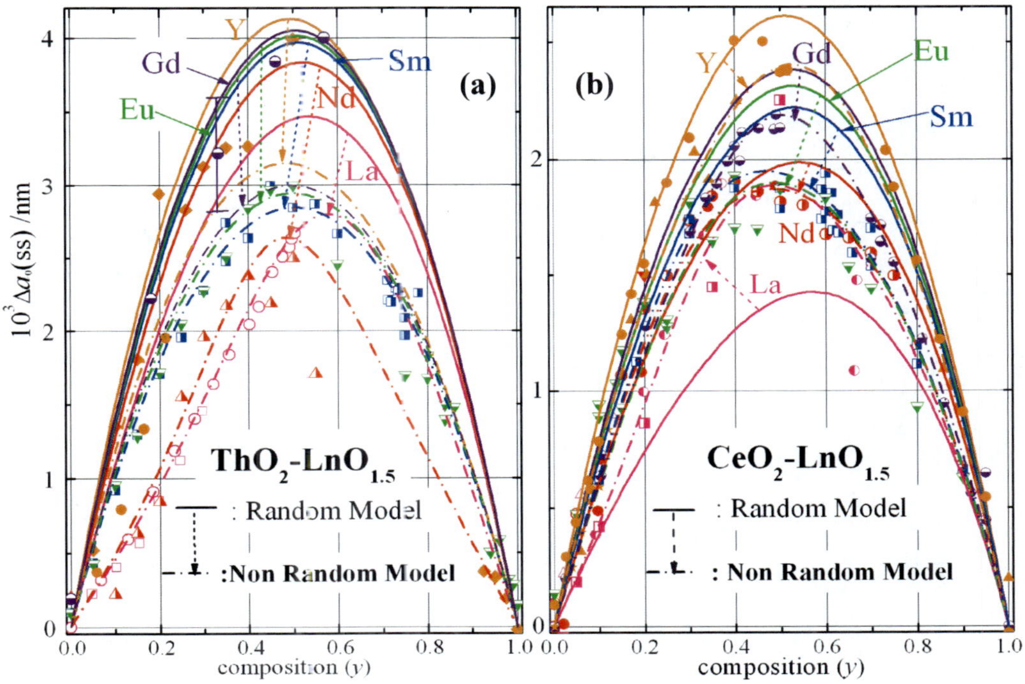

Figure 6. (a) Apparent non-Vegardianity ($\Delta a_0(ss)$) vs. composition (y) plots for Th-Lns, and Ce-Lns, (b) their DCC random and non-random model curves. (See text for details).

We thus know in Figs. 7a-b and 8a-b that Th-Lns are all basically strongly case (i) Ln^{3+}-V_O associative largely negatively generalized non-Vegardian $((\delta\Delta a_0(ss)) << 0)$, due to the highest-CN=Ⅷ adhering nature of the largest host Th^{4+}. In contrast, Ce-Lns are much more weakly and variably case (i) to (ii) non-random negatively to positively generalized non-Vegardian, because the smaller host Ce^{4+} can adopt the lower CN=Ⅶ and Ⅵ more flexibly, thus showing the grosser miscibility with many $LnO_{1.5}$s than the former. So, the cations / O^{2-} ionic-radii ratios defined by;

$$IRR(M/O) = r_C(M^{4+})(\text{VIII}) / r_C(O)(\text{IV}),\ IRR(Ln/O) = r_C(Ln^{3+})(\text{VI}) / r_C(O)(\text{IV}) \quad (12a\text{-}b),$$

especially the former is known to likely play the major role for both exhibiting such contrastive coupled $\delta CN(Ln^{3+})$ - $\delta(\Delta a_0(ss))$ behavior. The results of Ce-Lns also fairly agree with their σ(ion) data [84-88] that the highest $\sigma(ion)_{(max)}$ is found in Ce-Sm to Ce-Nd which have the most near random defect structure and therefore the smoothest V_O migration path. A recent ab-initio DFT study [27] indeed predicts this to occur at (the radioactive) Ln^{3+} = Pm just in between Sm and Nd, as confirmed experimentally with Sm+Nd co-doped ceria [88]. Note that just Shannon's $r_C(Pm^{3+})(VI) = r_C(Ce^{4+})(VIII) = 0.097$nm in Fig. 5. So, also the dopant / host ionic-radii ratio (IRR);

$$IRR(Ln/M) = r_C(Ln^{3+})(VI) / r_C(M^{4+})(VIII)\ (= IRR(Ln/O) / IRR(M/O)\ ) \quad (13),$$

is useful for describing their $\delta(\Delta a_0(ss))$ - $\delta CN(Ln^{3+})$ - σ(ion) correlation. Applying this scenario to Th-Lns, however, seems to contradict with rather the opposite trend in Fig. 7a that generally Th-Ln with the larger Ln^{3+} is much more non random and would be less conductive. To draw definite conclusion on this point more σ(ion) data for the latter seem necessary.

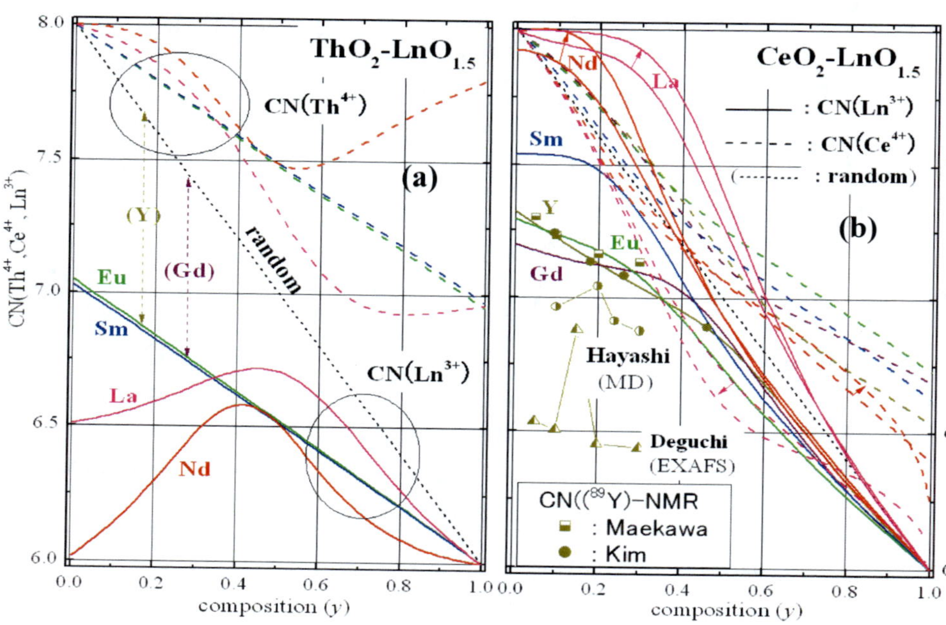

Figure 7. The best non-random $CN(M^{4+}, La^{3+})$ curves of Th-Lns (a) and Ce-Lns (b). In (b) $CN(Y^{3+})$ data of Ce-Y from NMR [31,33], EXAFS [55] and MD-simulations [22] are also plotted. NMR $CN(Y^{3+})$ data; filled circle at T_{prep} = 1500oC [33], half-filled rectangular at T_{prep} = 1650oC [31].

Another crucial factor here is the macroscopic F MO_2 - C $LnO_{1.5}$ lattice-match ratio (LMR) [1];

$$LMR(LnO_{1.5} / MO_2) = f_C(LnO_{1.5}) / f_F(MO_2) \quad (14),$$

which, as seen in Figs. 2 and 4, spans over a wide LMR range from largely positive (>> 1 for M^{4+} = Zr(Hf)) to largely negative (<< 1 for M^{4+} = Th). The global σ(ion) trend of Ce > Zr > Hf > Th accords with a general view that the most near-random and best lattice-match M^{4+} = Ce has the highest σ(ion), and increasing non-randomness (either to the case (i) or (ii) side) and lattice mismatch (either to the negative or positive side) (for M^{4+} = Th or Zr(Hf), respectively), degrades the σ(ion) of the system. Cooperation of all these IRR and LMR factors in Eqs. (12)-(14), covering mutually largely different range, makes the M^{4+} = Th, Ce and Zr(Hf) have mutually different unique δCN(Ln^{3+}) - δ(Δa_0(ss)) - σ(ion) correlations.

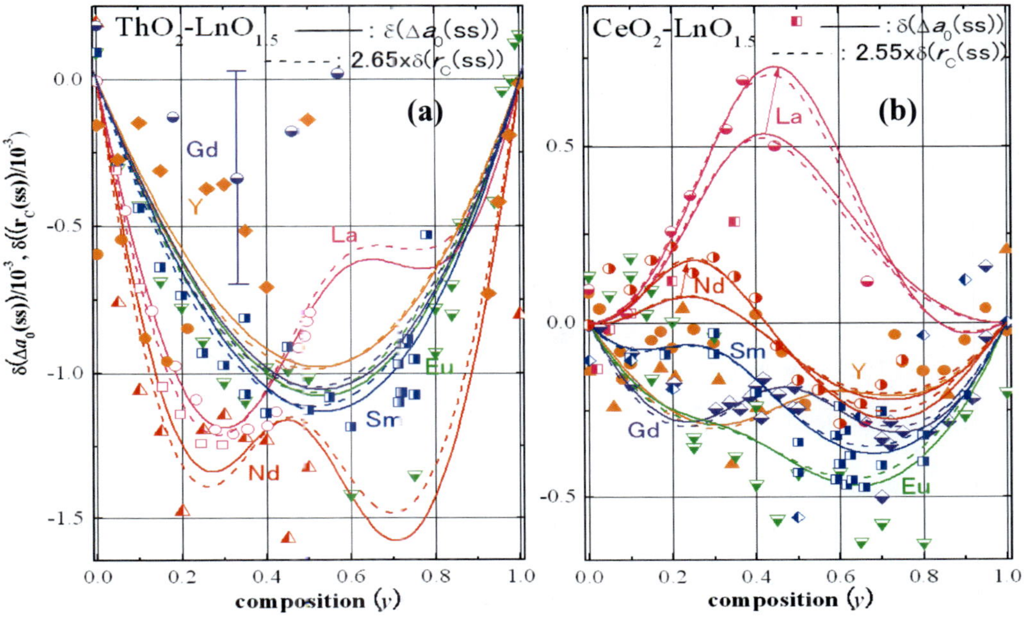

Figure 8. δ(Δa_0(ss)(exp)) -y plot of Th-Lns (a) and Ce-Lns (b) and their respective model curves [1].

2.2.2. Detailed Analysis of Th-La and Ce-Gd

Though the above Δa_0(ss) analyses based on each one representative a_0(ss) data set [1] may appear problematic, more complete analyses of almost all the reported a_0(ss) data of each system [3-5,89] have been in fact made and condensed here into these Figures for clarity. Such detailed situation is described below for each one typical example of both systems.

As to M^{4+} = Th, due both to the high sensitivity to δCN(Ln^{3+}) (large $\Delta\alpha_{Ln\text{-}Th} > 0$) and to accurate a_0(ss) data of Th-La and Th-Nd, the situation is simple. Fig. 9a-b show Δa_0(ss)-data analysis results for Th-La most sensitive to its δCN(La^{3+}) and its two a_0(ss) data sets [90,91] being in excellent agreement. So one can decide its highly non-random CN(La^{3+},Th^{4+}) curves accurately in a narrow range between the model III (pink) and IV (red) curves in Fig. 9a-b (the latter were drawn in Fig. 7a as the best). When the more non-random CN(La^{3+}, Th^{4+}) (model-II; green) that is the best non-random CN(Nd^{3+}, Th^{4+}) of Th-Nd in Fig. 7a is employed in Fig. 9b, the resultant Δa_0(ss) model curve in Fig. 9a in Th-La is definitely lower than both Δa_0(ss) data. Th-La is thus concluded to be less case (i) Ln^{3+}-V_O associative than Th-Nd. Why this is so is not clear at present. Its one possible scenario will be proposed and argued later in 3.2. One can thus conclude on M^{4+}= Th that Th-Nd is the most extremely case (i) Nd^{3+}-Vo associative most negatively δ(Δa_0(ss)), such extremity getting milder for either larger Ln^{3+} = La or smaller Sm to Y dopant. For their more complete understanding, more accurate a_0(ss) data of the samller-dopant Th-Gd and Th-Y systems (Figs. 6a and 8a), as well as their more detailed σ(ion) and local-structure data, are certainly needed.

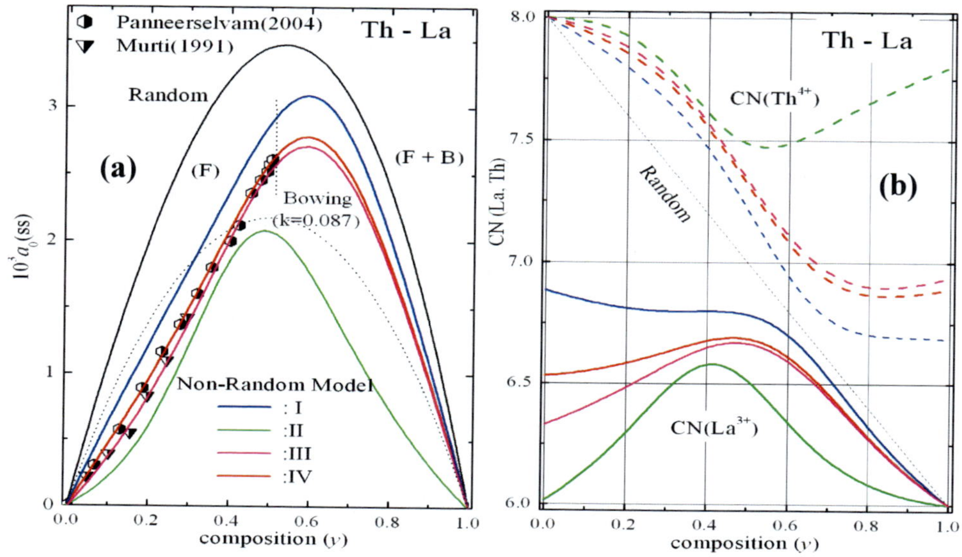

Figure 9. Δa_0(ss) analysis results of Th-La (a) and their respective model CN(La^{3+}, Th^{4+}) curves (b).

The situation in M^{4+} = Ce is generally much more delicate due both to their increasingly lower sensitivity to δCN(Ln^{3+}) for the smaller Ln^{3+} = La → Y with increasingly smaller $\Delta\alpha_{Ln\text{-}Ce}$ (>0) (Fig. 5) and to the smaller $|\delta(\Delta a_0(ss)|$ in Fig. 8b: Their $|\delta(\Delta a_0(ss)|$s except Ce-La are indeed quite small ($< 5\cdot 10^{-4}$ nm), making their meaningful a_0(ss) analysis difficult. Fig. 10a-b show such perplexing situation in Ce-Gd by plotting all the reported Δa_0(ss) data [6,14,16,92-101] and CN(Gd^{3+}, Ce^{4+}) data [21,22,55] for their several model curves. Besides Bevan's ref. data [6] (in blue) used in [1] and plotted in all these Figs. 2, 6b, 8b and 10a, more than ten sets of largely 'scattered' a_0(ss) data are reported, for which it is possible to make every description from strongly case (i) Gd^{3+}-V_O associative negatively $\delta(\Delta a_0(ss))$(<0) for the exceptionally lowest Seo's data [100] (in dark green) to just the opposite P-like extremely case (ii) Ce^{4+}-V_O associative positively $\delta(\Delta a_0(ss))$(>0) for the highest Grover's data [99] (in pink), indeed illustrating the difficulty in extracting reliable CN(Gd^{3+}, Ce^{4+}) data from this kind of a_0(ss) analysis alone, as seen in Fig. 10b. One certainly needs here other local-structure and σ(ion) etc data to judge which is the most reliable a_0(ss) data set. But, this difficulty of course means not the weakness but the strength of the DCC model in correctly predicting such only weakly $\delta(\Delta a_0(ss))$ behavior with relatively large uncertainty of these Ce-Lns. Apart from XRD-apparatus calibration and a_0-standard etc problem, one crucial material-side problem for such a_0(ss) data scatter [1] seems to be non-trivial a_0(ss) change depending on detailed heating / cooling / annealing etc condition (temperature and time) of each sample and besides possible long-term a_0(ss) drift occurring even at room temperature [92], both being likely to be most pronounced in these highest-σ(ion)(max) Ce-Lns with the most mobile O^{2-}s.

Our experience suggestes that less homogeneous the sample larger the (probably more distortion-dilated) a_0(ss) in these DF oxides. This appears to be indeed the case in most of these larger a_0(ss) data than the ref. Bevan's in Fig. 10a, except for Seo's alone-lowest data for nano-particle samples prepared by a spray-pyrolysis method [100]. On this knowledge we have recently prepared highly-homogeneous Ce-Gd samples in a deliberate Δy = 0.05 interval

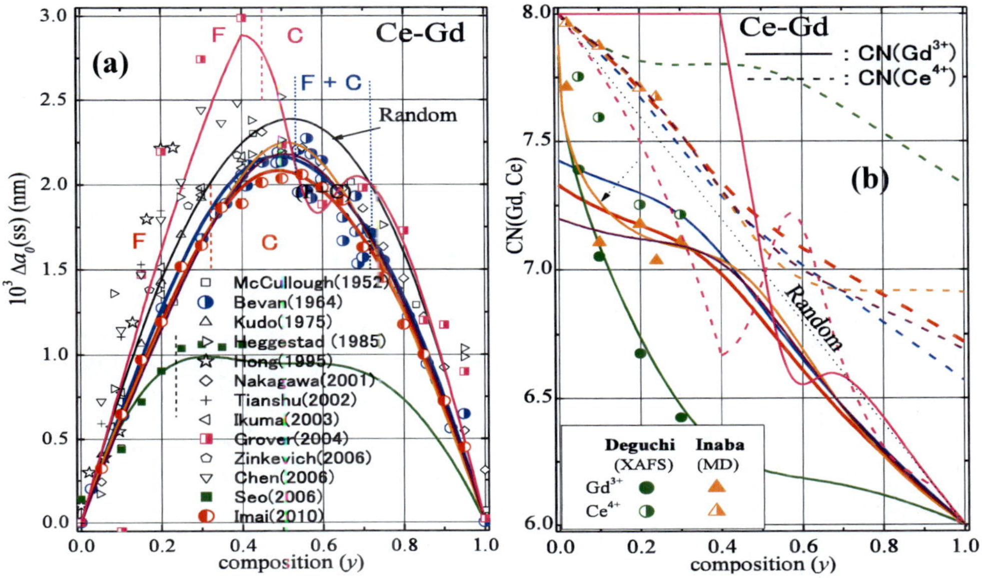

Figure 10. Detailed analysis results of Ce-Gd [89]; Δa_0(ss)(exp) and model curves (a) and their CN(Gd^{3+},Ce^{4+}) data and model curves (b). The former best Δa_0(ss) and CN(Gd^{3+},Ce^{4+}) curves of Bevan's data [1] in purple, and their here-revised one in blue. (See text for details).

covering the entire y range by a coupled wet-chemical and ceramic method at the final T_{prep} = 1450°C (< 1600°C in Bevan's), and plotted those Δa_0(ss) data (Imai (2010) in red color [3,101]). These two sets of most detailed data indeed show good agreement with each other, both lying on the lowest-a_0(ss) side, yet their a_0(ss), phase-diagram and CN(Gd^{3+},Ce^{4+}) behavior seem to exhibit some non-trivial differences related to their T_{prep} difference. These results show that even in Ce-Gd, which is one of the two most-throughly studied Ce-Lns with Ce-Y, much remains to be done to fully clarify its coupled a_0(ss), local-structure and (defect-) thermodynamic behavior, as summarized below.

(1) A relatively low T_{prep} = 1450 °C of our sample has almost firstly enabled Ce-Gd to be prepared as a complete solid solution, being disordered DF-type in the narrower $0 \leq y \leq 0.30$ range and ordered C-type in the much wider $0.30 < y \leq 1.0$ range, in contrast to Bevan's T_{prep} = 1600 °C data that reveal the existence of a wide miscibility gap over the $\sim 0.52 < y < \sim 0.73$. Grover's data for T_{prep} = 1400 °C (in pink) [99] too show the similar behavior to ours in that a smooth DF → C phase change occurs at $y \sim 0.45$, but the a_0(ss) data exhibit a peculiar largely-wavy variation with y over there. These data thus clarify that, unlike the general trend, this best lattice-match Ce-Gd (LMR~1; $a_0(CeO_2)$(nm) = 0.5411 ~ $a_0(GdO_{1.5})$ / 2 = 0.54063 in (Fig. 2) has the stronger tendency to form the ordered non-random C-type solid solution at the lower T_{prep} = 1450-1400 °C range rather than to form the disordered random DF-type one at the higher T_{prep} = 1600 °C. These also imply that the XRD 'apparently disordered' DF phase in the initial narrow $0 \leq y \leq 0.30$-0.40 range at the T_{oper} = 1450-1400 °C would most plausibly be also a short-range ordered C-type deeply penetrating into the lower $y < 0.30$-0.40 range, for unless otherwise one would here too observe the DF-C miscibility gap as in the Bevan's 1600°C data. Whether this is the more general unique phase-diagram and structural behavior of these DF oxides deserves further detailed study of this and related Ce-Lns in the future. Also note in this context that in our latest XRD and ^{89}Y MAS-NMR study of Ce-Y with

T_{prep} = 1400°C [102] the system has the ordered C-type phase in the ~0.30 (or 0.20) $\leq y \leq$ ~0.525 range in the XRD (or Y MAS-NMR) data also in the CeO_2-rich side of the wide, thus not DF-C but C-C miscibility gap extending over the ~0.525 $\leq y \leq$ ~0.675 range, again unlike Bevan's 1600°C data [6] with the usual DF-C miscibility gap existing over the similar y range.

(2) In Fig. 10a-b our T_{prep} = 1450°C data are a little but definitely less Δa_0(ss) and more case (i) Gd^{3+}-V_O associative than Bevan's. To my knowledge this is the third a_0(ss) evidence that the system moves to the more non-randomness with decreasing T_{prep}, the first and second ones being a Bevan's report that Ce-Y annealed at 1100 °C has lower a_0(ss) [6] and the two sets of ^{89}Y-MAS-NMR data in Ce-Y shown in Fig. 7b, respectively. Since even our more case (i) non-random lower CN(Gd^{3+}) curve in Fig. 10b is still slightly higher than (or near equal to) two sets of reliable MAS-NMR CN(Y^{3+}) data of Ce-Y (Fig. 7b) with the higher T_{prep} = 1500 or 1650 °C than 1450 °C of our Ce-Gd samples, revising the former results [1] (drawn in the same purple color both in Figs. 10b and 7b), it seems certain now that Ce-Gd and Ce-Eu (see [5a], as not detailed here) are more near random than Ce-Y rather than near equally non-random as in Fig. 7b. This is also more in line with their well-known σ(ion) trend. Since 'frozen-in non-randomness' fixed at T_{prep} due to the slow cation diffusivity is the basic premise of the DCC model [1,2], further data on such T_{prep}-dependent a_0(ss) and non-randomness behavior of various DF oxides are awaited.

(3) The ^{155}Gd-Mössbauer data [3,101] show that in the $GdO_{1.5}$-rich side C-type phase the outer more-distorted 8b-Gd doublet quickly disappears for y = 1.0 → 0.60, revealing that Ce^{4+} preferentially replaces the minor 8b Gd^{3+} rather than the less-distorted major 24d Gd^{3+}; QS (quadrupole-splitting) data; QS(8b) ~ 12 >> QS(24d) ~ 6 mm/s in C $GdO_{1.5}$. A steep QS increase of this minor 8b Gd^{3+} to the still higher ~14 at y → 0.60 (see Fig. 2 of [3]) further implies that increasing 8b Ce^{4+} is driving this decreasing 8b Gd^{3+} to the even more-distorted unstable state. Namely, the efficient elimination of more-distorted 8b Gd^{3+} by the preferential $Ce^{4+} \rightarrow Gd^{3+}$ replacement in C $GdO_{1.5}$ is an eloquent back-side view of the formation of the overall less-distorted stable C-type solid solution here. Now that the MO_2-rich side DF phase is found likely to have the more deeply C-type long- or short-range-ordered structure than thought before, its $LnO_{1.5}$-rich side C-type counterpart too should be more deliberately attented hereafter.

3. COMPATIBILITY WITH ION-PACKING MODEL

3.1. Ion-Packing (I-P) Model Lattice-Parameter Expression

The two f_F(I-P) curves of F MO_2 for Shannon's r_a(O)(nm) = 0.138 and a little smaller 0.1373 in:

$$f_F(\text{I-P}) = (4/\sqrt{3}) \cdot \{r_C(M^{4+}(\text{VIII})) + r_a(O)\} \quad (15),$$

are also drawn in Fig. 4, showing good agreement with Eq. (5a) for the $r_C(M^{4+}(\text{VIII})) > 0.08$nm range. Making substituting $r_C(\text{VIII}) \rightarrow r_C(ss)$ and $r_a(O) \rightarrow (1-y/4) \cdot r_a(O) + y/4 \cdot r(Vo)$ (where $r(Vo)$ is the effective Vo radius) for $MO_2 \rightarrow M_{1-y}Ln_yO_{2-y/2}(Vo_{y/2})$ in Eq. (15) gives;

$$a_0(ss)(\text{I-P}) = (4/\sqrt{3}) \cdot [r_C(ss) + r_a(O) + (y/4)\cdot\{r(Vo) - r_a(O)\}] \quad (16).$$

The correspondence of Eq. (16) to the modified Eq. (1), $a_0(ss) = f_F(r_C(ss)) + y\cdot\{f_C(r_C(ss)) - f_F(r_C(ss))\}$, gives the first below (= Eq. (15) itself of F MO_2 at y=0) and the second Eq. (17);

$$f_F(r_C(ss)) = (4/\sqrt{3}) \cdot \{r_C(ss) + r_a(O)\} \quad (15),$$

$$f_C(r_C(ss)) - f_F(r_C(ss)) = (1/\sqrt{3}) \cdot \{r(Vo) - r_a(O)\} \quad (17).$$

Inserting Eqs. (5a-b) for $f_F(r_C(ss))$ and $f_C(r_C(ss))$ into these, and then manipulating them finally results in:

$$r_a(O)=0.1373+1.7654\cdot\{r_C(ss)-0.0992\}^2 \text{ (nm)} \quad (18).$$

$$r(Vo)=0.1386+20.3335\cdot\{r_C(ss)-0.07096\}^2 \text{ (nm)} \quad (19).$$

Eq. (16) in combination with Eqs. (18)-(19) is the general $a_0(ss)$(I-P) formula used in the DCC model. Fig. 11 shows these $r_a(O)$ and $r(Vo)$ and the total Vo effect (Eq. (17)) vs. $r_C(ss)$ curves. While $r_a(O)$ shows only a minor variation with its min. ~0.1373nm at $r_C(ss)$ ~ 0.099nm (~ $r_C(U^{4+})$(VIII) = 0.100nm), $r(Vo)$ with a min. ~0.1386 nm at $r_C(ss)$ ~ 0.071nm shows a rapid increase, reaching ~0.16nm at the largest $r_C(La^{3+}(VI))$=0.1032nm, constituting most of the total distortion dilation for the full $2O^{2-} \rightarrow 2Vo$ in **F M'O_2 → C Ln'$O_{1.5}$**. $r(Vo) \geq r_a(O)$ always holds, for $LnO_{1.5} \rightarrow MO_2$ doping always decreases the system's major electrostatic cohesive energy. One can therefore expect only limited utilization of the simplified $a_0(ss)$(I-P) after assuming some constant $r(V_O)$ for each MO_2 [80-83]. $r(Vo)$ in Eq. (19) is not the real V_O radius due to various local lattice relaxations about V_O [73]. Yet, it is the only possible $r(Vo)$ formula, so far as Eq. (16) is adopted as the $a_0(ss)$(I-P) expression.

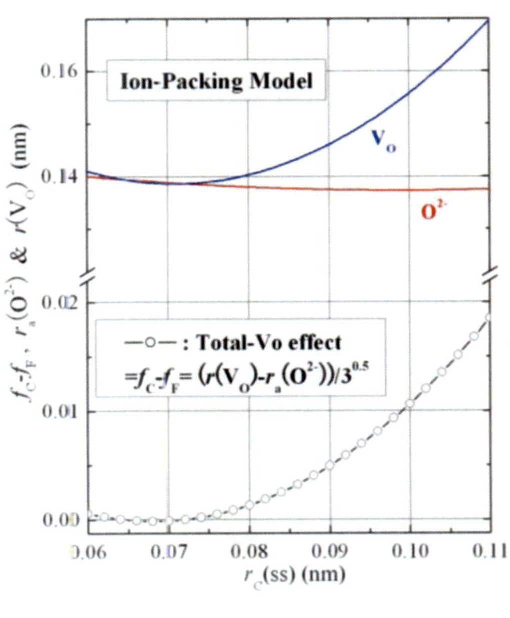

Figure 11. $r_a(O)$, $r(Vo)$ and (f_C - f_F) (the total Vo effect) vs. $r_C(ss)$ plot in the I-P model [1].

The M^{4+}-O^{2-} packing condition for F MO_2 to avoid direct O^{2-}-O^{2-} contact gives:

$$r_C(M^{4+})(\text{VIII}) \geq (\sqrt{3}-1) \cdot r_a(O) = 0.732 \cdot r_a(O) \quad (20).$$

Putting r_a(O)~0.1373nm in this again gives $r_C(M^{4+})$(VIII)~0.1005~$r_C(U^{4+})$(VIII)=0.100nm, i.e., UO_2 is the best $r_C(M^{4+})$-r_a(O) match. Only the largest Th^{4+} (r_C(VIII)=0.105nm) fulfills this condition. This is probably the most likely reason why the melting point (T_m) of AnO_2 lowers with the atomic number Z (An contraction); $T_m(An^{4+})$(°C)=3220(Th)→2880(U)→2730(Ce) →2570(Pu)→2500(Am) [103-105], a non-An CeO_2 also obeying this trend. Or, the increase in the O^{2-}-O^{2-} repulsion in the smaller MO_2 diminishes its structural and thermodynamic stability. So, such smaller M^{4+} will take the lower CN=VII and VI more easily than Th^{4+} in relaxing the O^{2-}-O^{2-} repulsion, consistently with the above a_0(ss)-analysis results of M^{4+}=Th and Ce systems. The smallest ZrO_2 and HfO_2 even transform to a distorted monoclinic (M) structure with the lower CN~7. While, 2Vo-containing O^{2-}-O^{2-} repulsion largely-relaxed $LnO_{1.5}$ (Fig. 1b) forms the most stable C-type structure (CN=VI) in such smallest-Ln^{3+} area where F MO_2 no more exists (Fig. 4), the T_m conversely increasing with Z (Ln contraction); $T_m(Ln^{3+})$(°C)=2320(La)→2350(Eu)→2410(Y)→2490(Lu)→2550(Sc) [106-110]. The least-distorted smallest $ScO_{1.5}$ with the smallest r(Vo)~r_a(O) (Fig. 11) thus attains the highest structural and thermodynamic stability. The lower T_ms of these $LnO_{1.5}$s than the MO_2s also accord with their defective nature with the electrostatic weaker Ln^{3+}-O^{2-} bond. The larger $LnO_{1.5}$ (Ln^{3+}=La-Gd) indeed transforms to a non-cubic A- or B-type structure [106,107,109], eliminating too large a void (Vo) in too much distortion-dilated C $LnO_{1.5}$ inside an overall less-distorted and more-dense higher CN~7 structure. This is a counter situation to the above M distortion of Zr(Hf)O_2.

The 'distortion-dilation' is thus rationally asserted to be a direct manifestation of the opposing structural and thermodynamic stabilities of C $LnO_{1.5}$ and F MO_2 with r_C: In Fig. 4, starting from the near null V_O effect area in the most-stable least-distorted smallest $ScO_{1.5}$ - the least stable most distorted smallest Zr(Hf)O_2 pair on the lower left side, it increases toward the largest V_O effect area in the least-stable most-distorted largest $LaO_{1.5}$ - the most stable least-distorted largest ThO_2 pair on the upper right side. This trend correlates well with the opposing trend in the $r_C(Ln^{3+}, M^{4+})$'s CN-slope in Fig. 5 (Eqs. (6a-b) and (10)); the smallest $r_C(Sc^{3+})$-$r_C(Zr^{4+})$ pair has near-equal $\alpha_{Sc} \sim \alpha_{Zr}$, i.e. near null V_O effect, and the largest $r_C(La^{3+})$-$r_C(Th^{4+})$ pair has the most different $\alpha_{La} \gg \alpha_{Th}$, i.e. the largest Vo effect. This cross linking of the opposing trend with the macroscopic thermodynamic and structural stability of C $LnO_{1.5}$ and F MO_2 with that in the microscopic $r_C(Ln^{3+}, M^{4+})$'s CN slope is the backbone physical reality of these DF oxides existing behind the above-clarified coupled macroscopic non-Vegardian a_0(ss) - microscopic non-randomness behavior. Or, one can more straight-forwardly mention that generalized non-Vegardianity, $\delta(\Delta a_0(ss))$, and non-random oxygen coordination, $\delta CN(Ln^{3+})$, in these DF oxides, are two sides of the same distortion-dilation phenomenon of defective C $LnO_{1.5}$ coupled between the macroscopic-a_0(ss) and microscopic-r_C(ss) level. This crucial point is worthwhile further thorough experimental and theoretical investigations by various crystal-chemical, thermodynamic and lattice-dynamic (and static) etc methods [78,111,112] in the future.

3.2. Metal – Oxygen Bond Length

Combining $r_a(O)$ (Eq. (18)) and $r_C(M^{4+}, Ln^{3+})$ (Eqs. (6a-b)) provides the metal-oxygen bond length (BL) data, e.g. as $BL(Ln^{3+}\text{-}O^{2-}) = r_C(Ln^{3+}) + r_a(O)$. EXAFS and MD-simulations $BL(Y^{3+}\text{-}O^{2-}, Ce^{4+}\text{-}O^{2-})$ data in Ce-Y are compared with the present model results in Fig. 12: These literature BL data plotted correspond to the respective $CN(Y^{3+}, Ce^{4+})$ data plotted in Fig. 7b. With the $CN(Y^{3+})$ Hayashi's (MD) is much closer to the model and Y MAS-NMR data than Deguchi's (EXAFS) in Fig. 7b. Conversely with the BL data here the former is too small compared with the latter. Fig. 12 also reveals another major trend difference of the model and reported data in that in the latter $BL(Ce^{4+}\text{-}O^{2-}) > BL(Y^{3+}\text{-}O^{2-})$ throughout the reported whole DF $0 \le y \le \sim 0.35$ range. It is possible in the present model if Ce-Y has more case (i) $Y^{3+}\text{-}V_O$ associative $CN(Y) < 7$ at $y \rightarrow 0$ in Fig. 7b. The conclusion is that the model $BL(Ce^{4+}\text{-}O^{2-}, Y^{3+}\text{-}O^{2-})$ are most reasonable on Ce-Y. The difficulty in determining both CN and BL with EXAFS is well known. Deguchi's data seem not free from this difficulty, giving the more reasonable BL than the CN, as is usually the case. The problem with classical MD and LS simulations [21-24] is the validity of the inter-atomic potentials used (see also [113,114]). In Hayashi's results the CN data seem more reasonable than the BL data.

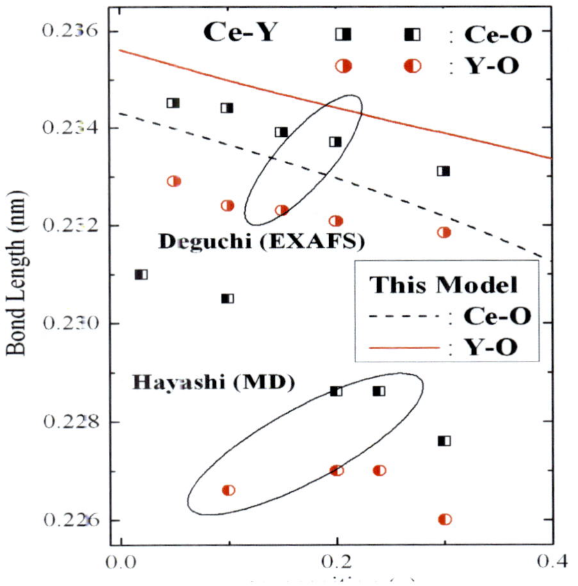

Figure 12. $BL(Ce^{4+}\text{-}O^{2-}, Y^{3+}\text{-}O^{2-})$ data in Ce-Y [22,55] and their DCC-model curves [1].

Of particular interest is the expected strongly non-liner $BL(Ln^{3+}\text{-}O^{2-})$ variation in Th-La and Th-Nd from the unique $CN(Ln^{3+})$ shape in Fig. 7a. Fig. 13a provides their $BL(Ln^{3+}\text{-}O^{2-}, Th^{4+}\text{-}O^{2-})$ model curves over the entire y range. Since the both $r_a(O)$ variation is small (Fig. 11), the $BL(Ln^{3+}\text{-}O^{2-})$ curves exhibit a steep initial rise similar to the $CN(Ln^{3+})$ curves in Fig. 7a, unlike the smoothly-decreasing BL curve with y as with Ce-Y (Fig. 12) and many other systems (Figs.7a-b). Besides, $BL(Th^{4+}\text{-}O^{2-})$ in Th-Nd is even slightly longer than that in Th-La, as expected from the $CN(Th^{4+})$ curve in Fig. 7a that is slightly higher in Th-Nd, despite the obvious fact in Fig. 13b that Th-Nd has naturally the shorter $a_0(ss)$ than Th-La.

Whether that eccentric BL(Ln^{3+}-O^{2-}, Th^{4+}-O^{2-}) behavior can indeed be pin-pointed in some suitable experiment and defect-simulations will be a stringent test of the model and a challenging experimental and computational study.

Fig. 13a also shows the both average-total BL(total metal - O^{2-}) = BL(Th+Ln - O^{2-}) = r_C(ss) + r_a(O) = (1-y)·BL(Th^{4+}-O^{2-}) + y·BL(Ln^{3+}-O^{2-}) curves. That of Th-La (in black) in Fig. 13a and the a_0(ss)(non-random) curve of Th-Nd (in red) in Fig. 13b exhibit almost no variation with y in both DF 0 ≤y < ~0.50 range, while the BL(total metal - O^{2-}) curve of Th-Nd in Fig. 13a and the a_0(ss) curve of Th-La in Fig. 13b monotonously decreases and increases, respectively. These results imply the possibility that BL(total metal - O^{2-})~constant in Th-La in Fig. 13a or a_0(ss)~constant in Th-Nd in Fig. 13b would be operating as each primary mechanism of DF solid solution formation, as if to manifest each strong will to retain its identity as 'ThO_2'-based solid solution in each possible way, enabling both to form extremely case (i) Ln^{3+}-Vo non-random unique defect structure (Fig. 7a). When such unique condition is lost in each y > ~0.50 range, both cease to form 'solid solution'.

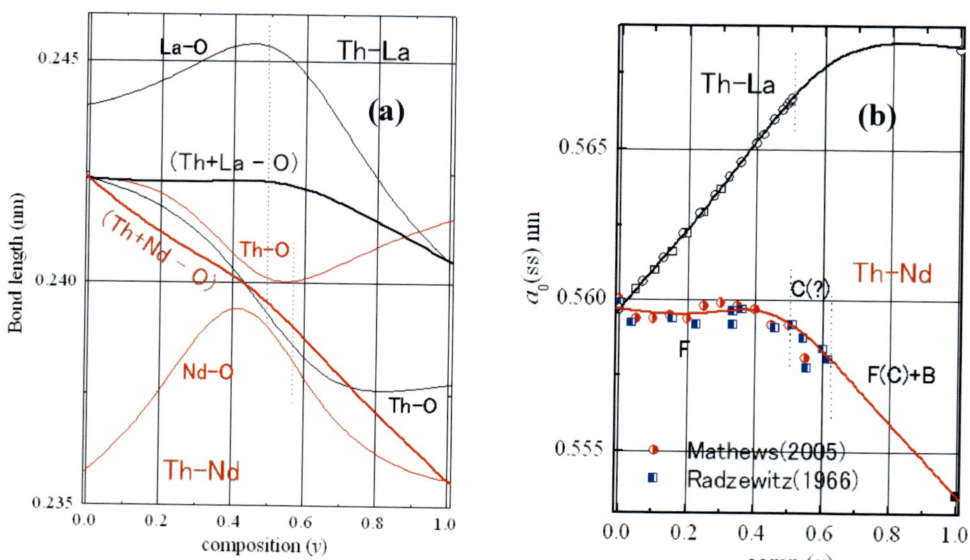

Figure 13. (a) The DCC-model BL(Th^{4+}-O^{2-}, Ln^{3+}-O^{2-}, total cation-O^{2-}) curves in Th-La and Th-Nd. (b) a_0(ss) data of Th-La and Th-Nd [115,116] and their best model curve.

4. 'From Overall to Detailed' Non-Random Local-Structure

We demonstrate here that full use of overall non-random CN(M^{4+}, Ln^{3+}) data (Figs. 7a-b) in combination with 'restricted (non-)randomness' of DF lattice briefly introduced in the last paragraph of 2.1 leads to a new consistent description of the controversial local and defect structure and σ(ion)(max) behavior of these DF oxides [2], opening a new horizon of defect crystal chemistry (DCC) of them. Before doing this, we once again confirm here that 'restricted (non-) randomness' in short means that cations (M^{4+}, Ln^{3+}) can take only the highest three oxygen-CN = Ⅷ, Ⅶ and Ⅵ states in these DF oxides, contrary to our naive expectation in the hitherto-usually-supposed completely-random O^{2-}+Vo mixing case in the anion sublattice. A decisive evidence of this crucial fact has obatined in recent less-than-ten years by several ^{89}Y-MAS-NMR local oxygen coordination number studies around ^{89}Y dopant in two representative Ce-Y and YSZ (Zr-Y) systems [30-36]. This fact is further reinforcrd by one other evidence which has been available for much longer time that in Fig. 5 Shannon's $r_C(M^{4+}, Ln^{3+})$ data for CN < Ⅵ(=6) are entirely absent for almost all these host M^{4+} (= Ce and Th) and dopant Ln^{3+} (= La-Lu, Y) cations, eloquently telling us that such lower oxygen-CN (< 6) compounds are rarely formed in these (relatively large) cation systems.

4.1. Restricted (Non-)Randomness of Defect-Fluorite Lattice - Exclusion of First Nearest -Neighbor (1NN) <100> Vo-Vo Configuration –

Figs. 14a-b depict in pair the proposed structural model for the $LnO_{1.5}$-rich side oxygen-excess ordered C-type solid solution, $M_{1-y}Ln_yO_{1.5}O_{i0.5-y/2}Vo_{y/2}$, and the MO_2-rich side oxygen-deficit disordered DF-type one, $M_{1-y}Ln_yO_{2-y/2}Vo_{y/2}$, respectively, embodying the above-asserted 'restricted (non-)randomness' of DF lattice. While the former is a simple oxygen-excess version of the well-defined Vo long-range ordered C $LnO_{1.5}$ with 24d and 8b Ln^{3+} site (both CN = Ⅵ) (Fig. 1b), the latter is posited to be a more disordered short-range ordered oxygen-excess hybrid-mixture structure phase of C-type $LnO_{1.5}$, disordered DF-type δ $BiO_{1.5}$ [117-120] and pseudo-pyrochlore-type P'-type $LnO_{1.5}$, etc. The last hypothetical P' $LnO_{1.5}$ is obtained in pyrochlore (P) structure depicted in Fig. 3 by simultaneous full $Zr(Hf)^{4+} \rightarrow Ln^{3+}$ and O(2) → Vo replacement (see Fig. 9 of [2]) wherein only 8b-like Ln^{3+} site with 3NN <111> Vo-Vo pair exists. Namely, as seen in Fig. 14b, the latter can take various three-body <110>-<110>, <110>-<111> and <111>-<111> Vo-Vo-Vo configurations and connections much more flexibly and randomly than the former as a microscopic-heterogeneous composite mixture or micro-domain phase of short-range ordered C-type and P'-type and disordered DF-type δ $BiO_{1.5}$ based structures, yet both fulfilling the basic restricted-(non)random CN ≥ 6 condition by having at most one either <110> or <111> Vo-Vo pair about each Ln^{3+} and M^{4+}.

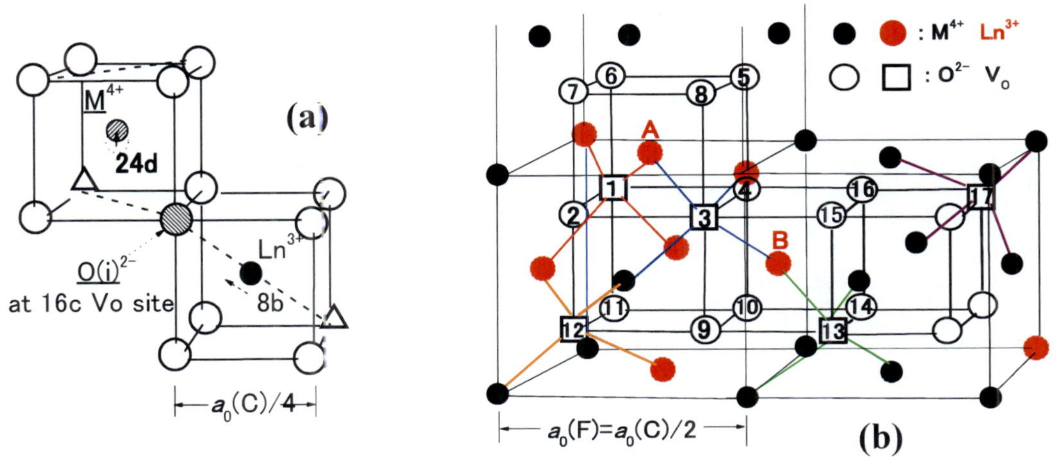

Figure 14. Structure model for the ordered C-type (a) and disordered DF-type (b) solid solution.

In the former simpler case (Fig. 14a) $MO_2 \rightarrow LnO_{1.5}$ doping introduces a fixed amount of O^{2-} interstitials ({Oi} = (1-y)/2) into the built-in 16c Vo site (= 1/2), so that the CN=VI, VII or VIII cation site is identified as that with each one Vo-Vo, Vo-O_i or O_i-O_i pair, respectively. In the ref. random mixing case of O_i and Vo each site fraction is provided by the second-order binomial theorem, ${}_2C_n(1-y)^n y^{2-n}$ with n = 0, 1 or 2, respectively, as follows;

$$[CN(VI)] = y^2,\ [CN(VII)] = 2y\,(1 - y),\ [CN(VIII)] = (1 - y)^2 \qquad (21a).$$

It is denoted as the ref. three-site model (a): Its [CN(VI,VII,VIII)] profile drawn in Fig. 15a in dashed line is symmetric about y=0.50, at which 1/2 is the middle CN(VII) and the remaining 1/2 is equally divided into the CN = VIII and VI. Fig. 15a also shows the limiting two-site model (a') solid curves in which V_O-O_i pair formation is enhanced to 1 at y = 0.50;

$$[CN(VIII)] = 1\text{-}2y,\ [CN(VII)] = 2y \ \ vs.\ \ [CN(VII)] = 2(1\text{-}y),\ [CN(VI)] = 2y\text{-}1 \qquad (21b).$$
$$(0 \le y \le 0.50) \qquad\qquad (0.5 \le y \le 1.0)$$

The reason for considering this limiting two-site model (a)' becomes clear soon in 4.2.

Once the 1NN <100> Vo-Vo pair formation (// a_0) were allowed in Figs. 14a-b, such 'restricted random' model (a) [CN(VI,VII,VIII)] profile in Fig. 15a immediately reduces to the usual (but in fact non-existing) 'completely-random' model-(b)-type one depicted in Fig. 15b, in which the [CN(VIII,VII,VI, V ,IV,III, --, 0)] is given by the eighth-order binomial theorem of ${}_8C_n(1-y)^n y^{8-n}$ with n=8, 7, 6, 5, 4, 3, ···, 0, respectively. It directly follows from this that the exclusion of the 1NN <100> Vo-Vo pair formation is the necessary condition for 'restricted (non-)randomness' in DF solid solutions. Recent several MD and Monte-Carlo (MC) σ(ion)-simulation studies of Ce-Pr [69], YSZ [70,71] and Ce-Y [72] the all have reached a similar conclusion that strong Vo-Vo repulsion (possibly 1NN <100> Vo-Vo pair exclusion) greatly

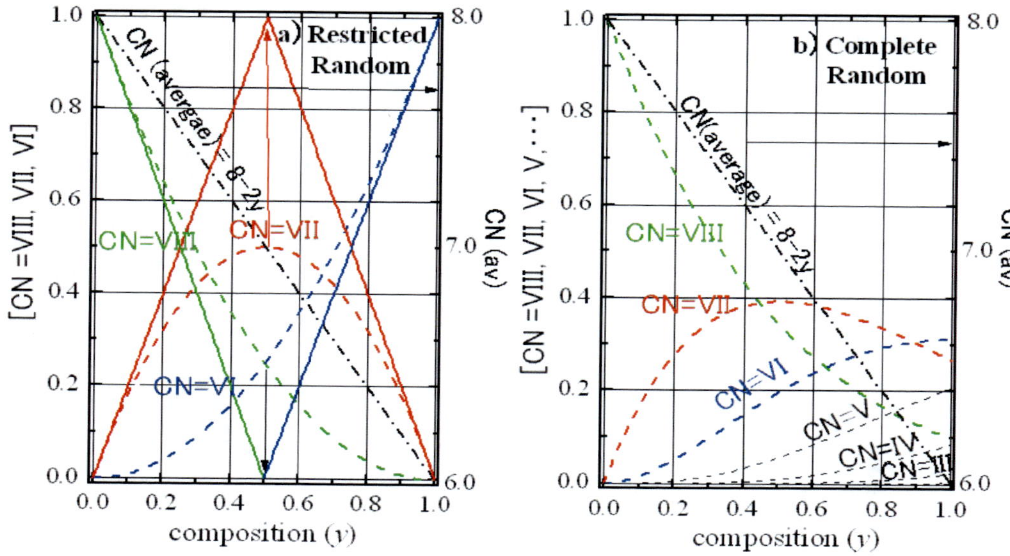

Figure 15. Restricted random (a) and completely-random (b) total-cation site-fraction profile.

reduces the σ(ion) of the system in the higher y range by more severely narrowing the Vo migration path there and hence giving rise to the σ(ion)(max) onset in the low-y range. This topic will be discussed in detail in 4. 3.

As argued in [2], however, the 1NN <100> V_O-V_O pair exclusion is not the sufficient condition for the restricted (non-)randomness, for, as drawn on the top-left horizontal oxygen plane in Fig. 14a in dashed line, putting one more <110> V_O-V_O pair about the 24d cation (M^{4+} or Ln^{3+}) in orthogonal direction to that in the bottom plane can make the lower CN = V and even IV site without making the 1NN <100> V_O-V_O pair. (This is not the case for the 8b site.) Indeed, the presence of such lower CN = V Bi^{3+} site in δ Bi_2O_3 is recently suggested in its coupled ND and MC simulations study [117-120] due to existence of the Bi^{3+} $6s^2$ lone-pair electrons favoring such irregular CN = V state. But, as argued there, it is rather an exceptional structural feature of 'disordered DF-type' highest-σ(ion) δ Bi_2O_3 different from 'ordered C-type' $LnO_{1.5}$ (Ln^{3+}=Y-La); e.g., it alone has a very limited high-temperature stability range of $729 \leq T \leq 824°C = T_{mp}$), i.e., an exceptionally low T_{mp} below 1000°C, and moreover, as plotted in Figs. 4-5 (in red), albeit Bi^{3+} and La^{3+} have near equal largest $r_C(Bi^{3+}) \sim r_C(La^{3+})$, Bi^{3+} alone has Shannon's $r_C(Bi^{3+})$ datum at CN = 5 and δ Bi_2O_3 alone has even shorter f_F = 0.5655nm at 778°C than the latter f_C(C $LaO_{1.5}$) = 0.5683nm at room temperature. It is thus assumed here that these DF M-Lns (M^{4+} = Th and Ce; Ln^{3+} = La-Y) have only the restricted (non-)random CN ≥ 6 cation sites, such energetically high-cost disproportionation reaction, 2CN(Ⅵ) ↔ CN(Ⅴ) + CN(Ⅶ), possible to some extent in δ Bi_2O_3, hardly occurring in either of their C- or DF-type solid solution phase.

4.2. Mutually Non-Random Detailed Cation ↔ Anion Coordination

With combination of overall CN(M^{4+}, Ln^{3+}) data (Fig. 7a-b) of these M-Lns and the above-introduced restricted (non-)random framework (Figs. 14a-b and 15a) of their DF lattice we can now embark on the derivation of their detailed mutually non-random cation ↔ anion concentrations of {Ln^{3+}(Ⅷ,Ⅶ,Ⅵ)} and {M^{4+}(Ⅷ,Ⅶ,Ⅵ)} and the counter Ln^{3+} n(=0-4)-fold coordinated {O^{2-}(n)} and {Vo(n)} in a consecutive manner, surmounting the major obstacle to previous studies of DF oxides that such key detailed cation and anion concentration data have been almost entirely lacking.

4.2.1. Two to Three Site Model Approach to CN(Y^{3+}, Ce^{4+}) in Ce-Y - Detailed Non-Random Anion (V_O, O^{2-}) → Cation (Y^{3+}, Ce^{4+}) Coordination -

Figs. 16a-b show in pair the individual site-fraction [Y^{3+}(Ⅷ,Ⅶ,Ⅵ)] - [Ce^{4+}(Ⅷ,Ⅶ,Ⅵ)] curves in Ce-Y derived in the two to three-site model approach to the CN(Y^{3+}, Ce^{4+}) pair curves shown in Fig. 7b. Ce-Y is chosen here as the best system to illustrate their derivation process [1], for it has two sets of reliable direct MAS-NMR [^{89}Y(Ⅷ,Ⅶ,Ⅵ)] data in the DF-type $0 \leq y \leq \sim 0.50$ range in good mutual agreement and to be compared with the present model results. Their combined Y^{3+}+Ce^{4+} mole-fraction curves in Fig. 16c reveal that Ce-Y has enhanced [CN(Ⅶ)] (> [CN(Ⅶ)(random)]) toward the limiting two-site model (a') side in Fig. 15a.

First, the derivation procedure of these model [Y^{3+}(Ⅷ,Ⅶ,Ⅵ)] curves in Figs. 16a is described; the total Y^{3+} sub-site fraction and the overall average CN(Y^{3+}) conditions give the following Eqs. (22a-b), respectively;

$[Y^{3+}(\mathrm{VIII})]+[Y^{3+}(\mathrm{VII})]+[Y^{3+}(\mathrm{VI})]=1,\ 8\cdot[Y^{3+}(\mathrm{VIII})]+7\cdot[Y^{3+}(\mathrm{VII})]+6\cdot[Y^{3+}(\mathrm{VI})]=\mathrm{CN}(Y^{3+})$ (22a-b).

In the $0\leq y\leq 0.33$ range over which $8>\mathrm{CN}(Y^{3+})\geq 7$, the limiting two-site $Y^{3+}(\mathrm{VIII,VII})$ state by setting $[Y^{3+}(\mathrm{VI})]=0$ in Eqs. (22a-b) gives Eq. (23a-b) and their solution, Eq. (24a-b);

$[Y^{3+}(\mathrm{VIII})]+[Y^{3+}(\mathrm{VII})] = 1,\ 8\cdot[Y^{3+}(\mathrm{VIII})]+7\cdot[Y^{3+}(\mathrm{VII})] = \mathrm{CN}(Y^{3+})\ (0\leq y\leq 0.33)$ (23a-b).

$[Y^{3+}(\mathrm{VII})] = 8 - \mathrm{CN}(Y^{3+}),\ [Y^{3+}(\mathrm{VIII})] = \mathrm{CN}(Y^{3+}) - 7\ (0\leq y\leq 0.33)$ (24a-b).

The similar two-site $Y^{3+}(\mathrm{VII,VI})$ state for the $0.33\leq y\leq 1.0$ range where $7\geq \mathrm{CN}(Y^{3+})\geq 6$ gives:

$[Y^{3+}(\mathrm{VII})]+[Y^{3+}(\mathrm{VI})] = 1,\ 7\cdot[Y^{3+}(\mathrm{VII})]+ 6\cdot[Y^{3+}(\mathrm{VI})] = \mathrm{CN}(Y^{3+})\ (0.33 \leq y\leq 1.0)$ (25a-b).

$[Y^{3+}(\mathrm{VII})] = \mathrm{CN}(Y^{3+}) - 6\ [Y^{3+}(\mathrm{VI})] = 7 - \mathrm{CN}(Y^{3+})\ (0.33 \leq y\leq 1.0)$ (26a-b).

Thus derived limiting two-site model (a') $[Y^{3+}(\mathrm{VIII,VII,VI})]$ curves in Eqs. (24a-b) and (26a-b) in each region are drawn in Fig. 16a in thin solid color line, for which $[Y^{3+}(\mathrm{VII})]=1$ and $\mathrm{CN}(Y^{3+})=7$ at $y=0.33$. However, such $[Y^{3+}(\mathrm{VII})]=1$ state is a perfectly ordered dopant Y^{3+}-V_O configuration hardly realizable in these disordered DF oxides. So, around this $y\sim 0.33$ (or already before reaching there) a fraction of $[Y^{3+}(\mathrm{VII})]$ decomposes into $[Y^{3+}(\mathrm{VIII})]$ and $[Y^{3+}(\mathrm{VI})]$ according to the following intra-Y^{3+} disproportionation reaction, Eq. (27):

$2\ Y^{3+}(\mathrm{VII}) \leftrightarrow Y^{3+}(\mathrm{VIII}) + Y^{3+}(\mathrm{VI})$ (27)

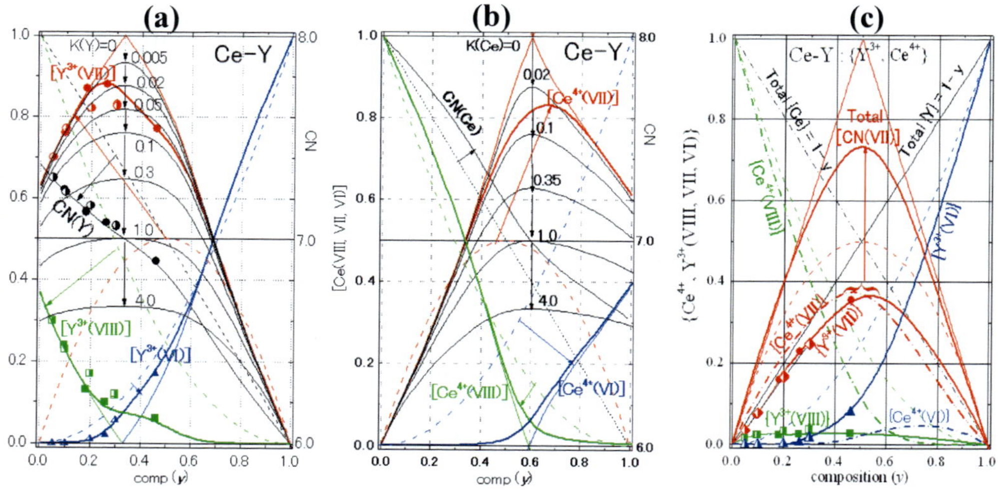

Figure 16. The model site-fraction profile of Y^{3+} (a) and Ce^{4+} (b) and their combined total-cation mole-fraction data (c) in Ce-Y. In (a) experimental ^{89}Y-MAS-NMR data are also plotted for comparison.

Namely, as the two sets of ^{89}Y-NMR data (in filled and half-filled circles) and the derived final model curves (in thick solid color line) indicate, already from the starting $y = 0$ point a fraction of the most abundant middle $[Y^{3+}(\mathrm{VII})]$ in fact starts disproportionating gradually following Eq. (27), deviating all these [Y^{3+}(VIII,VII,VI)] curves from each limiting two-site model (a') ones. As a result, the increasing [Y^{3+}(VI)] (from ~0 at y=0) and the decreasing [Y^{3+}(VIII)] (from ~0.36 at y=0) attain a coincidence point, [Y^{3+}(VIII)]=[Y^{3+}(VI)] at y=0.33, meanwhile the major middle [Y^{3+}(VII)] has already turned to diminish at y~0.25 fairly before y =0.33. Using Eqs. (22a-b) in such actual three-site situation [Y3+(VIII,VI)] are solved for a given [Y3+(VII)] and CN(Y) as follows:

$$[Y^{3+}(\mathrm{VIII})] = \{(CN(Y)\text{-}6) - [Y^{3+}(\mathrm{VII})]\}/2,\ [Y^{3+}(\mathrm{VI})] = \{(8\text{-}CN(Y)) - [Y^{3+}(\mathrm{VII})]\}/2 \quad (28a\text{-}b).$$

The [Y^{3+}(VIII,VI)] curves obtained by inserting the most reasonable [Y^{3+}(VII)] curve in thick solid red line into Eq. (28a-b) are drawn in thick solid green and blue line, respectively, in Fig. 16a,. all these being fully consistent with the individual ^{89}Y-MAS-NMR concentration data over the reported entire DF-type 0≤y≤0.5 range. If we start with the slightly more near-random Maekawa's CN(^{89}Y) data [31] in half-filled symbol (T_{prep}=1650°C) instead of Kim's [33] in filled symbol (T_{prep}=1500°C), we will of course reach the more near-random [Y^{3+}(VIII, VII, VI)] profile better fit to the former. These two MAS-NMR [$^{89}Y^{3+}$(VIII, VII, VI)] data sets thus provide a direct evidence that DF oxides are shifting to the more random direction with increasing T_{prep}, as expected.

Applying the Guggenheim's quasi-chemical equilibrium (QCE) approach to the intar-Y disproportional reaction in Eq. (27), we obtain;

$$K(Y) = \exp(-\Delta G°(Y^{3+}) / RT) = 4 \cdot [\, Y^{3+}(\mathrm{VIII})] \cdot [\, Y^{3+}(\mathrm{VI})] / [\, Y^{3+}(\mathrm{VII})]^2 \quad (27),$$

where $\Delta G°(Y^{3+})$ is the standard Gibbs free-energy change for Eq. (27), K the equilibrium constant, R the gas constant, and T the temperature in Kelvin (K). Factor 4 (=2^2) in the rhs accounts for the fact that Y^{3+}(VII) has nominally two equivalent O^{2-}-Y-Vo and Vo-Y-O^{2-} configurations (being not essential, for it only alters the K(Y) by some constant factor). Inserting Eq. (28a-b) into Eq. (27)' gives Eq. (29) quadratic in [Y^{3+}(VII)], and its solutions, Eqs. (30a-b);

$$\{K(Y) - 1\} \cdot [Y^{3+}(\mathrm{VII})]^2 + 2 \cdot [Y^{3+}(\mathrm{VII})] - (CN(Y)\text{-}6) \cdot (8\text{-}CN(Y)) = 0 \quad (29).$$

$$[Y^{3+}(\mathrm{VII})] = \{CN(Y)\text{-}6\} \cdot \{8\text{-}CN(Y)\} / 2 \ \ (\text{for } K(Y) = 1) \quad (30a),$$

$$[Y^{3+}(\mathrm{VII})] = [-1+\{1+(K(Y)\text{-}1)\cdot(CN(Y)\text{-}6)\cdot(8\text{-}CN(Y))\}^{0.5}] /\{K(Y)\text{-}1\} \ (\text{for } K(Y) \neq 1) \quad (30b).$$

The calculated QCE [Y^{3+}(VII)] curves in Eqs. 30a-b for several K(Y) values are drawn (in thin solid black line) in Fig. 16a and indicate that as K(Y)=0→1 they smoothly shift from the limiting two-site model (a') to the ref. three-site model (a), the smallest K(Y) ~ 0.005 giving a closest fit to Kim's [^{89}Y(VII)] data in the low y≤~0.20 range. But, unlike the MAS-NMR data and the above-derived best model curve, the QCE model cannot shift the [Y^{3+}(VII)] peak position (y_{peak}) to the lower y side from the original y=0.33 for any K(Y). This deficiency of the QCE model most plausibly implies that K(Y) (i.e. $\Delta G°$(Y)) is not constant here but varying with y. This is a different situation either from (i) molecular gas reaction; $2AB(g) \leftrightarrow A_2(g) + B_2(g)$, or (ii) its A-B binary alloy analogue supposed by Guggenheim; $2AB \leftrightarrow AA + BB$ [122,123], for which one can reasonably assume some constant values of A-B, A-A and B-B molecular-bonding energies or the corresponding 1NN bond-pair energies independent of the mixture composition (y) as a good first approximation. However, in these DF oxides, it seems more reasonable to suppose not the constant but variable Y^{3+}(Ⅷ, Ⅶ, Ⅵ) cluster energies with y, for, as seen in Fig. 16a, these, as well as Ce^{4+}(Ⅷ,Ⅶ,Ⅵ) clusters in Fig. 16b, are altogether forming here a highly-defective and highly non-random microscopic-heterogeneous composite hybrid mixture and/or multiple micro-domain state (Figs. 14a-b) accompanying various macroscopic and microscopic distortion-dilations, structural clustering and relaxations, lattice-parameter (a_0(ss)) and mlar-volume etc (V_m) changes whose global mode and degree greatly varying with y from those either in the original F MO_2 at y=0 (Fig. 1a) or C $LnO_{1.5}$ at y=1.0 (Fig. 1b). For example, the largely-distorted Ce^{4+}(Ⅷ) cluster at $y \neq 0$ in the highly-defective DF-type solid solution is more naturally expected to have significantly different energy (enthalpy) and (vibrational etc) entropy from that of the ideally regular non-defective oxygen cube in pure F CeO_2 at y=0 (Fig. 1a). Not to mention, this seems to be more consistent also with the above analysis results of largely distortion-dilated markedly non-Vegardian a_0(ss) data of these systems. Fig. 16a shows that K(Y)~0.005 for the $y \leq$~0.20 range increases steeply to a much larger ~0.05 level in the relatively narrow ~0.20<y<~0.40 range traversing the y_{peak}~0.25, i.e. the system is undergoing a steep non-randomness shift around there.

Applying the same treatment to the host Ce^{4+} after the whole $Y^{3+} \rightarrow Ce^{4+}$ replacement in Eqs. (22)-(30), we obtain the QCE [Ce^{4+}(Ⅷ,Ⅶ,Ⅵ)] model curves (Fig. 16b) reinforcing the above view of variable K(Ce) with y too. Though here exist no direct experimental Ce^{4+} site-fraction data, we can derive their approximate curves, for the minor [Ce^{4+}(Ⅷ)] = [Ce^{4+}(Ⅵ)] at the coincidence point of y_{peak}=0.60 where CN(Ce^{4+})=7 resulting from the corresponding intra-Ce^{4+} disproportionation reaction;

$$2\, Ce^{4+}(\text{Ⅶ}) \leftrightarrow Ce^{4+}(\text{Ⅷ}) + Ce^{4+}(\text{Ⅵ}) \qquad (31),$$

has after all a narrow allowance range between the min. zero (=0) in the limiting two-site model (a') and the max. ~0.16 in the ref. three-site model (a), albeit some ambiguity certainly exists here unlike the above Y^{3+} case. One finds a striking mirror-symmetry between the largely-slanted highly non-random [Y^{3+}(Ⅷ,Ⅶ,Ⅵ)] - [Ce^{4+}(Ⅷ,Ⅶ,Ⅵ)] profiles about y~0.50, and particularly between the most abundant [Y^{3+}(Ⅶ)] vs. [Ce^{4+}(Ⅶ)], revealing that the dopant Y^{3+} in the CeO_2-rich DF phase and the host Ce^{4+} in the $YO_{1.5}$-rich C-type phase form quite similar V_O 1-fold associated (CN=Ⅶ) major non-random local defect structures which are new for both cations and are prevalent in each dopant region.

The corresponding mol-fraction curves of the total cation (Fig. 16c) are obtained by multiplying the [Y^{3+}] and [Ce^{4+}] curves in Figs. 16a-b by the respective mol fractions, y and (1-y) per $Ce_{1-y}Y_yO_{2-y/2}$:

$$\{Y^{3+}(\text{Ⅷ},\text{Ⅶ},\text{Ⅵ})\}=y\cdot[Y^{3+}(\text{Ⅷ},\text{Ⅶ},\text{Ⅵ})],\ \{Ce^{4+}(\text{Ⅷ},\text{Ⅶ},\text{Ⅵ})\}=(1-y)\cdot[Ce^{4+}(\text{Ⅷ},\text{Ⅶ},\text{Ⅵ})] \qquad (32a\text{-}b),$$

and reveal some intriguing new features of Ce-Y: First, the $\{Y^{3+}(VII)\}$-$\{Ce^{4+}(VII)\}$ form in pair two most abundant bowing curves which are close to overlapping in the central area. Yet, the most abundant on both side are naturally the host $\{Ce^{4+}(VIII)\}$ in the CeO_2-rich DF $0 \leq y \leq \sim 0.35$ area and the host $\{Y^{3+}(VI)\}$ in the $YO_{1.5}$-rich C-type $\sim 0.65 \leq y \leq 1.0$ area, the respective most alien$\{Y^{3+}(VIII)\}$ and $\{Ce^{4+}(VI)\}$ both remaining below ~5 % over the entire y range.

Second, as indicated by a vertical red arrow, the [CN(VII)] for all cations provided by:

$$[CN(VII)] = \{Y^{3+}(VII)\} + \{Ce^{4+}(VII)\} = y \cdot [Y^{3+}(VII)] + (1-y) \cdot [Ce^{4+}(VII)] \quad (33),$$

is largely enhanced up to ~0.72 at $y \sim 0.50$ from 0.50 in the ref. three-site model (a), keeping its near-symmetric bowingness. Such significant near-bowing [CN(VII)] enhancement indicates that Ce-Y has in fact inherent tendency toward the limiting two-site model (a') in Fig. 15a. Namely, Vo and O^{2-} are here undergoing a partial Vo-O^{2-} ordering, (Vo-Vo) + (O^{2-} - O^{2-}) → 2(Vo-O^{2-}), avoiding not only the 1NN <100> but also more-distant 2NN <110> and 3NN <111> V_O-V_O configurations, i.e., confirming the afore-mentioned Vo-Vo repulsion scheme found in several computation simulation studies. Fig. 16c thus establishes the validity of the model (a) + (a)' hybrid restricted (non-)random framework (Fig. 15a) for Ce-Y. Since the formation of this new middle [CN(VII)] state both for M^{4+} and Ln^{3+} can be regarded as evidence of their mixing compatibility to form a common DF phase, it is natural that Ce-Lns which generally mix much better than Th-Lns have an enhanced middle [CN(VII)]>0.50, as seen here. The model application to other eleven CN(Ln^{3+}, M^{4+}) data of Ce-Lns and Th-Lns in Figs. 7a-b is straightforward, and will provide a complete set of their approximate non-random cation (VIII, VII, VI) concentration data.

Putting the random CN(Y^{3+})=8-2y in Eq. (30a) gives $[Y^{3+}(VII)]=2y \cdot (1-y)=[CN(VII)]$ in Eq. (21a) in the ref. three-site model (a). Namely, only in the ideally restricted-random limit the individual $[Y^{3+}(VIII,VII,VI)]$ and $[Ce^{4+}(VIII,VII,VI)]$ and all-cations' [CN(VIII,VII,VI)] converge to the same limiting complete restricted-random values in Eq. (21a). The last question to ask here is then 'what enables it to form such highly non-random coupled Y^{3+}-Ce^{4+} defect structure (Figs. 16a-c)? The answer is the heterogeneous inter $Y^{3+}(VII)$-$Ce^{4+}(VII)$ disproportionation reactions that couple the both sub-sites:

$$Y^{3+}(VI) + Ce^{4+}(VIII) \leftarrow Y^{3+}(VII) + Ce^{4+}(VII) \rightarrow Y^{3+}(VIII) + Ce^{4+}(VI) \quad (34a\text{-}b),$$

which is modestly shifted to the case (1) Y^{3+}-Vo associative lhs Eq. (34a) side in Ce-Y. We have above derived the non-random cation concentrations using the CN(Y^{3+},Ce^{4+}) data (Fig. 7b) obtained in the a_0(ss) analysis. Consequently, the above procedure involved only the intra Y^{3+} and Ce^{4+} disproportionation reactions, Eqs. (27) and (31). However, such overall non-random CN(Y^{3+}, Ce^{4+}) structure itselfis is in fact produced by the intervention of the inter Y^{3+}-Ce^{4+} ones, Eqs. (34a-b). Thus, in the higher-level combined QCE and DCC model approach, one needs to treat all these intra and inter non-randomness-generation processes in an integrated manner. Although some such attempts [4] have already been initiated by us, they would more proprly be described elsewhere in the future.

4.2.2. Overall Non-Random Cation → Anion Coordination

In DF $M_{1-y}Ln_yO_{2-y/2}Vo_{y/2}$ anions (Vo, O^{2-}) are constantly cation 4-fold coordinated (CN= Ⅳ) for all *y* (Figs. 1a-c, 3 and 14a-b). To distinguish this cation → anion coordination from the anion → cation one, the average Ln^{3+} CN of Vo or O^{2-} is denoted as $N(Ln^{3+})_{Vo\ or\ O2-}$, respectively. The overall average $N(Ln^{3+})$ condition corresponding to Eq. (7) for the converse $CN(Ln^{3+}, M^{4+})$ is:

$$N(Ln^{3+})_{Vo}\cdot (y/4) + N(Ln^{3+})_{O2-}\cdot (1-y/4) = 4 \cdot y \qquad (35).$$

The random $CN(Ln^{3+}, M^{4+}) = 8 - 2y$ (Eq. (4)) corresponds to the random $N(Ln^{3+})_{Vo,\ O2-}$ of

$$N(Ln^{3+})_{Vo}(\text{random}) = N(Ln^{3+})_{O2-}(\text{random}) = N(\text{random}) = 4 \cdot y \qquad (36).$$

One can easily derive the general relationship between $N(Ln^{3+})_{Vo,\ O2-}$ and $CN(Ln^{3+})$ given by;

$$N(Ln^{3+})_{Vo} = 2 \cdot \{8 - CN(Ln^{3+})\},\ \ N(Ln^{3+})_{O}{}^{2-} = 2\,y \cdot CN(Ln^{3+}) / (4 - y) \qquad (37a\text{-}b).$$

Here Eq. (37b) for $N(Ln^{3+})_{O2-}$ is obtained by combining Eq. (37a) for $N(Ln^{3+})_{Vo}$ with Eq. (35). If the system is case (i) Ln^{3+}-Vo associative non random and therefore Ln^{3+} has more 1NN Vos (=8-$CN(Ln^{3+})$) than the random value of 2*y*, 8-$CN(Ln^{3+}) > 2y$ ($CN(Ln^{3+}) < (8-2y)$), the V_O will also have more 1NN Ln^{3+}s (=$N(Ln^{3+})_{Vo}$) than the random value of $N = 4y$ in Eq. (36). That is, a simple linear relationship between them, $N(Ln^{3+})_{Vo} = C\{8-CN(Ln^{3+})\}$ with the proportional constant C = 2 in Eq. (37a), readily follows. For instance, in the limiting C $LnO_{1.5}$ at *y*=1.0 (Fig. 1b) the max. 2Vo coordination about Ln^{3+} (8-$CN(Ln^{3+})$ = 8-6 = 2) corresponds to the full 4Ln^{3+} coordination of Vo; $N(Ln^{3+})_{Vo} = 4 = 2\cdot\{8-CN(Ln^{3+})\} = 2\cdot 2$. Eq. (37a) can also describe another extreme case of P $Zr(Hf)_2Ln_2O_7$ at *y*=0.50 (Fig. 3), where CN = 8 for Ln^{3+}, so 8-$CN(Ln^{3+})$ = 8–8 = 0 and therefore $N(Ln^{3+})_{Vo} = 0$ in Eq. (37a). Namely, as the Ln^{3+} has no 1NN Vo, the Vo has no 1NN Ln^{3+} either. In this sense assuming random cation coordination about O^{2-} and Vo is a self-contradiction in either parent F-type M^{4+} = Ce and Th or P(δ)-type SZ(SH), unless the system is a completely random ideal solid solution with both CN = 8-2*y* (Eq. (4)) and N = 4*y* (Eq. (36)). In other words mutually non-random anion ↔ cation coordination must always have 1:1 correspondence given by Eqs. (37a-b).

Figs. 17a-b show the non-random $N(Ln^{3+})_{Vo}$ - $N(Ln^{3+})_{O2-}$ pair curves of both systems calculated with Eqs. (37a-b) using each $CN(Ln^{3+})$ data (Figs. 7a-b). Both $N(Ln^{3+})_{Vo}$ curves are just the re-normalized upside-down curves of both $CN(Ln^{3+})$ curves. Therefore, the M^{4+} = Th (Fig. 17a) reveal their extreme Ln^{3+} association to V_O. For the much more abundant host O^{2-}, which decreases only to (1-*y*/4)=0.75 at *y*=1.0 from 1 at *y*=0, the random → non-random $N(Ln^{3+})_{O2-}$ shift is quite small, albeit it is certain that $N(Ln^{3+})_{O2-} < 4y$ slightly in all the six Th-Lns. That is, Ln^{3+} does not readily dissociate from O^{2-}. The M^{4+} = Ce (Fig. 17b) again reveal a much milder steady case (i) → (ii) non-randomness shift with increasing $r_C(Ln^{3+})$, the crossover occurring at the Ln^{3+} = Sm → Nd. Here, the non-random $N(Ln^{3+})_{O2-}$ shift is even more minor. All these non-random $CN(Ln^{3+}, M^{4+})$, [(Ln^{3+}, M^{4+})(Ⅷ, Ⅶ, Ⅵ)] and $N(Ln^{3+}, M^{4+})_{Vo}$ are likely to be fixed by the T_{prep} of the system.

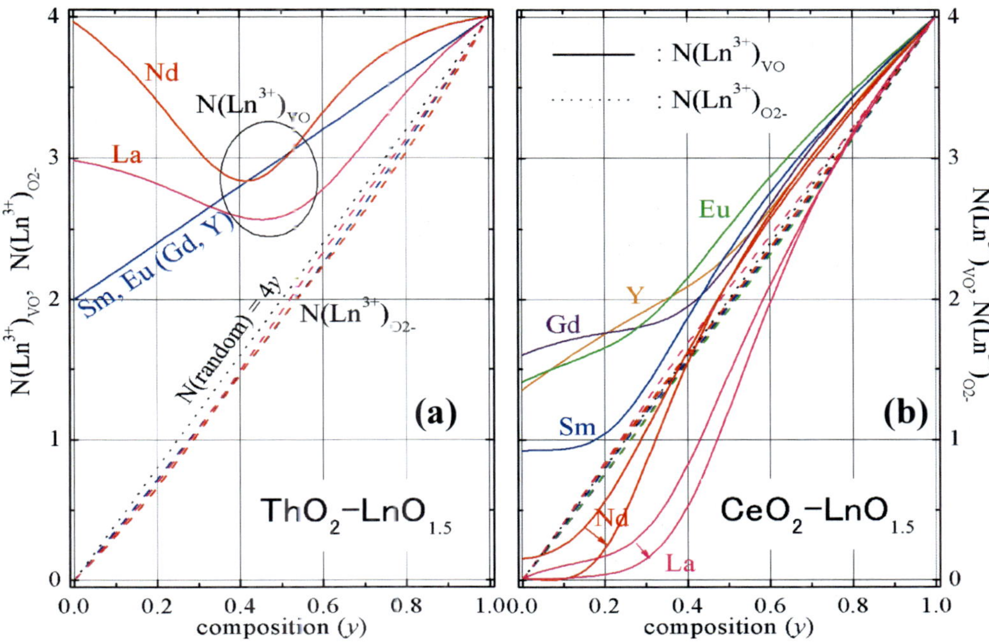

Figure. 17. The overall $N(Ln^{3+})Vo$ and $N(Ln^{3+})O2-$ in Th-Lns (a) and Ce-Lns (b).

4. 2. 3. Detailed Non-Random Cation Coordination about Anion in Ce-Y

As illustrated for cations in Figs. 16a-b, the overall non-random $N(Ln^{3+})_{Vo,\ O2-}$ data of anions in these systems in Figs. 17a-b can be used to derive detailed non-randomly Ln^{3+}-n (= 0-4)-fold coordinated Vo and O^{2-} concentrations. But, to attain this we need some additional local-structure and **σ**(ion) etc data of the system. This is illustrated again for Ce-Y [2].

Fig. 18a depicts the most likely non-random [Vo(n=0-4)] curves in Ce-Y derived from its $N(Y^{3+})_{Vo}$ data in Fig. 17b with recourse to its most reliable low-temperature **σ(ion) data** reported by Nowick et al. [64,124] (see Fig. 19 for brevity). Fig. 18b depicts the ref. random [Vo(0-4)] profile with its ref. random $N(Ln^{3+})_{Vo}/4 = \{\sum n_4C_n(1-y)^{1-n}y^n\}/4 = y$ diagonal line in black solid line. For comparison, the direct MAS-NMR $[^{17}O^{2-}(0\text{-}4)]$ data in the DF $0 \leq y \leq 0.25$ range [32] are also plotted with its non-random $N(Ln^{3+})_{O2-}/4$ curve in black dashed line. Fig. 18c further depicts one other less-probable half-way [Vo(n=0-4)] profile. The random cation coordination profile about V_C (Fig. 18b) has been often used for **σ(ion)** analysis of these M^{4+} = Ce and Zr [61,64,67-69,125]. Since their $N(Ln^{3+})_{Vo}$ is mostly non-random (Figs. 17a-b), Fig. 18b is not the accurate [Vo(0-4)] profile. However, it is a good first-approximation $[O^{2-}(0\text{-}4)]$ profile in both M^{4+} = Ce and Th from their close to random $N(Ln^{3+})_{O2-}$ curves, though the actual ^{17}O-NMR $[O^{2-}(0\text{-}4)]$ data show a definite but slight non-randomness. Besides, as seen below, it can give a valuable hint to derive the non-random [Vo(0-4)] profile in Fig. 18a.

As to the two non-random $[V_O(0\text{-}4)]$ profiles in Figs. 18a and c, since $CN(Y^{3+})$=8, 7 and 6 correspond to $N(Y^{3+})_{Vo} = 2\cdot\{8\text{-}CN(Y^{3+})\}$ = 0, 2 and 4 in Eq. (37a), respectively, the $[Y^{3+}$(Ⅷ,Ⅶ,Ⅵ)] in Fig. 16a can be directly transferred as [Vo(0,2,4)] as drawn there in dot-dashed line. This is of course the non-final profile, rather like the limiting two-site model (a') state in the cation case (Figs. 16a-b). The final one should include two odd-n [Vo(1,3)] too, for which the full Vo sub-site and overall average $N(Y^{3+})_{Vo}$ conditions are given by;

$$\sum_{n=0}^{4} [Vo(n)] = 1, \quad \sum_{n=0}^{4} n \cdot [Vo(n)] = N(Y^{3+})_{Vo} \qquad (38a\text{-}b).$$

In analogy to the above cation case we could call this the three- to five-site model approach to Vo(n). The incorporation of the two Vo(1,3) can be made mainly with the following three (more effective) combination / dissociation reactions (due to the large mutual overlap over the common y range), Eqs. (39a-c), together with other (less-effective) ones (due to the smaller mutual overlap), Eqs. (40a-b):

$$Vo(0) + Vo(2) \leftrightarrow 2\,Vo(1),\ Vo(2) + Vo(4) \leftrightarrow 2\,Vo(3),\ Vo(1) + Vo(3) \leftrightarrow 2Vo(2) \qquad (39a\text{-}c),$$

$$Vo(0) + Vo(4) \leftrightarrow 2Vo(2),\ Vo(1)+Vo(4) \leftrightarrow Vo(2)+Vo(3) \qquad (40a\text{-}b).$$

Starting from the initial [Vo(0,2,4)] state in dot-dashed line, by transformation among all these [Vo(0-4)] by using Eqs. (39a-c) and (40a-b), their most likely final state has been sought so as to provide a consistent description of the σ(ion) behavior (Fig. 19) described in 4.3, and drawn in solid line in Figs 18a. Fig. 18c shows one other less probable extreme half-way one.

In doing this, we have utilized a valuable hint obtained from the ref. random [Vo(n)] profile in Fig. 18b that there is a full correspondence between the overall $N(Ln^{3+})_{Vo} = 4y$ curve and the individual [Vo(n)] peaks that each [Vo(n)] (n = 0,1,2,3,4) always is most abundant at $y_{peak} = 0$, 0.25, 0.50, 0.75 and 1.0, precisely at which $N(Ln^{3+})_{Vo} = 4y = 0$, 1, 2, 3 and 4, respectively.

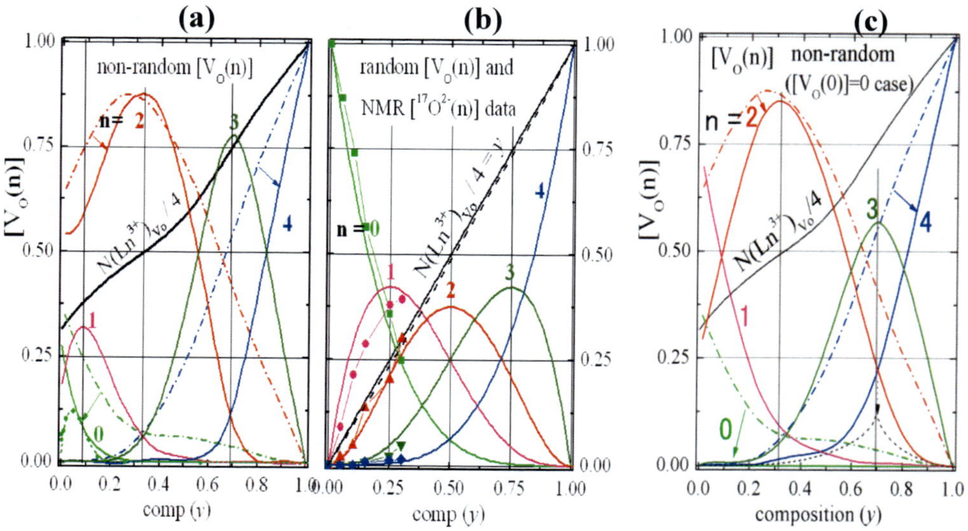

Figure 18. Possible Y^{3+} n-fold -associated non-random [Vo(n)] profiles (a) and (c) and the ref. random anion profile and NMR $[^{17}O^{2-}(n)]$ data (b) in Ce-Y. (a) and (c) also include the initial [Vo(0,2,4)] (dot-dashed line) and final calculated [Vo(0-4)]s (in solid line).

By applying this rule tc the non-random $N(Y^{3+})_{Vo}$ curve, it is no doubt that the non-random [Vo(2,3,4)] have their respective peaks at y_{peak} = 0.33, 0.69 and 1.0 at which the $N(Ln^{3+})_{Vo}$ = 2, 3 and 4, as with each derived final curve in Fig 18a. But, this is only approximately so in the half-way [Vo(2)] in Fig. 18c. In the latter the [Vo(0)] is set zero, transferring all $[Vo(0)]_{initial}$ as [Vo(1)] by combining with $[Vo(2)]_{initial}$ in Eq. (39a); $[V_O(1)] = 2[Vo(0)]_{initial}$.

As these two [Vo(0-4)] profiles in Figs. 18a and c illustrate, most important but most problematic is to fix unambiguously the final states of the two lowest-n [Vo(0,1)] in the lowest-y < ~0.30 range, for the $N(Ln^{3+})_{Vo}$ > 1 already at y = 0 outside the application range of the above rule from the random case. There are many possibilities here, such as (i) y_{peak} = 0 for both [Vo(0,1)], (ii) y_{peak} = 0 for [Vo(0)] but $y_{peak} \neq 0$ for [Vo(1)] as in Fig. 18a or other way around, *etc*, which to choose being hard to decide from this $N(Ln^{3+})_{Vo}$ curve alone. Judging from the extreme abundance of $[Vo(2)]_{initial}$ which is largely shifted to the lower-y side from the ref. random [Vo(2)] curve with its peak just at y = 0.50 in Fig. 18b, it seems highly likely that [Vo(0)] and [Vo(1)] are the least and the second-least abundant Vo site, respectively, both largely diminished and shifted to the lower-y side from the ref. random [Vo(0,1)], i.e., [Vo(0)] < [Vo(1)] < [Vo(2)], as indeed the case in Figs. 18a and c. To go beyond this, however, one seems to need some additional relevant defect and σ(ion) etc data of this and related systems.

The [Vo(0,1)] in Fig. 18a are derived with recourse to its low-temperature σ(ion) data (Fig. 19); the [Vo(0)] (= $0.275(1-y)^{13}$) is greatly reduced and more steeply diminishing with y than the random [Vo(0)] (= $(1-y)^4$) in Fig. 18b. This choice of [Vo(0)] results in the [Vo(1)] with a small peak ~0.31 at y_{peak} ~ 0.08. It has the largest random → non-random y_{peak} shift of Δy_{peak} = -0.17 (0.25 → 0.08), in line with the trend along with the larger-n [Vo(n=2,3,4)]. Though these [Vo(0,1)] are largely diminished and rammed into a narrow low-y range, both still retain the respective original random-case shapes in Fig. 18b to some extent, Whereas, in Fig. 18c an extreme [Vo(0)] = 0 choice gives the $[Vo(1)] = 2[Vo(0)]_{initial}$ with its y_{peak} = 0 and most abundant for the y < ~0.10 range. In both (and any) cases the peak [Vo(2)] value does not seem to vary much from its initial $[Vo(2)]_{initial}$ ~ 0.33. It is a paradox that, while the actual peak position of the middle [Y(VII)] deviates from the original y_{peak} = 0.33 in the limiting two-site model (a') as seen in Fig. 16a, this y_{peak} = 0.33 is retained as that of the middle [Vo(2)] both in Fig. 18a and c.

One support for the [Vo(0,1)] profile in Fig. 18a comes from the $N(Sm^{3+})_{Vo}$ curve of less case (i) Sm^{3+}-Vo associative non-random Ce-Sm in Fig. 17b; it has y_{peak} ~ 0.18 for $[V_O(1)]$ at which $N(Sm^{3+})_{Vo}$ = 1. This y_{peak} is much higher than the y_{peak} ~ 0.08 for [Vo(1)] in Ce-Y in Figs. 18a. The strongest support comes from σ(ion) data at 182°C of this Ce-Y itself (Fig. 19) with a characteristic maximum at such lowest y (< 0.10) range as described in 4.3. Fig. 18a is thus proposed as a possible most likely 'schematic' non-random [Vo(0-4)] profile in Ce-Y in such low temperature range. 'Schematic' here means that the derived [Vo(0,1)] curves have relatively large uncertainty in absolute value and relative [Vo(0)] / [Vo(1)] ratio, one possibly approaching more to those much smaller in Fig. 18c, as discussed in 4.3.

As to the $[O^{2-}(0\text{-}4)]$ profile, in fair accordance with the model result, within experimental uncertainty the direct ^{17}O MAS-NMR $[O^{2-}(0\text{-}4)]$ data [32] plotted in Fig. 18b indeed confirm their close to random but a little shifted profile to the higher-y side opposite to the above Vo case. This marked difference between the strongly non-random [Vo(0-4)] profile (Figs. 18a and c) and the near random $[O^{2-}(0\text{-}4)]$ one (Fig. 18b) in this and many related DF oxides directly mirrors the difference between their overall $N(Ln^{3+})_{Vo}$ and $N(Ln^{3+})_{O2-}$ in Figs. 17a-b, which in turn originates from the obvious fact that DF oxides have the 'host' O^{2-} ions far-more abundant than the doped Vos; $[O^{2-}] = (1-y/4) >> y/4 = [Vo]$ up to y=1.0. This is different from what occurs in the cation sublattice, wherein as y=0→1.0 the host M^{4+} is totally replaced with the dopant Ln^{3+}, and the two form a near-symmetric non-random defect structure revealed in Figs. 16a-c. In DF oxides the minor V_Os mostly shoulder the non-randomness generated in the anion sublattice.

The above results indicate that even in a given (fixed) overall non-random $N(Ln^{3+})_{Vo,\,O2-}$ state in Figs. 17a-b, both extremely non-random [Vo(0-4)] and near-random $[O^{2-}(0\text{-}4)]$ profiles, in particular the former, cannot be uniquely fixed, i.e., many different [Vo(0-4)] profiles other than that in Fig. 18a (e.g. that in Fig. 18c) are in principle possible depending on the temperature of the system. In other words, as detailed in [2], the anions seem to be able to adjust their detailed non-randomness rather freely to the system's operation temperature (T_{oper} < 1000°C) well below T_{prep} without disturbing its overall non-random $N(Ln^{3+})_{Vo}$ by virtue of the high Vo (i.e. O^{2-}) mobility as oxide-ion conductor in such much lower T_{oper} range than T_{prep}.

4.3. Ionic-Conductivity Analysis of Ce-Y

As summarized in Fig. 19, the derived [Vo(0-4)] and $[O^{2-}(0\text{-}4)]$ profiles (Figs. 18a-b) in Ce-Y can combine to give a straightforward description of its intriguing $\sigma(ion)_{(max)}$ behavior observed in the low y range. To compare with the model curves, we have plotted the raw $\sigma(ion)_{(exp)}$ $((\Omega\ cm)^{-1})$ data at the reported lowest and a higher temperatures of 182 and 560°C [124] multiplied by $5 \cdot 10^3$ and 1.4, respectively, so that the both give nearly the same normalized (norm) $\sigma(ion)_{(max)} \sim 1.4 \cdot 10^{-2}$ at the common $y(max)$~0.077. Or, it should be rather oppositely stated that the non-random $[V_O(0\text{-}4)]$ profile in Fig. 18a has been fixed so as to give a most consistent σ(ion) vs. y isotherm curve at 182°C as a result of multiple Vo(n) ↔ O^{2-}(n) site-exchange process through the Y^{3+}-n(=0-4)-fold associated anion site with each Vo mobility, $\mu_m(n)$;

$$\sigma(total) \sim \sum_{n=0}^{4} \sigma(n) \propto \sum_{n=0}^{4} \mu_m(n) \cdot \{Vo(n)\} \cdot \{O^{2-}(n)\} = (y/4) \cdot (1-y/4) \sum_{n=0}^{4} \mu_m(n) \cdot [Vo(n)] \cdot [O^{2-}(n)] \tag{41}$$

Fig. 19 shows that the lowest-n=0 term, σ(0) in Eq. (41) alone (in dashed green) can already reproduce reasonably well the basic feature of its $\sigma(ion)_{(max)}$ behavior, with its concentration product, $\{Vo(0)\} \cdot \{O^{2-}(0)\} = (y/4) \cdot (1-y/4) \cdot [Vo(0)] \cdot [O^{2-}(0)]$ (in thin green line) taking the max. at y~0.06 and its Vo(0) mobility taking by far the highest value; $\mu_m(0)=10 \gg \mu_m(1)=1.25 \gg \mu_m(2)=0.025$ as listed in the Figure legend. Though it certainly exhibits a much steeper initial increase and a little lower $y_{(peak)}$~0.055-0.06 (< 0.077 = $y_{(max)}$) than both σ(ion) data, it has a reasonable subsequent decrease after the $\sigma(ion)_{(max)}$ lying just in between both σ(ion) data. It appears therefore enough valid to conclude that the non-dopant (Y^{3+})-associated lowest n = 0 state provides the most freely-mobile Vo(0) transport path through the thickest $[O^{2-}(0)]$ channel (Fig. 18b) in the lowest y < 0.20 range, i.e., the most favourable dopant-free anion channel for Vo motion, giving rise to its characteristic σ(ion)(max) behavior. Though not detailed here, our more recent σ(ion) analyses show that this is basically also the case for many other related M^{4+} = Ce and even SZ(SH) systems.

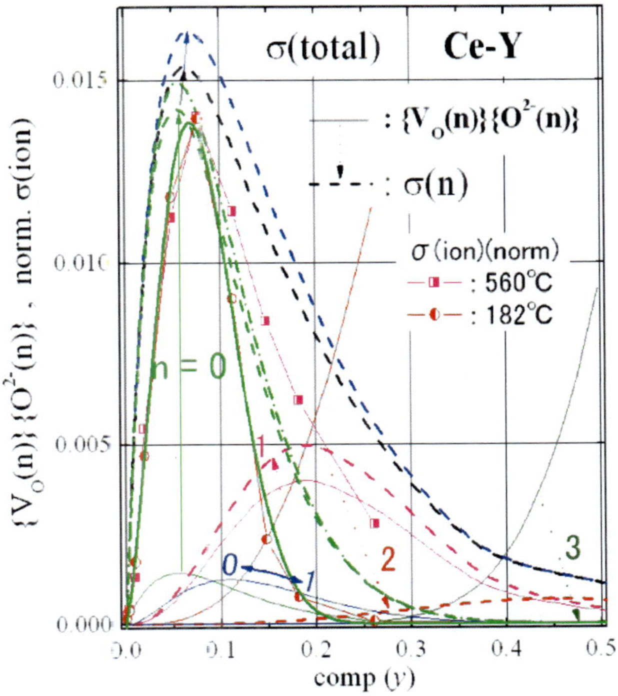

Figure 19. σ(ion) data [124] and model curves in Ce-Y. Each σ(n) and σ(total) curves (dashed line) are calculated using [Vo(n)]-[O^{2-}(n)] curves in Figs. 18a-b. and μ_m(n) values; $\mu_m(0) = 10 >> \mu_m(1) = \mu_m(1\leftrightarrow0) = 1.25 >> \mu_m(2) = 0.025 >> \mu_m(3) = 0.005 >> \mu_m(4) = 0$. σ(0)(green), σ(0) +σ(1) (black), σ(0)+σ(1)+2σ(0↔1) (blue). Here-derived final σ(0) curve (thick solid green) (See text for details).

Possible some further elaborations of this base σ(0) curve are discussed below: The most abundant concentration product changes from {Vo(0)}{O^{2-}(0)} for the initial $0 \le y \le 0.075$ range to {Vo(1)}{O^{2-}(1)} for the $0.075 \le y \le 0.175$ range and further to {Vo(2)}{O^{2-}(2)} for the y>~0.175 range. But, due to their much lower mobilities (μ_m(1 or 2) / μ_m(0) = 1/8 or 1/400), such higher n ≥ 1 states make only a little contribution to the σ(ion)(total) (that is ~0 for the n ≥2). Unless otherwise the onset of σ(ion)(max) should shift to the much higher y>0.10 range. The fact that both σ(ion) data take the similar sharp σ(ion)(max) at the same y(max)=0.077 indicates that at least in such low temperature range ($182 \le T \le 560$°C) the contribution of the higher n ≥ 1 states remains minor. The present μ_m(1)=1.25 gives the σ(ion)(total) (= σ(0)+ σ(1)) curve (in dashed black line) with its max. being shifted to a little higher y~0.07 and decreasing more gradually in the higher-y > 0.10 range. This qualitatively accords with the observed experimental trend that higher the T_{oper} (180→560°C) broader the σ(ion)(max) curve in the higher-y (> 0.077) side after the σ(ion)(max); this is because with increasing T_{oper} both [Vo(0-4)] and μ_m(0-4) are thermally more equiponderated (averaged), so that the relative contribution of the higher-n σ(1) will increase especially in the higher-y > 0.77 range, though the degree and effect of such thermal equiponderation (averaging) is judged to be quite small in this low-temperature range.

Obviously, the more complete σ(ion)(total) formula than Eq. (41) will be the next Eq. (41)' including all the zig-zag $Vo(i) \leftrightarrow O^{2-}(j)$ site-exchange processes ($i \neq j$) with $\mu_m(i\text{-}j)$

$$\sigma(\text{total})' \sim \sum_{i,j=0}^{4} \sigma(i\text{-}j) \propto \sum_{i,j=0}^{4} \mu_m(i\text{-}j)\cdot\{Vo(i)\}\cdot\{O^{2-}(j)\} = (y/4)\cdot(1\text{-}y/4)\sum_{i,j=0}^{4} \mu_m(i\text{-}j)\cdot[Vo(i)]\cdot[O^{2-}(j)] \quad (41)',$$

which correctly reduces to the usual high-temperature random-limit σ(ion)(total) expression for hopping ionic conduction: $\sigma(\text{total})(\text{random}) \propto (y/4)\cdot(1\text{-}y/4)\cdot\mu_m$. In that case all the $\mu_m(i\text{-}j)$s become the same μ_m and are removed from the summation term, so that it follows that;

$$\sigma(\text{total})'(\text{random})/\{(y/4)\cdot(1\text{-}y/4)\cdot\mu_m\} \sim \sum_{i,j=0}^{4} [Vo(i)]\cdot[O^{2-}(j)] = \sum_{i,=0}^{4} [Vo(i)]\sum_{i=0}^{4} [O^{2-}(j)] = 1\cdot 1 = 1 \quad (41)''.$$

It corresponds to the thermally fully-equiponderated high-temperature random limit σ(ion)(total) formula. However, in reality, these DF oxides including Ce-Y are unlikely to reach such high-temperature random limit below ~2000°C, for all these prepared by the usual ceramic method (T_{prep}~1400-1800°C) still have largely restricted non-random defect structure, as above clarified in this chapter.

The only exception to this appears to be a disordered-DF type δ Bi_2O_3 (and its-based complex oxides) with the highest **σ(ion)** known to these DF oxides: An ^{17}O-MAS-NMR study of δ Bi_2O_3 [126] clarified that its O^{2-} has a low migration activation energy $E_m \sim 0.46$ eV comparable with that of the **σ(ion)**$\cdot T$ itself, showing no discontinuity between the solid δ phase in the $729 \leq T \leq 824$°C (= T_{mp}) range and the liquid phase above T_{mp}. This means that anions i.e. oxygens in δ Bi_2O_3 are very in the so-called 'sublattice-melt' state, unlike the present DF oxides. Namely, δ Bi_2O_3 alone has much more (restricted-) random highly-disordered liquid-like anions (O^{2-} and Vo) all fully moving around much more freely to make its highest **σ(ion)**. Yet, as argued briefly in 3.1 its anion sublattice is judged to be still far from being 'completely random' (Fig. 15b), restricting the Bi^{3+} site only to the CN ≥ 5 state by excluding the 1NN <100> Vo-Vo pair formation as well.

Eq. (41) for σ(ion)(total) has in fact singled out only the n = i = j = 0 - 4 terms by setting $\sigma(n\text{-}n) = \sigma(n) = \{V_O(n)\}\{O^{2-}(n)\}\mu_m(n)$ in the following homogeneous $Vo\leftrightarrow O^{2-}$ site exchange processes;

$$Vo(n) + O^{2-}(n) \leftrightarrow O^{2-}(n) + Vo(n) \text{ (where n = 0-4)} \quad (42),$$

neglecting all the $i \neq j$ terms in Eq. (41)'. However, in reality, what makes it possible for the Vo(n=0-4) to produce such highly non-random profile (Fig. 18a) is again the intervention of these heterogeneous inter $Vo(i)\leftrightarrow O^{2-}(j)$ site-exchange processes ($i \neq j$) in Eq. (41)';

$$Vo(i) + O^{2-}(j) \leftrightarrow O^{2-}(i) + Vo(j) \text{ (where } 0 \leq i < j \leq 4 \text{)} \quad (43).$$

and the dynamic equilibrium between them that always proceeds as a pair; σ(i-j) = σ(j-i);

$$\sigma(i\text{-}j) \propto \{V_O(i)\}\{O^{2-}(j)\}\ \mu_m(i\text{-}j) = \{V_O(j)\}\{O^{2-}(i)\}\ \mu_m(j\text{-}i) \propto \sigma(j\text{-}i)\ (0 \leq i < j \leq 4) \qquad (44),$$

In this sense the heterogeneous inter Vo(i)↔O^{2-}(j) site-exchange reaction expressed by Eqs. (43)-(44) is the most important internal non-randomness re-equilibriation (readjustment) process for the anions in response to temperature (T_{oper}) and composition (y) change of the system that actually builds the individual non-random [Vo(n)] and [O^{2-}(n)] curves in Figs. 18a-c. In parallel to the cation case in 3.2, these most important heterogeneous Vo(i)↔O^{2-}(j) site-exchange reactions in Eq. (43) did not appear at all in the above derivation process of the non-random individual [Vo(n)] curves in Figs. 18a-c (see Eqs. (39a-c) and (40a-b)), again because we have used there the overall non-random $N(Ln^{3+})_{Vo,O2-}$ curves derived from another a_0(ss)-analysis route. The most crucial fact here is that anions in these DF oxides have an additional freedom of adjusting their non-randomness rather freely in the much lower T_{oper} (< ~1000°C) range than the usual T_{prep}~1400-1800°C by virtue of their high oxide-ion conductivity expressed in Eqs. (41-43). Indeed σ(ion) even at 182°C is measured in Ce-Y in Fig. 19.

As a matter of fact, as drawn in Fig. 19 in blue dashed line, the contribution of such heterogeneous zig-zag i ↔ j processes to the σ(ion)(total) in Eq. (41)' is small in this low temperature range: From both the mobility and concentration factors, this comes only from the lowest-two-n (i, j = 0, 1) states in Eqs. (43)-(44). To estimate the possible max. contribution from the σ(1-0) + σ(0-1) = 2σ(1-0) = 2σ(0↔1) in Eq. (41)', it is set that $\mu_m(1\text{-}0) = \mu_m(1) = 1.25$ and adding this 2σ(0↔1) (the lowest-side thin solid blue curve) to the above σ(ion)(total) (the dashed black curve) results in its only slight upward shift (the dashed blue curve).

To gain further insight into the σ(ion)(max) behavior from a different point of view, consider the three lowest-n = 0-2 in Eq. (43);

$$Vo(0) + O^{2-}(1) \leftrightarrow O^{2-}(0) + Vo(1),\ Vo(0) + O^{2-}(2) \leftrightarrow O^{2-}(0) + Vo(2), \qquad (45a\text{-}b).$$

$$Vo(1) + O^{2-}(2) \leftrightarrow O^{2-}(1) + Vo(2). \qquad (45c).$$

Eqs. (45a-c) indicates that at lower T_{oper} more and more Vo(0)s are trapped and immobilized in the more stable Vo(1) and further in the most-stable Vo(2), making the least abundant Vo(0) increasingly more dominant almost sole contributor to σ(ion)(total) of the system, ultimately reaching the lowest-limit [Vo(0)]~0 state supposed in Fig. 18c. Oppositely speaking, with increasing T_{oper} the more [Vo(0,1)] produced by the thermal excitations from the most-stable [Vo(2)] ground state and their thermally more equiponderated mobilities ($\mu_m(0)$, $\mu_m(1)$, $\mu_m(0\leftrightarrow1)$) lead to the increasing relative contribution of the latter Vo(1). As seen in Fig. 19 this simultaneously causes the σ(ion)(max) broadening in the higher y > 0.077 range and its $y_{(max)}$ shift to the higher-y side.

Though the above analysis [2] is thus a good first-approximation σ(ion)(max) description of Ce-Y, Fig. 19 still reveals that, compared with the σ(ion)(exp) data at 182°C, all the derived σ(ion) model curves (in dashed green, black and blue lines) tend to increase initially much more steeply, attain the milder (i.e., more round) σ(ion)(max) at a slightly lower y ~ 0.05-0.07

and then decrease more gradually in the higher $y > 0.08$ range. To obtain the better fit to σ(ion) data at 182°C with the sharpest σ(ion)(max), some attempts have been made. The first one is to replace the random $[O^{2-}(n)]$ profile to the slightly non-random smooth-curved MAS-NMR $[^{17}O^{2-}(n=0,1)]$ profile in Fig. 18b, but, this (dash-dotted green curve there) is found to bring little improvement of the σ(0) curve. Therefore, this discrepancy is judged to stem from another fact that the present $[Vo(0)] = 0.275(1-y)^{13}$ profile in Fig. 18a is not enough accurate to reproduce the σ(ion)(max)(182°C) data, for the mobility $\mu_m(0)$ (and its migration activation energy H_m) in the lowest $y < 0.070$-0.08 (~ y(max)) range is reported to be nearly constant independent of y [124,126,127]. So, adopting the same $\mu_m(0)=10$ and the random $[O^{2-}(0)]$ in Fig. 18b, the [Vo(0)] was calculated backwards from the relationship; σ(ion)(norm)(182°C) ~ $\sigma(0) = \mu_m(0)\cdot\{Vo(0)\}\cdot\{O^{2-}(0)\} = 10\cdot(y/4)\cdot(1-y/4)\cdot(1-y)^4[Vo(0)]$, as;

$$[Vo(0)] = \sigma(ion)(norm)(182°C) / \{10 \cdot (y/4) \cdot (1-y/4) \cdot (1-y)^4\} \qquad (46),$$

and is plotted in Fig. 18a in filled green symbol. Unlike the above steeply-decreasing [Vo(0)] $(= 0.275(1-y)^{13})$ with y, thus-calculated [Vo(0)] is found to exhibit a peak at $y \sim 0.005$, being well fit by the following expression as drawn there;

$$[Vo(0)] = 0.413\cdot y^{0.35}\cdot(1-y)^{0.75}\cdot\exp(-(0.01986-y)^2/0.007) \qquad (47).$$

Using thus-calculated [Vo(0)] curve in Eq. (47) the σ(ion)(norm)(182°C) curve is re-calculated and drawn in Fig. 19 in thick solid green line. It is apparent that the calculated σ(ion)(norm) curve closely follows the σ(ion)(182°C) data over the entire $0 \le y \le 0.30$ range. So, it appears very likely that at this lowest $T_{oper} = 182°C$ the lowest-n σ(0) is indeed almost the sole contributor to the σ(ion)(total) in the system. The calculated [Vo(0)] curve with its peak at $y \sim 0.005$ in Fig. 18a also seems to accord with the experimental finding that the overall activation energy E_a for σ(ion)·T takes a min. ~0.77eV at y~0.038-0.078 (~0.005 in this [Vo(0)] peak) in the vicinity of the σ(ion)(max), decreasing initially from ~0.90eV at y~ 0.001 and then increasing rapidly up to ~1.43eV at y=0.57 after the $y_{(max)}$ [64,125]. This is shortly because this minimum in E_a (= H_m+ H_A) means the minimum in H_A, the association enthalpy for $Y^{3+} + V_O \leftrightarrow (Y^{3+}\text{-}V_O)$ in Eq. (45a), (for H_m is constant for the initial $y \lesssim 0.77$ range, as above noted), consequently enhancing the [Vo(0)] to have a peak at y~0.005. The more thorough σ(ion)-data analyses including their full T dependence over the reported wide $182 \le T \le 1400°C$ range is certainly necessary, in order to determine more precisely the absolute value of the [Vo(0)] curve in Eqs. (46-47), for, if the assumed $\mu_m(0) = 10$ is changed to e.g. 30 in Eq. (46), the resulting [Vo(0)] is changed to the one-third (10/30 = 1/3) of the present one, approaching to the [Vo(0,1)] profile drawn in Fig. 18c (as afore-mentioned). Accordingly, what one can say at most here seems to be that; the above proposed is a 'schematic' basic scenario for the onset of σ(ion)(max) in Ce-Y as a typical example of such almost commonly-observed intriguing σ(ion)(max) phenomenon in these DF oxides.

One fascinating future issue of such σ(ion)(max) behavior in DF oxides is finally addressed in brief. As alluded to in [2] and also in the above 4.1 in this chapter, the σ(ion) in these systems is considered to be a complicated cooperative and correlated process in which Vos are moving around dynamically avoiding the mutual 1NN Vo-Vo configuration in Fig. 14a-b, as partially revealed and substantiated in a MD-simulations study of Okazaki et al.

for YSZ [65,66]. It appears far from clear at present 'how and to which extent such dynamical cooperative simultaneous motions of two or even three Vos are affecting their actual σ(ion) behavior'. As afore-cited in 4.1, most recent several Monte-Carlo σ(ion)-simulation studies of these DF oxides [69-72] have indeed arrived at the similar conclusion to the present DCC model that strong 1NN <100> Vo-Vo repulsion greatly enhances the onset of their σ(ion)(max) in the low *y* range by more severely hindering the free Vo motion in the higher *y* range. To resolve this basic problem more quantitatively and unambiguously we seem to need to have really microscopic understanding on the static as well as dynamic local-structure and σ(ion) behavior of these systems at high temperatures based on ab-initio DFT and related defect-property calculations.

CONCLUSION

As was shown above, the proposed new defect-crystal-chemistry approach can provide a possible unified picture of non-Vegardianity and non-random defect structure of parent fluorite (F)-type defect-fluorite MO_2-$LnO_{1.5}$ solid solutions (ss) with M^{4+} = Ce and Th as inherently-coupled two sides of 'distortion-dilation' phenomenon of C-type $LnO_{1.5}$ [1-5]. As detailed in **1-3** of this chapter, its capability as (i) a new quantitative generalized Vegard-law model compatible with ion-packing model enables the extraction not only of their metal-oxygen bond-length data but also of key overall non-random oxygen coordination number data of both cations (CN(Ln^{3+}, M^{4+})) from the lattice-parameter data analysis, clarifying a strikingly contrastive coupled non-Vegardian - non-random behavior of these M^{4+} = Ce and Th systems. Once these overall non-random CN(Ln^{3+}, M^{4+}) data are obtained in such lattice-parameter data analysis and/or by lanthanide (Ln) MAS-NMR method etc, making full use of them in combination with 'restricted (non-) randomness of defect fluorite lattice, as detailed in 4, this shows its further capability as (ii) a new direct link to their controversial defect structure; as illustrated here for the representative CeO_2-$YO_{1.5}$ system; the overall non-random CN(Ln^{3+}, M^{4+}) data are successively converted to the detailed mutually non-randomly coordinated cation ↔ anion concentration data of oxygen 8-, 7- and 6-fold coordinated {Y^{3+}(Ⅷ, Ⅶ, Ⅵ)} and {M^{4+}(Ⅷ, Ⅶ, Ⅵ)} and the converse Ln^{3+} n(=0-4)-fold coordinated {Vo(n)} and {O^{2-}(n)}, and the last two anionic ones can further combine to provide a new straightforward description of their intriguing ionic-conductivity maximum behavior in the low dopant (Ln^{3+}) content range. As allued to in brief, one major model extension/elaboration direction is to integrate the non-randomness derived from the lattice-parameter analysis in the above (i) with that from the more general defect-thermodynamics, i.e., to develop a new defect-thermodynamic formalism of DF oxides that can further pave the way to their detailed defect-thermodynamic (non-random defect structure, phase-diagram and calorimetric mixing *etc*) property description.

The above significant success of this defect-crystal chemistry lattice-parameter model means that these M^{4+} = Ce and Th are indeed the F MO_2 - C $LnO_{1.5}$ binary systems describable by the F-C binary model, and because of this very nature it is not applicable to stabilized M^{4+} = Zr and Hf (SZ and SH) involving pyrochlore (P) (and/or δ)-type intermediate ordering and its-associated broad lattice-parameter hump (i.e. extra distortional dilation) over the extended central area. So, how to extend and/or revise it to such F-P(δ)-C ternary (or other-type) one applicable to the latter is another major model-extension direction in the future. Due to the limited page of this chapter this important problem could not be covered and therefore I should ask the readers to refer to [5] for such more-recent model extension / elaboration attempts.

REFERENCES

[1] A. Nakamura, *Solid State Ionics* (2010) 181, 1543–1564.

[2] A. Nakamura, *Solid State Ionics* (2010) 181, 1631-1653.

[3] A. Nakamura, K. Imai, N. Igawa, Y. Okamoto, E. Yamamoto, S. Matsukawa, M. Takahashi, *Hyperfine Interact.* (2012) 207, 67–71

[4] A. Nakamura, invited lectures at ICCT2010 (Tsukuba, Japan) Aug. 2010, and at ISMC3 (Mumbai, India) Dec. 2010 (unpublished work).

[5] a) A Nakamura, K. Imai, N. Igawa, Y. Okamoto, E. Yamamoto, S., matsukawa, M. Takahashi, *Hyperfine Interact.* (2012) 207, 67-71,
b)A. Nakamura, N. Igawa, Y. Okamoto, Y. Hinatsu, J. Wang, M. Takahashi and M.Takeda, in *'Mossbauer Spectroscopy: Applications in Chemistry, Biology, Nanotechnology, Industry, and Environment'*, Chap.4, V.K. Sharma, G. Klingelhoefer and T. Nishida eds, John-Wiley & Sons (2013) pp.73-94.
c) A Nakamura, Y. Okamoto, N. Igawa, J. Wang, Y. Hinatsu, M. Takahashi and M. Takeda, *Hyperfine Interact.* (2013) 217, 17-26

[6] D.J.M. Bevan, W.W. Barker, R.L. Martin, T.C. Parks, *Proc. 4th Conf. on Rare Earth Research,* Phoenix, AZ (1964) pp. 411-468.

[7] M.R. Thornber, D.J.M. Bevan, E. Summerville, *J. Solid State Chem.* (1970) 1, 545-533.

[8] M. Morinaga, J.B. Cohen, *Acta. Cryst.* (1980) A36, 520.

[9] D. Steele, B. E. Fender, *J. Phys. C. Solid St. Phys.* (1974) 7, 1-11.

[10] J.P. Goff, W. Hayes, S. Hull, M.T. Hutchings, K.N. Clausen, *Phys. Rev.* (1999) B59, 14202-14219.

[11] F. Frey, H. Boysen, I. Kaiser-Bischoff, *Z. Kristallogr.* (2005) 220, 1017-1026.

[12] S. Suzuki, M. Tanaka, M. Ishigame, *Jap. J. Appl. Phys.* (1985) 24, 401-410.

[13] A. Gallardo-Lopez, J. Martınez-Fernandez, A. Domınguez-Rodrıguez, *J. Eur. Ceram. Soc.* (2002) 22, 2821–2825.

[14] M. Zinkevich, D. Djurovic, F. Aldinger, in *Proc. 7th Eur. SOFC Forum 2006*, J. Kilner ed., Lucerie, Switzerland (2006) P0512 (CD ROM).

[15] W. Chen, T.A. Lee, A. Navrotsky, *J. Mater. Res.* (2005) 20, 144-150.

[16] W. Chen, A. Navrotsky, *J. Mater. Res.* (2006) 21, 3242-3252.

[17] P. Simoncic, A. Navrotsky, *J. Am. Ceram. Soc.* (2007) 90, 2143-2150.

[18] A. Navrotsky, P. Simoncic, H. Yokokawa, W. Chen, T. Lee, *Faraday Discuss.* (2007) 134, 171-180.

[19] A. Navrotsky, *J. Mater. Chem.* (2010) 20, 10577–10587.

[20] G. Ceder, P.D. Tepesch, A.F. Kohan, A. Van der Ven, *J. Electroceram.* (1997) 1, 15-26

[21] H. Inaba, R. Sagawa, H. Hayashi, K. Kawamura, *Solid State Ionics* (1999) 122, 95-103.

[22] H. Hayashi, R. Sagawa, H. Inaba, K. Kawamura, *Solid State Ionics* (2000) 131, 281-290.

[23] L. Minervini, M.O. Zacate, R.W. Grimes, *Solid State Ionics* (1999) 116, 339-349.

[24] M.O. Zacate, L. Minervini, D.J. Bradfield, R.W. Grimes, K.E. Sickafus, *Solid State Ionics* (2000) 128, 243-254.

[25] A. Bogicevic, C. Wolverton, G.M. Crosbie, E.B. Stechel, *Phys. Rev.* (2001) B64, 014106.

[26] X. Wei, W. Pan, L. Cheng, B. Li, *Solid State Ionics* (2009) 180, 13-17.

[27] D.A. Andersson, S.I. Simak, N.V. Skorodumova, I.A. Abrikosov, B. Johansson, *PNAS* (2006) 103, 3518-3521

[28] A. Predith, G. Ceder, C. Wolverton, K. Persson, T. Mueller, *Phys. Rev.* (2008) B77, 144104.

[29] P. Dalach, D.E. Ellis, A. van de Walle, *Phys. Rev.* (2010) B 82, 144117.

[30] K. Kawata, H. Maekawa, T. Nemoto, T. Yamamura, *Solid State Ionics* (2006) 177, 1687-1690.

[31] H. Maekawa, K. Kawata, T. Yamamura, J. Kawamura, Y. P. Xiong, N. Sakai, H. Yokokawa, *Solid State Ionics* (2009) 180, 313-319.

[32] M. Ando, I. Oikawa, Y. Noda, S. Ohki, M. Tansho, T. Shimizu, H. Kiyono, H. Maekawa, *Solid State Ionics* (2011) 192, 576–579.

[33] N. Kim, J. F. Stebbins, *Chem. Mater.* (2007) 19, 5742-5747.

[34] N. Kim, C-H Hsieh, J. F. Stebbins, *Chem. Mater.* (2006) 19, 5855-5859.

[35] N. Kim, C-H. Hsieh, H Huang, F.B. Prinz, J.F. Stebbins, *Solid State Ionics* (2007) 178, 1499–1506

[36] W.M. O'Neill, M.A. Morris, *Chem. Phys. Lett.* (1999) 305, 389–394.

[37] N.M. Masaki, N.R.D. Guillermo, H. Otobe, A. Nakamura, Y. Izumiyama, Y. Hinatsu, in *Mass and Charge Transport in Inorganic Materials: Fundamentals to Devices, Part B*, P. Vincenzini, V. Buscaglia eds., TECHNA, Faenza (2000) pp.1233-1240.

[38] N.M. Masaki, H. Otobe, A. Nakamura, D. Harada, K. Ito, Y. Sasaki, Y. Hinatsu, *J. Nucl. Sci. Technol.* Suppl. (2002) 3, 217-220.

[39] D. Harada, Y. Hinatsu, N. Masaki, A. Nakamura, *J. Am. Ceram. Soc.* (2002) 85, 647-52.

[40] N.M. Masaki, H. Otobe, A. Nakamura, N.R.D. Guillermo, Y. Izumiyama, D. Harada, Y. Hinatsu, *Hyperfine Int.* (2003) *C*305, 305-308.

[41] N. Masaki, H. Otobe, A. Nakamura, D. Harada, Y. Hinatsu, *Mater. Res. Bull.* (2005) 40, 650-655.

[42] N.M. Masaki, A. Nakamura, F. Furuuchi, Y. Hinatsu, *J. Phys. Chem. Solids* (2005) 66, 312-317.

[43] J. Wang, A. Nakamura, M. Takeda, *Solid State Ionics* (2003) 164, 185-191.

[44] J. Wang, H. Otobe, A. Nakamura, M. Takeda, *J. Solid State Chem.* (2003) 176, 105-110.

[45] J. Wang, M. Takeda, A. Nakamura, *J. Nucl. Mater.* (2005) 340, 33-38.

[46] A. Nakamura, H. Otobe, J. Wang, M. Takeda, *J. Phys. Chem. Solids* (2005) 66, 356-363.

[47] A. Nakamura, N. Masaki, H. Otobe, Y. Hinatsu, J. Wang, M. Takeda, *Pure and Appl. Chem.* (2007) *79*, 1691-1729.

[48] C.R.A. Catlow, A.V. Chadwick, G.N. Greaves, L.M. Moroney, *J. Am.Ceram. Soc.* (1986) 69, 272-77.

[49] M.H. Tuilier, J. D-Ghys, H. Dexpert, P. Lagarde, *J. Solid State Chem.* (1987) 69, 153-161.

[50] T. Uehara, K. Koto, S. Emura, F. Kanamaru, *Solid State Ionics* (1987) 23, 331-337.

[51] D. Komyoji, A. Yoshiasa, T. Moriga, S. Emura, F. Kanamaru, K. Koto, *Solid State Ionics* (1992) 50, 291-301.

[52] G. Antonioli, P.P. Lottici, I. Manzini, G. Gnappi, A. Montenero, F. Paloschi, P. Parent, *J. Non-Cryst. Solids* (1994) 177, 179-186.

[53] N. Ishizawa, Y. Matsushima, M. Hayashi, M. Ueki, *Acta Cryst.* (1999) B55, 726-735.

[54] G.E. Rush, A.V. Chadwick, I. Kosacki, H.U. Anderson, *J. Phys. Chem.* (2000) B104, 9597-9606.

[55] H. Deguchi, H. Yoshida, T. Inagaki, M. Horiuchi, *Solid State Ionics* (2005) 176, 1817-1825.

[56] T. Moriga, S. Emura, A. Yoshiasa, S. Kikkawa, F. Kanamaru, K. Koto, *Solid State Ionics* (1990) 40/41, 357-361.

[57] M. Yashima, H. Arashi, M. Kakihana, and M. Yoshimura, *J. Am. Ceram. Soc.* (1994) 77, 1067-1071.

[58] J.M. Calderon-Moreno, M. Yoshimura, *Solid State Ionics* (2002) 154/155, 125-133.

[59] B.P. Mandel, N. Garg, S. M. Sharma, A.K. Tyagi, *J. Solid. State Chem.* (2006) 179, 1990-1994.

[60] R.C. Maher, L.F. Cohen, P. Lohsoontorn, D.J.L. Brett, N.P. Brandon, *J. Phys. Chem.* (2008) A112, 1497-1501.

[61] O.T. Sørensen (ed.), *'Nonstoichiometric Oxides'*, Academic Press (New York, 1981); see many chapters and references therein.

[62] A. Nakamura, J.B. Wagner, Jr, *J. Electrochem. Soc.* (1980) 127, 2325-33.

[63] A. Nakamura, J.B. Wagner, Jr, *J. Electrochem. Soc.* (1986)133, 1542-48.

[64] A.S. Nowick, in *'Diffusion in Crystalline Solids'* G.E. Murch and A.S. Nowick eds, Academic Press (New York, 1984) pp.143-188. (See also many reference therein).

[65] H. Shimojo, H. Okazaki, *J. Phys. Soc. Jpn.* (1992) 61, 2848-2857

[66] H. Shimojo, H. Okazaki, *J. Phys. Soc. Jpn.* (1992) 61, 4106-4118

[67] M. Martin, *J Electroceram.* (2006) 17, 765–773.

[68] M. Nakayama, M. Martin, *Phys. Chem. Chem. Phys.* (2009) 11, 3241-3249.

[69] P.P. Dholabhai, S. Anwar, J.B. Adams, P. Crozier, R. Sharma, *J. Solid State Chem.* (2011) 184, 811–817.

[70] S.T. Norberg, S. Hull, I. Ahmed, S.G. Eriksson, D. Marrocchelli, P.A. Madden, P. Li, J.T. S. Irvine, *Chem. Mater.* (2011) 23, 1356–1364

[71] D. Marrocchelli, P.A. Madden, S.T. Norberg, and S Hull, *Chem. Mater.* (2011) 23, 1365-1373.

[72] M. Burbano, S.T. Norberg, S. Hull, S.G. Eriksson, D. Marrocchelli, P.A. Madden, G.W. Watson, *Chem. Mater.* (2012) 24, 222-229.

[73] L. Pauling, M.D. Shappel, *Zeitschrif. Krystallographie* (1930) A75, 128-140.

[74] M.A. Subramanian, G. Aravamudan, and G.V. Subba Rao, "Oxide Pyrochlores - A Review", *Prog. Solid State Chem.* (1983) 15, 55-143.

[75] F.A. Kröger and H.J. Vink, in *'Solid state physics'* Vol. 3, F. Seitz and D. Turnbull eds., Academic Press, New York (1957) pp.307-405.

[76] H.G. Scott, *Acta. Cryst.* (1977) B33, 281-282.

[77] H. Otobe, A. Nakamura, *SOFC-VI*, The Electrochem. Soc. Proc. (1999) 99-19, 463-473.
[78] R.D. Shannon, *Acta Crystallogr.* (1976) 32, 751.
[79] D.J. Kim, *J. Am. Ceram. Soc.* (1989) 72, 1415-1421.
[80] T. Ohmichi, S. Fukushima, A. Maeda, H. Watanabe, *J. Nucl .Mater.* (1981) 102, 40-46.
[81] M. Yashima, N. Ishizawa, M. Yoshimura, *J. Amer. Ceram. Soc.* (1992) 75, 1541-1549.
[82] M. Yashima, N. Ishizawa, M. Yoshimura, *J. Amer. Ceram. Soc.* (1992) 75, 1550-1557.
[83] S.J. Hong, A.V. Virkar, *J. Am. Ceram. Soc.* (1995) 78, 433-439.
[84] J. Faber, C. Geoffroy, A. Roux, A. Aylvestre, P. Abelard, *Appl. Phys.* (1989) A49, 225-232.
[85] K. Eguchi, T. Setoguchi, T. Inoue, H. Arai, *Solid State Ionics* (1992) 52, 165-172.
[86] M. Mogensen, N.M. Sammes, G.A. Tompsett, *Solid State Ionics* (2000) 129, 63-94.
[87] M. Mogensen, D. Lybye, N. Bonanos, P.V. Hendriksen, and F.W. Poulsen, *Solid State Ionics* (2004) 174, 279-286.
[88] S. Omar, E.D. Wachsman, J.C. Nino, *Appl. Phys. Lett.* (2007) 91, 144106.
[89] A Nakamura, unpublished lattice-parameter data compilation and their DCC model analysis for various DF MO_2-$LnO_{1.5}$ systems (M^{4+}= Ce, Th, Zr, Hf) (2010-2013).
[90] G. Panneerselvam, M.P. Antony, T. Vasudevan, *Mater. Lett.* (2004) 58, 3192-3196.
[91] S.P. Murti, C.K. Mathews, *J. Phys. D: Appl. Phys.* (1991) 24, 2202-2209.
[92] A. Kossoy, Y. Feldman, R. Korobko, E. Wachtel, I. Lubomirsky, J. Maier, *Adv. Funct. Mater.* (2009) 19, 634-641.
[93] J.D. MucCullough and J.D. Britton, *J. Am. Chem. Soc.* (1952) 74, 5225-5227.
[94] T. Kudo and H. Obayashi, *J. Electrochem. Soc.* (1975) 122, 142-147.
[95] K. Heggestad, J.L. Holm, O.T. Soerensen, Lab. Electrochem. Thermdyn. Mater. Uni. Paris, France, Report, *RISOM-2478*, 1985.
[96] T. Nakagawa, T. Osuki, T.A. Yamamoto, Y. Kitauji, M. Kano, M. Katsura, S. Emura, *J. Synchrotron Rad.* (2001) 8, 740-742.
[97] Z. Tianshu, P. Hing, H. Huang, J. Kilner, *Solid State Ionics* (2002) 148, 567– 573.
[98] Y. Ikuma, K. Takao, M Kamiya, E. Shimada, *Mat. Sci. Eng.* (2003) B99, 48–51,.
[99] V. Grover, A.K. Tyagi, *J. Solid State Chem.* (2004) 177, 4197-4204.
[100] D.J. Seo, K.O. Ryu, S.B. Park, K.Y. Kim, R-H. Song, *Mater. Res. Bull.* (2006) 41, 359-366.
[101] K. Imai, MS-thesis, 2010, Dept. Chem., Toho Univ. (Chiba, Japan).
[102] M. Andoh and H. Maekawa, private communication for JAEA - Tohoku Univ. Joint Research Project (2011).
[103] J.J. Katz, G. Seaborg, L.S. Morss, *The Chemistry of the Actinide Elements*, Vol. 2, Chapman and Hall, New York (1986).
[104] M. Kato, K. Morimoto, H. Sugata, K. Konashi, M. Kashimura, T. Abe, *J. Nucl. Mater.* (2008) 373, 237-245.
[105] M. Zinkevich, D. Djurovic, F. Aldinger, *Solid State Ioncs* (2006) 177, 989-1001.
[106] L. Eyring, in *Handbook on the Physics and Chemistry of the Actinides*, K. A. Gschneidner, Jr, L. Eyring, eds., Vol. 3, Chap. 27, North-Holland Pub. (Amsterdam, 1979) pp.337-399.
[107] D.J.M Bevan, E. Summerville, in *Handbook on the Physics and Chemistry of the Actinides*, K. A. Gschneider, Jr, and L. Eyring, eds., Chap. 28, North-Holland Publ. Amsterdam, 1979) pp.401-524.

[108] G. Adachi, N. Imanaka, *Chem. Rev.* (1998) 98, 1479-1514.
[109] M. Zinkevich, *Prog. Mater. Science* (2007) 52, 597-647.
[110] T. Masui, Y.W. Kim, N. Imanaka, G. Adachi, *J. Alloys and Compd.* (2004) 374, 97-100.
[111] R. Kittel, *Introduction to Solid State Physics*, 5th ed., Wiley, New York (1976).
[112] L. Pauling, *The Nature of the Chemical Bond*, Cornell Univ. Press, Ithaca, N. Y. (1960).
[113] K. Yamada, K. Kurosaki, M. Uno, S. Yamanaka, *J. Alloys and Compds.* (2000) 307, 10-16.
[114] N.-D. Morelon, D. Ghaleb, J.-M Delave, L. Van Brutzel, *Phil. Mag.* (2003) 83, 1533-1550.
[115] H. Radzewitz, *KFK 433* (July 1966) (in Germany).
[116] M.D. Mathews, B.R. Ambekar, A.K. Tyagi, *J. Alloys and Compds*, (2005) 386, 234-237.
[117] P.D. Battle, B. Montez, E. Oldfield, *J. Chem. Soc., Chem. Commun.* (1988) 584-585
[118] M. Yashima and D. Ishimura, *Chem. Phys. Lett.* (2003) 378, 395-399.
[119] X. Liu, *PhD thesis*, Univ. London (UK, 2009).
[120] C.E. Mohn, S. Stølen, S.T. Norberg, S. Hull, *Phys. Rev. Lett.* (2009) 102, 155502.
[121] P.D. Battle, B. Montez, E. Oldfield, *J. Chem. Soc., Chem. Commun.* (1988) 584-565
[122] E.A. Guggenheim, *Mixture,* Oxford Univ. Press, Oxford, UK (1952).
[123] R.H. Fowler, E.A. Guggenheim, *Statistical Thermodynamics,* Cambridge Univ. Press, Cambrigde, UK (1956).
[124] D,Y. Wang, D.S. Park, J. Griffith, A.S. Nowick, *Solid State Ionics* (1981) 2, 95-105.
[125] M. Meyer and N. Nicoloso, *Ber. Bunsenges. Phys. Chem.* (1997) 101, 1393-1398.
[126] K. Fuda, K. Kishio, S. Yamauchi and K. Fueki, *Solid State Comm.* (1985) 53, 83-85.
[127] K. Fuda, K. Kishio, S. Yamauchi, K. Fueki, Y. Onoda, *J. Phys. Chem. Solids* (1984) 45, 1253-1257.
[128] K. Fuda, K. Kishio, S. Yamauchi, K. Fueki, *J. Phys. Chem. Solids* (1985) 46, 1141-1146.

In: New Research Trends of Fluorite-Based Oxide Materials
Editors: Akio Nakamura and Junichiro Mizusaki
ISBN: 978-1-63117-350-9

Chapter 5

YSZ AND DOPED CERIA: DIFFERENCE IN SOFC ELECTRODE REACTION MECHANISM

Haruo Kishimoto[1*] and Harumi Yokokawa[2]
[1]National Institute of Advanced Industrial Science and Technology (AIST), Tsukuba, Japan
[2]Institute of Industrial Science, The University of Tokyo, Japan

ABSTRACT

For solid oxide fuel cells, CaF_2-type high temperature oxide ion conductors such as cubic stabilized zirconia and rare-earth doped ceria are used as an oxide component of composite electrodes as well as electrolyte materials. Chemical and electrochemical reaction processes are different depending on whether cubic stabilized zirconia or rare-earth doped ceria is used for the oxide component of electrode. In this chapter, first the properties of oxide ion conductivity, electronic conductivity, proton solubility and oxygen surface exchange reactivity are compared between cubic stabilized zirconia and rare-earth doped ceria. The differences in chemical and electrochemical reaction processes of the SOFC electrodes are then discussed based on those differences in the physicochemical properties. Interactions between oxide ion conductor and water vapor are finally focused as one of the key factors of providing such different electrode characteristics.

1. INTRODUCTION

Solid Oxide fuel cells (SOFCs) are one of the most promising new power generation systems. SOFCs have some advantages compared with other power generation systems and other fuel cells as follows;

- all components are solid (ceramics and metals);

[*] E-mail address: haruo-kishimoto@aist.go.jp, National Institute of Advanced Industrial Science and Technology (AIST), Tsukuba Central No.5, Higashi 1-1-1, Tsukuba, 305-8565 Japan.

- high temperature operation(700~1000°C);
- high electrical conversion efficiency;
- fuel flexibility;
- no use of precious metals;
- wide applicability (from small residential combined heat and powergeneration (CHP) system tolarge-scale power generation plant).

Most of these advantages are provided by using high temperature oxide ion conductor as electrolyte materials. Typical electrolytes for SOFCs are cubic stabilized zirconia, such as yttria (Y_2O_3) or scandia (Sc_2O_3) stabilized zirconia (ZrO_2) (YSZ and ScSZ, respectively), and rare-earth doped ceria, such as gadolinia (Gd_2O_3) or samaria (Sm_2O_3) doped ceria (CeO_2) (GDC and SDC, respectively); those have the CaF_2-type crystal structure. In addition, these materials are also used as an oxide component of composite electrodes consisting of oxide ion conductive materials and electrochemically active electrode materials. For metal electrodes, such composite electrodes are called "*cermet*" just because this consists of *cer*amics and *met*als. Main reasons for adding such oxides are to provide oxide ion conduction path inside the electrodes and to adjust the effective thermal expansion coefficient to that of the electrolyte materials from the mechanical stability point of view. In this chapter, electrode properties important for SOFC applications are described by comparing between cubic stabilized zirconia and rare-earth doped ceria. The impacts of such differences on the chemical and electrochemical reaction processes in the SOFC electrodes are discussed in details.

2. CaF_2-Type High Temperature Oxide Ion Conductors in SOFCs

For SOFCs, high temperature oxide ion conductor is used as electrolyte. Therefore, oxygen reduction reaction and fuel oxidation reaction proceed at the cathode (air electrode) side and at the anode (fuel electrode) side, respectively. Electrode reactions with hydrogen fuel are expressed as follows:

Oxygen reduction reaction at the cathode (air electrode): $O_2(g) + 4e' = 2O^{2-}$ (1)

Hydrogen oxidation reaction at the anode (fuel electrode):$2H_2(g) + 2O^{2-} = 2H_2O(g) + 4e'$ (2)

Overall reaction : hydrogenoxidation reaction: $2H_2(g) + O_2(g) = 2H_2O(g)$ (3)

Operation principle of SOFC is schematically shown in Fig. 1. Because fuels are oxidized on the anode, in principle, CO and hydrocarbons can be directly used as fuel for SOFCs. There are some requirements for the respective electrode materials:

- High catalytic activity for reactions (1) and (2)
- Porous microstructure for gas diffusion
- Stable (less sinterable) microstructure for electrical path and gaseous channel

- High mechanical strength
- Matching thermal expansion coefficient to the electrolyte material.

To meet these requirements, composite materials consisting of active (catalytic) materials and electrolyte materials are used for electrode material.

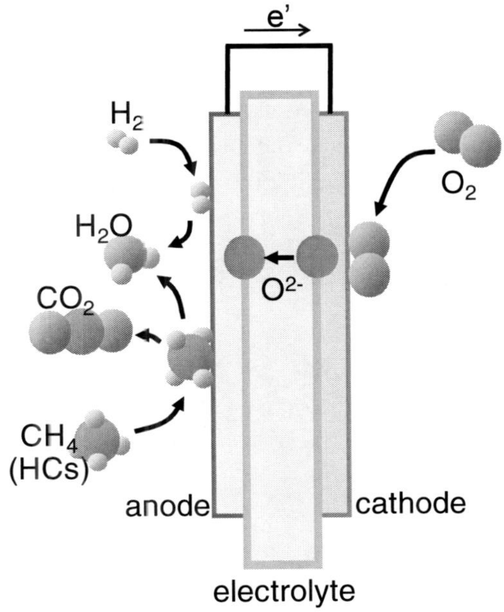

Figure1.Schematic drawing of operating principle for SOFC.

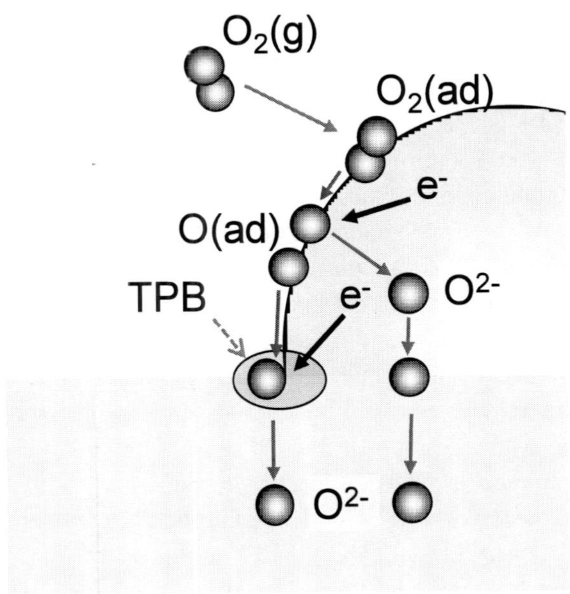

Figure 2. Schematic drawing of reaction pathways of oxygen incorporation [1].

The actual reaction process on the electrode is not as simple as Eqs. (1) and (2). The reaction process in the vicinity of cathode can be, as schematically summarized in Fig. 2 [1], divided into several elemental steps; for example, adsorption of gaseous molecules on electrode, dissociation of the molecules to atoms, diffusion of dissociated atoms on electrode to electrochemically active sites, and finally ionization (charge transfer) and incorporation; some of them can be influenced by the property of electrolytematerial component in the composite electrodes. In general, gas / electrode / electrolyte oxide triple phase boundaries (TPBs) play a crucial role.

For electrode reaction, not only the major charge carrier in the material but also other minor charge carriers must make contributions to those steps because all charge carriers are mobile according to their electrochemical potential gradient. Some important properties are first compared between cubic stabilized zirconia and rare-earth doped ceria, and then plausible effects originated from the difference in those properties on the electrode reaction process will be discussed.

3. Charge Carriers in Cubic Stabilized Zirconia and Rare-Earth Oxide Doped Ceria and Their Properties

3.1. Oxide Ion Conductivity

In both cubic stabilized zirconia and rare-earth doped ceria, oxygen vacancy, $V_O^{\cdot\cdot}$, in Kröger-Vink notation, is introduced into the CaF_2 type lattice by lower-valent cation doping:

$$R_2O_3 \text{ (in } MO_2 \text{ lattice)} = 2R_M' + V_O^{\cdot\cdot} + 3O_O^x \quad (4)$$

Oxide ions can move in the lattice by oxygen vacancy mechanism.

The total electrical conductivities of some representative high-temperature oxide-ion conductors are summarized in Fig. 3 [2-7]. Generally, the total electrical conductivity of GDC is about a-half to one-order-of-magnitude as high as that of YSZ. That of ScSZ is about twice as high as that of YSZ, and is the highest among cubic stabilized zirconia. Oxygen partial pressure dependence of total conductivities of YSZ [8] and SDC [9] are shown in Figs. 4 and 5, respectively. If the mobility of oxideions does not depend on oxygen partial pressure, oxide ion conductivity of YSZ and GDC is constant under the condition where oxygen vacancy concentration is determined by dopant cation concentration as in Eq. (4). "Pure oxide ion conductor" is defined as the material that has an oxide ion transport number, t_{ion}, of more than 0.99. In other words, we can call this region "electrolytedomain". The "pure oxide ion conductor" region is that total conductivity shows little or no oxygen partial pressure dependence in such a logarithmic plot as in Figs. 4 and 5. For YSZ, pure oxide ion conductor region is enough wide for use as SOFC's electrolyte material. That extends down to about $\sim 10^{-40}$ atm at 700°C and becomes about $\sim 10^{-30}$ even at the high temperature of 900°C.On the other hand, that of SDC is very limited, about $\sim 10^{-12}$ atm at 700°C and becoming about $\sim 10^{-7}$ atm at 900°C. Therefore, care must be taken in utilizing rare earth doped ceria as SOFC's electrolyte materials. This is due to its high electronic conductivity under reducing condition; details will be described in the next section.

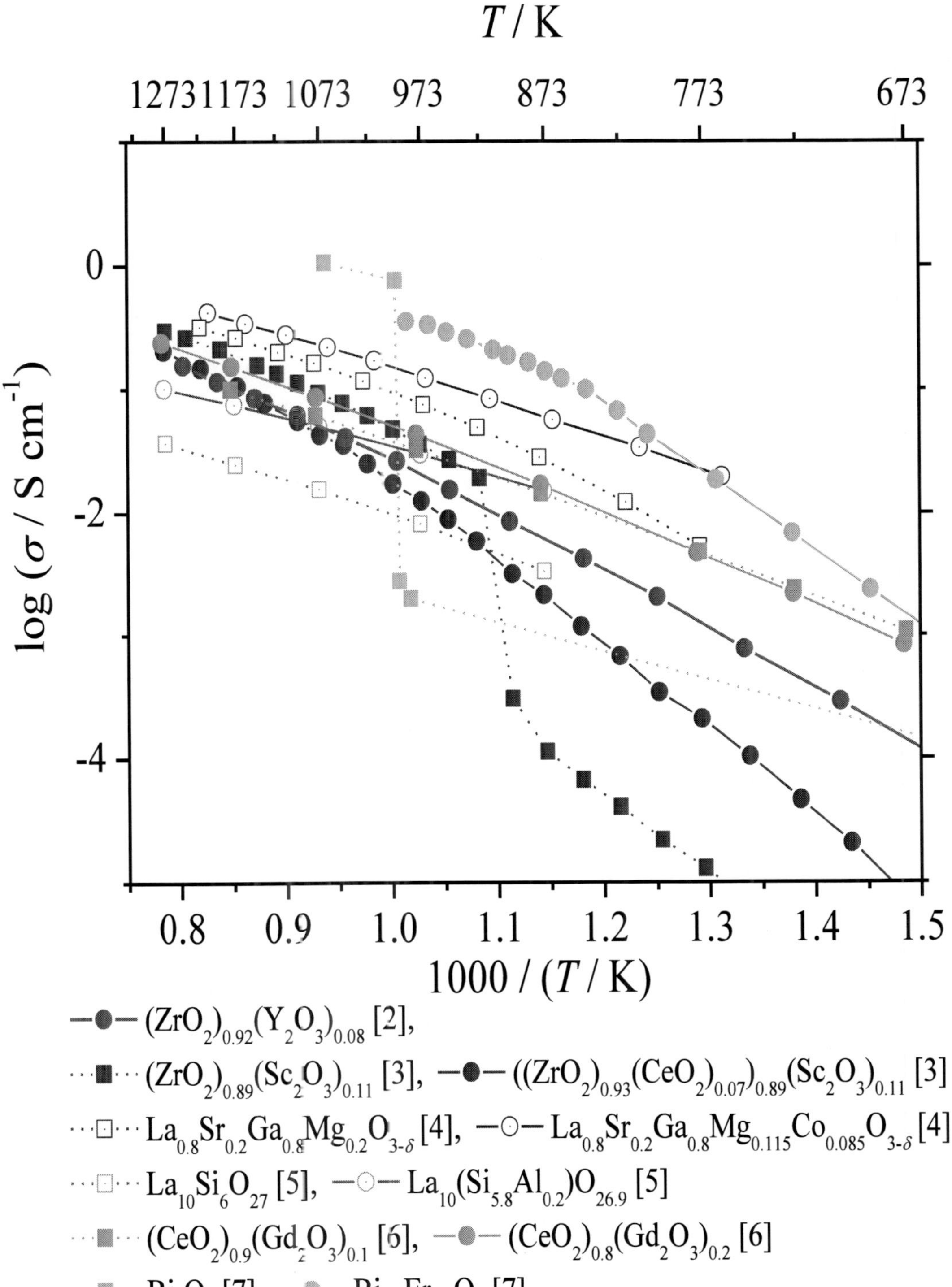

Figure 3. Temperature dependence of total electrical conductivity of several representative high temperature oxide ion conductors [2-7].

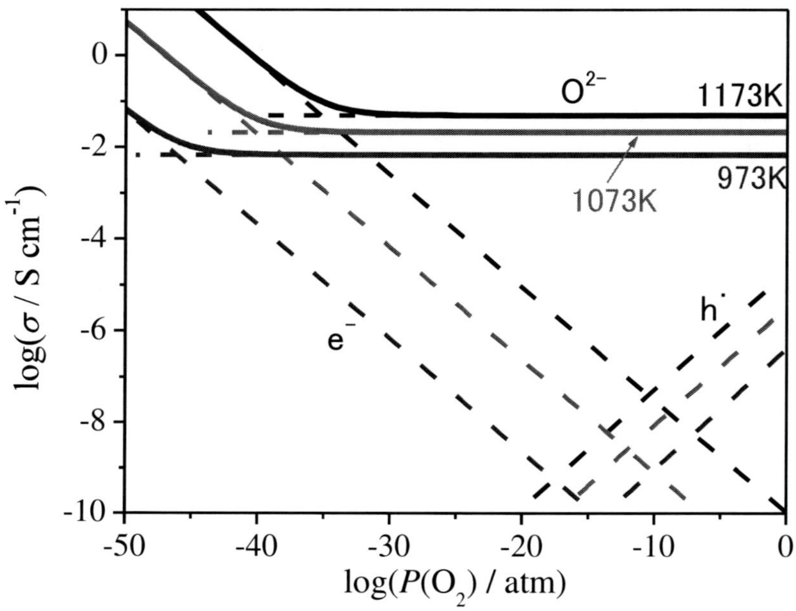

Figure 4. Electrical conductivity of YSZ ($Zr_{0.9}Y_{0.2}O_{2.1}$) as a function of $p(O_2)(g)$ [8].

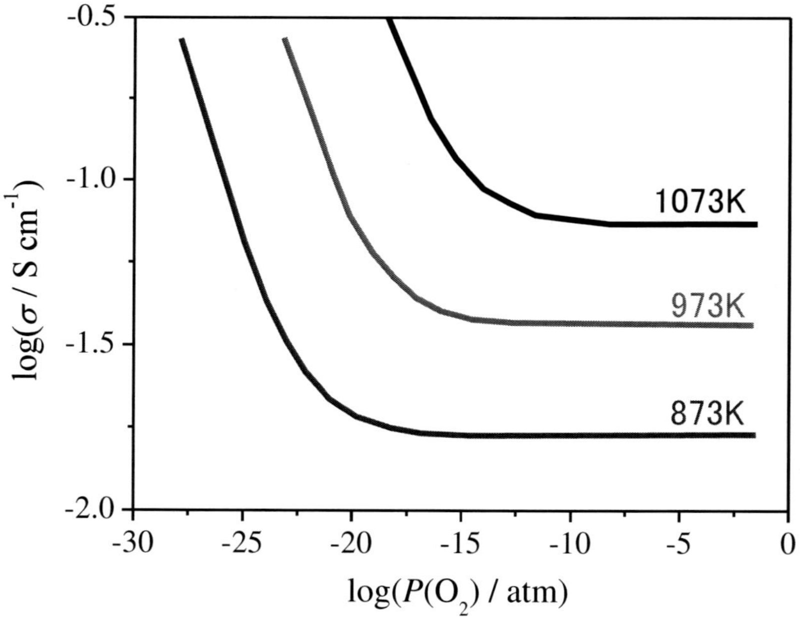

Figure 5. Electrical conductivity of SDC ($Ce_{0.8}Sm_{0.2}O_{1.9-x}$) as a function of $p(O_2)(g)$ [9].

3.2. Electronic Conductivity

Electronic conductivity of cubic stabilized zirconia and rare earth doped ceria is related to the electron or hole formation associated with oxygen nonstoichiometry change in the CaF_2-type lattice.

Electronic charge carrier formation reactions:

Holes: $V_O^{\cdot\cdot} + 1/2O_2(g) = O_O^x + 2h^{\cdot}$ (5)

Electrons: $O_O^x = V_O^{\cdot\cdot} + 1/2O_2(g) + 2e'$ (6)

It is well known that electronic conduction in CeO_2proceeds via small polar on (hopping) mechanism [10,11]. When the change in concentration of oxide ion vacancies in the above electron/hole formation reactions (Eqs. (5) or (6)) is negligibly small compared with the originally introduced concentration (as in YSZ and doped ceria), the concentrations of hole and electron increase with $p(O_2)^{1/4}$ and $p(O_2)^{-1/4}$, respectively, as expressed by the following mass action law equations:

Hole concentration:$[h^{\cdot}] = K_1\,[V_O^{\cdot\cdot}]^{1/2}p(O_2)^{1/4}$ (7)

Electron concentration: $[e'] = K_2\,[V_O^{\cdot\cdot}]^{-1/2}p(O_2)^{-1/4}$ (8)

where, K_1 and K_2 are the equilibrium constants for Eqs. (5) and (6), respectively. If the mobility of these electronic charge carriers does not depend on the oxygen partial pressure (as is usually assumed) the electronic conductivities of holes and electrons increases with $p(O_2)^{1/4}$ and $p(O_2)^{-1/4}$, respectively. Electronic conductivity in oxide ion conductors is generally measured by the Hebb-Wagner-type ion blocking (DC-polarization) method [12-22].

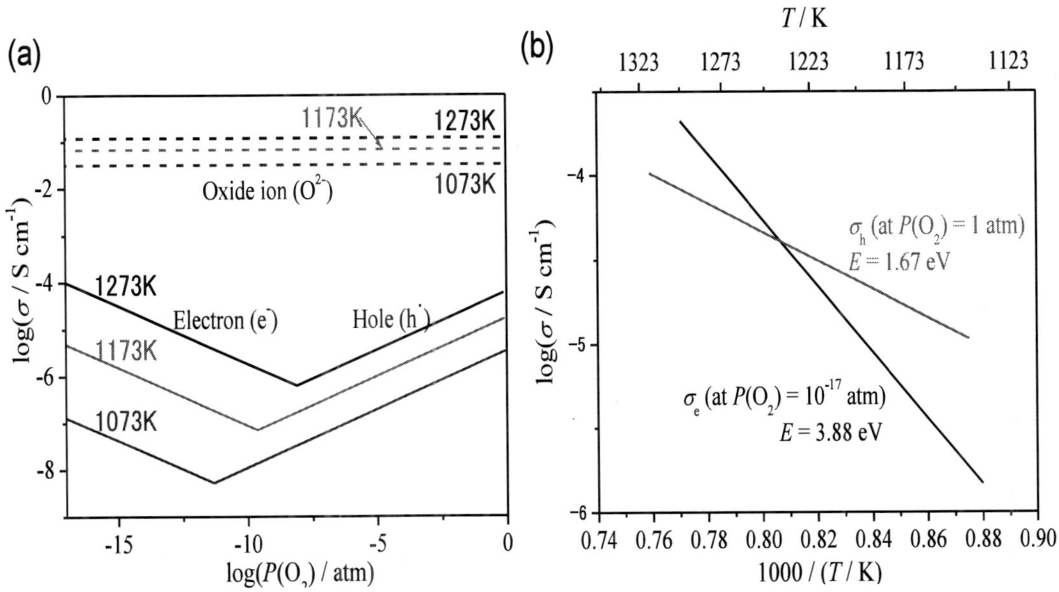

Figure 6. Electronic conductivity of8YSZ (8 mol% Y_2O_3-ZrO_2) (a) as a function of oxygen partial pressure and(b) as a function of temperature[14].

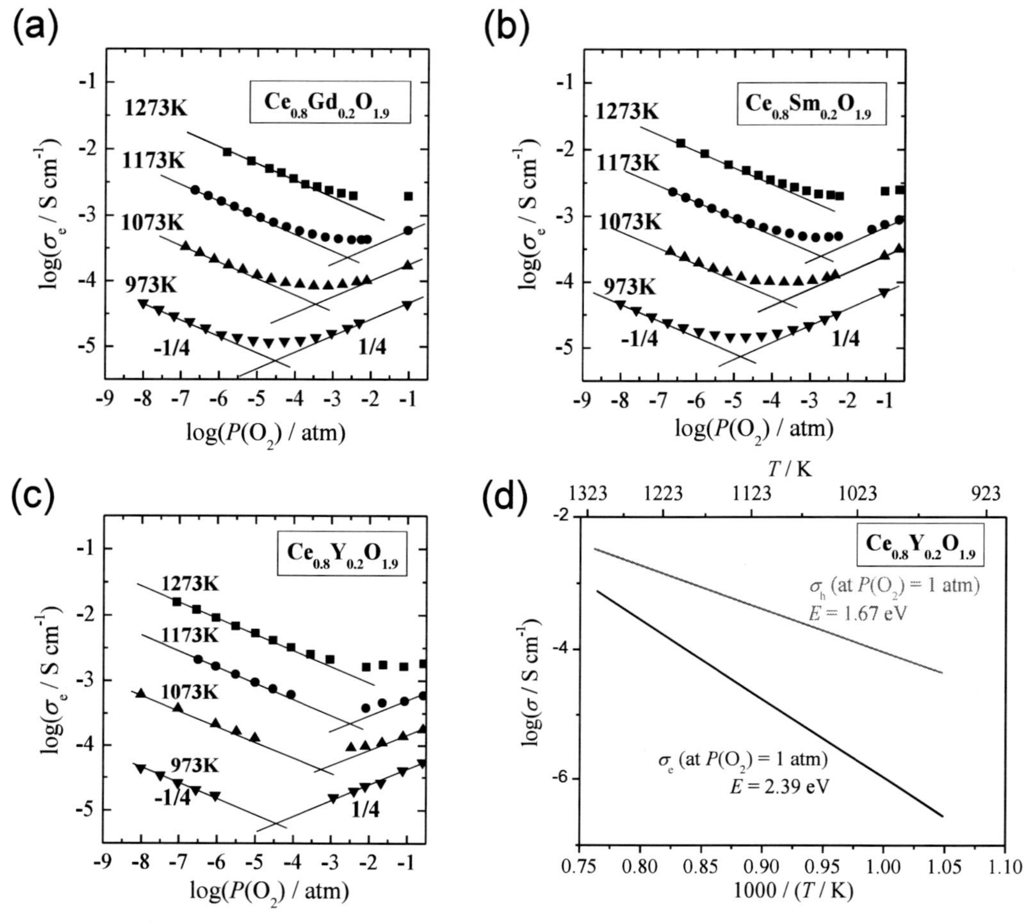

Figure 7. Electronic conductivity of rare-earth doped ceria as a function of oxygen partial pressure, (a) $Ce_{0.8}Gd_{0.2}O_{1.9}$, (b) $Ce_{0.8}Sm_{0.2}O_{1.9}$ and (c) $Ce_{0.8}Y_{0.2}O_{1.9}$, and (d) as a function of temperature for $Ce_{0.8}Y_{0.2}O_{1.9}$ [21,22].

The electronic conductivities for YSZ [14] and rare-earth doped ceria [21,22] determined by the ion-blocking cell are shown in Figs. 6 and 7, respectively. In both cases, the electronic conductivity changes with $p(O_2)^{1/4}$ and $p(O_2)^{-1/4}$ under high and low $p(O_2)$ range, respectively; these two regions are identified as the p-type hole conductive region and the *n*-type electron conductive region, respectively. Rare-earth doped ceria is generally regarded as electron conducting material especially under the reducing condition. But the hole conduction region is clearly observed even for rare-earth doped ceria. It is an interesting and important finding.

Fig. 8 compares the oxide-ion and electronic conductivities between YSZ and GDC at 800°C [23]. Electron and hole conductivities of GDC is about 5 to 6 order-of-magnitude and about 2 order-of-magnitude as high as those of YSZ, respectively, while the difference in the oxide ion conductivity is only in the range of a half order-of-magnitude. YSZ shows a very wide "pure oxide ion conductor" region because of its quite low electronic conductivity. On the other hand, rare-earth doped ceria shows much higher electronic conductivity especially under reducing condition. Therefore, the "pure oxide ion conductor" region is much narrower compared with YSZ especially under reducing condition. This makes it very limited to use rare-earth doped ceria as electrolyte material. But, for electrochemical reactions, its high electronic conductivity is expected to show interesting advantageous properties.

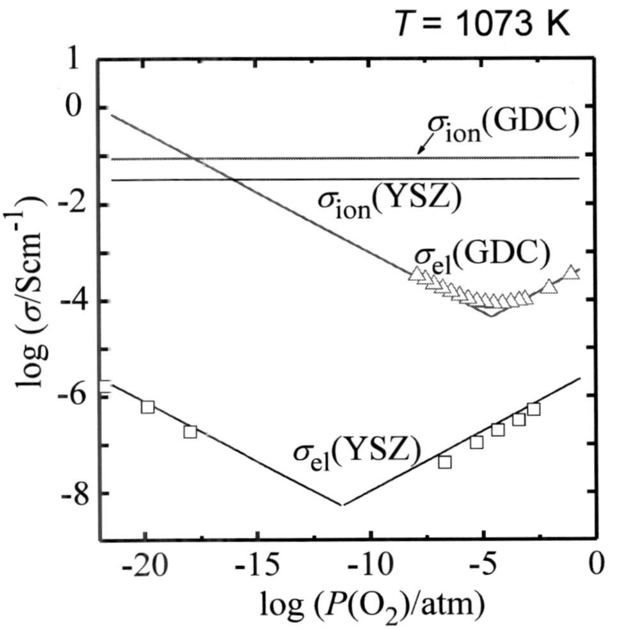

Figure 8. Comparison of oxide ion and electronic conductivities between YSZ and GDC [23].

3.3. Proton Solubility

It is well known for perovskite type proton conductors, such as $SrZrO_3$ and $BaCeO_3$, that proton in the oxide lattice is formed by water vapor incorporation mediated by oxygen vacancy [24-27]. One of the water-incorporation related phenomena for CaF_2-related zirconia based oxides is hydrothermal degradation of tetragonal zirconia (partially stabilized zirconia) [28-30]. Water vapor incorporates into the lattice and diminishes the oxygen vacancy concentration, and as a result, embrittlement occurs with phase transformation from the tetragonal phase to the monoclinic phase. This phenomenon occurs at an intermediate temperature range (400~700 K) under a water vapor containing condition. Protons are formed even for rare-earth doped ceria in the CaF_2-type crystal structure by water vapor incorporation [31,32]. However, the solubility is quite small and its influence on the overall electrical conductivity of the system is negligible.

Generally speaking, protons are formed upon water incorporation mediated by oxygen vacancy, as follows:

$$\text{Proton formation with water vapor: } V_O^{\cdot\cdot} + H_2O(g) = O_O^{x} + 2H^{\cdot} \quad (9)$$

In addition, proton formation mediated by holes is expressed as follows:

$$\text{Proton formation with hydrogen: } 2h^{\cdot} + H_2(g) = 2H^{\cdot} \quad (10)$$

Fig. 9[33] shows the concentration of dissolved protons into rare-earth doped ceria determined by secondary ion mass spectroscopy (SIMS) analysis and isotope labeling technique with heavy water (D2O) [31] together with that in YSZ determined by Carl Wagner [34].The proton solubility in rare-earth doped ceria is as low as about 10-3 mol/1 mol oxide depending on dopant cations. YSZ shows even lower value of 10-5 mol / 1 mol oxide; it is almost comparable with that in non-doped pure ceria and is about two order of magnitude lower than rare-earth doped ceria. It is clear from comparison between Eqs. (5) and (9), the proton formation reaction is very similar to the hole formation reaction which is well known for perovskite-type proton conducting oxides. The relationship between proton solubility and electronic conductivity (electron and hole conductivities) for Y2O3 doped ZrO2 - CeO2 solid solution is shown in Fig. 10[23].For the Y2O3 doped ZrO2 - CeO2 solid solutions with the CaF2-type structure, proton solubility is apparently related to the hole conductivity, as Eq. (10) indicates.

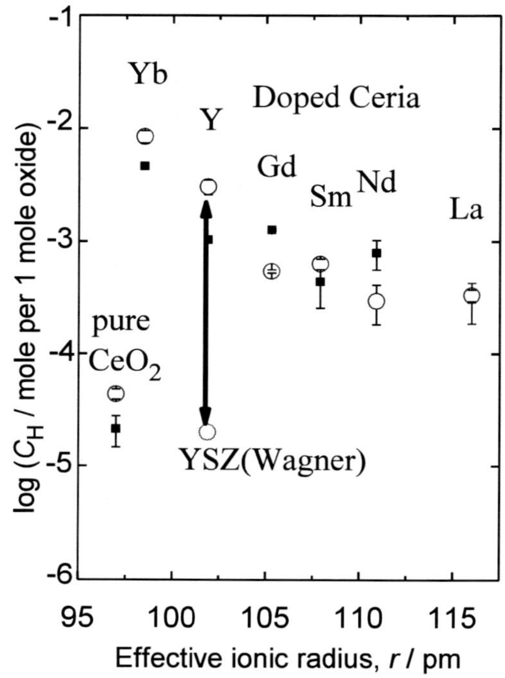

Figure 9. Experimental proton solubility for rare-earth doped ceria [31] and YSZ [34]. For ceria: at 1153 K (solid squares) and 1263 K (open circles) in 3% D_2O condition [33].

While Wagner's value in Fig. 9 is quite small, many researchers have investigated water uptake on YSZ or rare-earth doped ceria surface. The adsorption equations in different form are derived for ceria and zirconia. These indicate that the coverage with water will decrease with increasing temperature [35-38]. Since the proton solubility determined by SIMS can be interpreted as surface or grain boundary effect, the bulk proton solubility is measured using single crystals. Fig. 11 shows the normalized deuterium intensity for poly- and single crystals of Nd_2O_3 doped ceria as a function of Nd concentration determined by SIMS and D_2O labeling method [39]. It is clear that deuterium actually dissolves into bulk of doped ceria under the wet condition, whereas the solubility under dry condition is about two-order of magnitude as low as that of the former; it is near the detection limit of SIMS analysis.

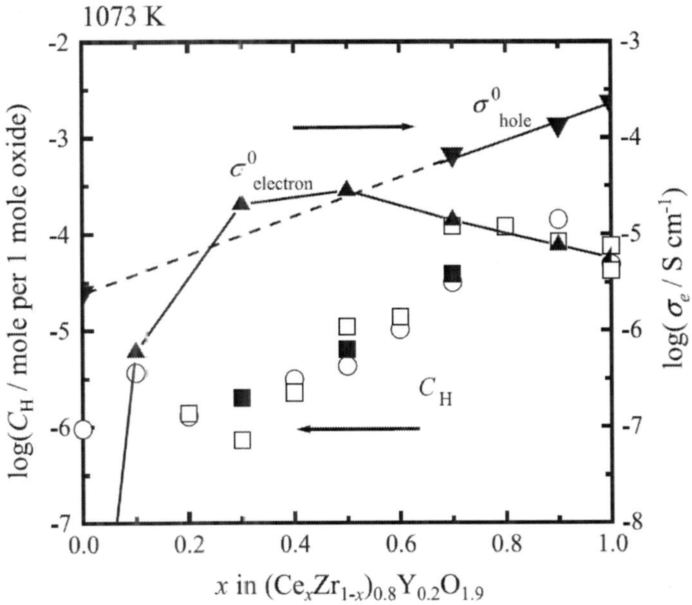

Figure 10. The proton solubility, hole and electron conductivities of 20 mol% of Y_2O_3 doped ZrO_2-CeO_2 solid solution plotted against the CeO_2 content [23].Electron and hole conductivity are extrapolated values to $p(O_2)$=1 atm.

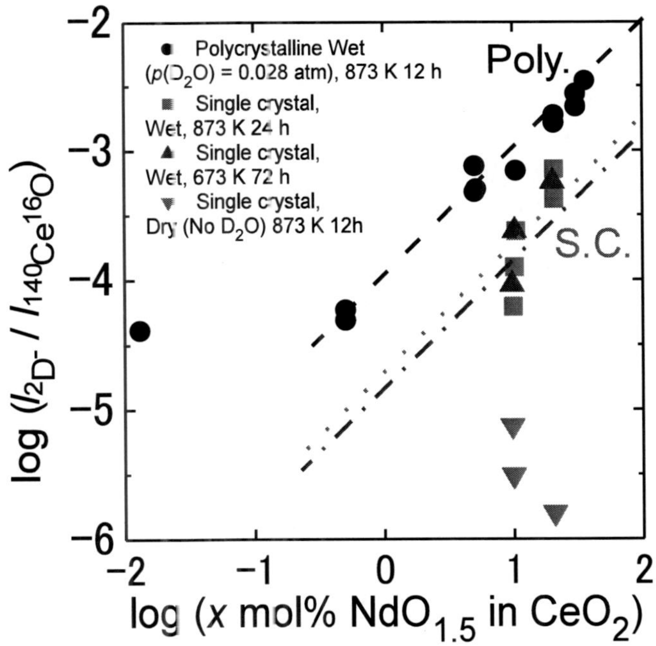

Figure 11. Experimental proton solubility for single crystal (S.C) and poly crystal NDC (Nd_2O_3 doped ceria) as a function of dopant concentration[39].

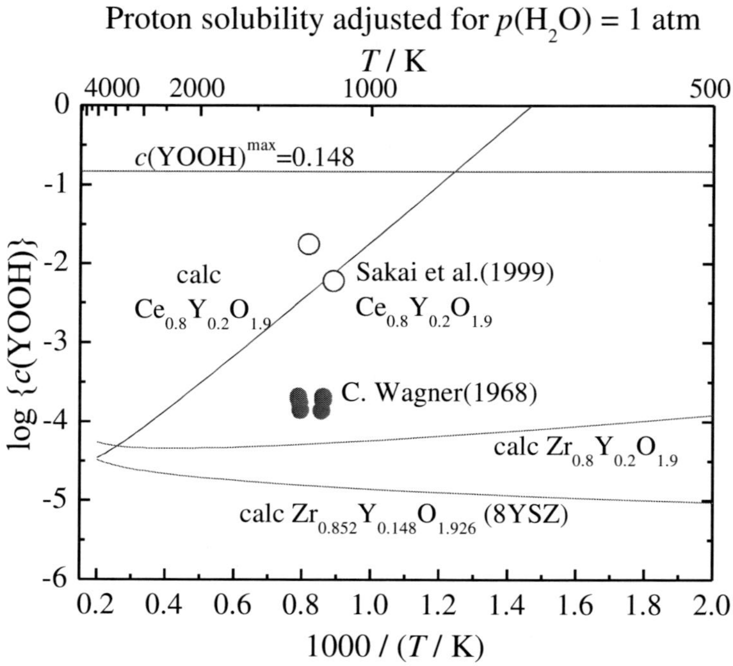

Figure 12. Calculated proton solubility from the chemical potential of $YO_{1.5}$ and the hydroxidation energy, δ(YOOH) [33]. Experimental data by Sakai et al. [31] and by Wagner [34] are adjusted for values at $p(H_2O)$=1 atm.

Since the information of proton behavior in the fluorite (CaF_2) type oxides is limited, it is useful to make some estimates on the solubility using the thermodynamic properties. In the usual thermodynamic description, proton in rare earth doped CaF_2-type oxides can be treated as a hypothetical hydroxide constituent, MOOH, in the fluorite lattice. Water incorporation reaction can be written as follows;

$$H_2O(g) + 2MO_{1.5} = 2MOOH \tag{11}$$

where $MO_{1.5}$ means the normal lattice occupation of dopant oxide with one half of oxide ion vacancies and its thermodynamic properties were estimated so as to reproduce phase relations in CeO_2-$MO_{1.5}$ system. In a similar manner, the thermodynamic properties of constituent MOOH were estimated from the thermodynamic properties of compound MOOH and $M(OH)_3$. Calculated proton solubility at $p(H_2O)$=1 atm is plotted in Fig. 12 [33]. The experimental values are also adjusted to those values at $p(H_2O)$=1 atm from values obtained at $p(H_2O)$=0.03 atm. Interesting difference can be seen in Fig. 12 in the temperature dependence of the proton solubility; that is, the solubility in doped ceria increases with decreasing temperature, whereas almost no change is expected for YSZ. Thermodynamically, this difference originates from difference in the enthalpy change for the MOOH formation between them. The stabilization of rare earth oxides in ceria is not so large that it is easy for them to react with water to form MOOH, whereas yttria in YSZ is well stabilized so that there YOOH formation is not favored compared with in doped ceria. As shown in Fig. 12, calculated concentration of YOOH in 8YSZ ($Zr_{0.852}Y_{0.148}O_{1.926}$) is about in four order of magnitude smaller than maximum concentration, $c(YOOH)^{max} = 0.148$.

From these differences, interesting different interaction behavior of these oxides with water vapors can be expected; When YSZ is cooled down in a given water vapor pressure, water will be deposited on the surface but does not dissolve, whereas, when doped ceria is cooled, the solubility will increase, so that deposited water on the surface can be dissolved into the bulk. Since water vapor cannot be removed completely from air, there is always some possibility that water vapor in air tends to dissolve into doped ceria during thermal cycles. In other words, it must be difficult to remove dissolved water out of doped ceria. Such water component exists in bulk as well as grain boundaries or surfaces.

3.4. Oxygen Surface Exchange Rate

Oxygen exchange reaction between gas phase and oxide surface occurs even under equilibrium condition in a microscopically reversible way. When $^{18}O_2$ is introduced as oxygen source in gas phase, the oxygen incorporation can be seen from the change in isotope concentration. Measurement of such ^{18}O concentration inside the oxide by SIMS provides isotope oxygen exchange reaction rate as well as oxygen diffusion coefficient. The oxygen exchange reaction rates for YSZ and GDC under both *dry* and *wet* conditions are shown in Fig.13 [40-43]; strictly speaking, dry condition means quite a low dew-point condition. Under such a dry condition, where GDC shows about one-order-of-magnitude higher value than YSZ. Under a wet condition, oxygen exchange reaction rate drastically changes for YSZ; it increases to about two order-of-magnitude higher value. This strongly suggests that the presence of water molecules or OH groups on the surface of YSZ enhances the oxygen exchange reaction significantly. The surface exchange reaction mechanism on YSZ surface is described as follows;

$$^{18}O_2(g) = {}^{18}O_2(ad) \text{ and/or } {}^{18}O_2(ad) = 2\,{}^{18}O(ad) \qquad (12,13)$$

$$^{18}O(ad) + V_O^{\cdot\cdot} = {}^{18}O_O^{x} + 2h^{\bullet} \qquad (14)$$

$$^{16}O_O^{x} + 2h^{\bullet} = {}^{16}O(ad) + V_O^{\cdot\cdot} \qquad (15)$$

$$2\,{}^{16}O(ad) = {}^{16}O_2(ad) \text{ and/or } {}^{16}O_2(ad) = {}^{16}O_2(g) \qquad (16,17)$$

When the water vapor exists in the atmosphere, surface oxygen exchange reaction is enhanced for YSZ by adsorbed water vapor on the YSZ surface;

$$H_2{}^{18}O(g) = H_2{}^{18}O(ad) \qquad (18)$$

$$H_2{}^{18}O(ad) + {}^{16}O_O^{x}(\text{in YSZ}) = V_O^{\cdot\cdot} + {}^{18}OH'(ad) + {}^{16}OH'(ad) \qquad (19)$$

$$^{18}OH'(ad) + {}^{16}OH'(ad) + V_O^{\cdot\cdot} = H_2{}^{16}O(ad) + {}^{18}O_O^{x}(\text{in YSZ}) \qquad (20)$$

$$H_2{}^{16}O(ad) = H_2{}^{16}O(g) \qquad (21)$$

On the other hand, for GDC, oxygen exchange reaction rate increases by about one order at a low temperature as 873 K, but it is not different between wet and dry conditions at higher temperature range above 1000K. Namely, surface oxygen exchange reaction rate for GDC under dry condition at a high temperature as 1000K is essentially the same as that for both YSZ and GDC under a wet condition. From these facts, it is believed that the GDC surface should be covered with water vapor related species, such as adsorbed OH, even under *dry* condition at least at high temperature.

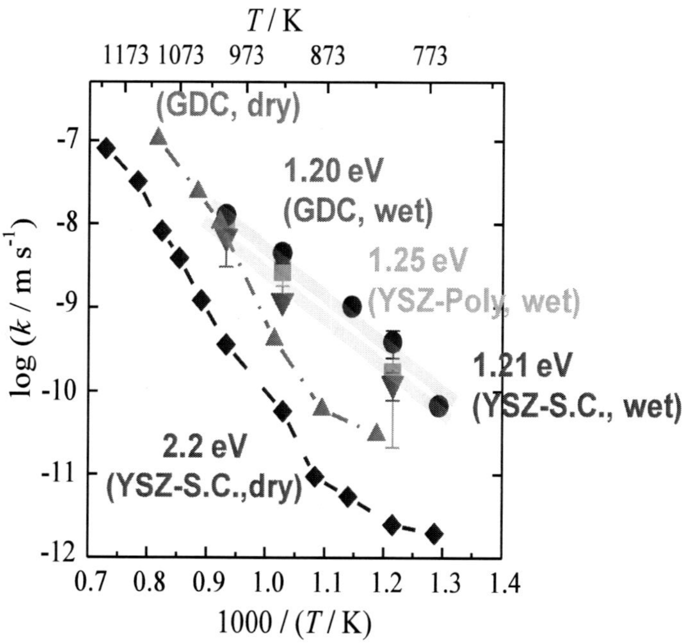

Figure 13. Surface oxygen exchange reaction rates for YSZ and GDC under dry and wet conditions [40-43]. Activation energies are also noted in this figure. S.C. is an abbreviation of single crystal.

4. Oxygen Exchange and Incorporation Reaction Around Metal / Oxide Interface

Oxygen incorporation reaction around metal electrode / oxide electrolyte is schematically depicted in Fig. 2. This reaction step is essentially unaltered whether the cathodic potential is applied (i.e., for oxygen incorporation reaction) or not (i.e., for oxygen exchange reaction). Oxygen incorporation process leading to oxygen reduction can be divided into the following several elemental steps:

Adsorption on the electrode: $O_2(g) = O_2(ad)$ (22)

Dissociation: $O_2(ad) = 2O(ad)$ (23)

Surface diffusion: O(ad, surface) = O(ad, TPB) (24)

These steps proceed on the electrode surface, and then;

Ionization (charge transfer): O (ad, TPB) + 2e' = O^{2-}(TPB) (25)

Incorporation (exchange): O^{2-}(TPB) = O^{2-}(in oxide) (26)

Where, as mentioned formerly, TPB is abbreviation of gas / electrode / oxide triple phase boundary (see Fig. 2). When we use the metal electrode, the reaction step in which oxygen incorporates into electrode material can be ignored because oxygen incorporation and exchange reactions into oxide electrolyte via inside of the metal electrode is supposed to be negligibly small. In general, effects of oxide electrolyte, which is oxide ion conductor, on these reaction steps are very limited; it is mostly reactions (25) and (26) which occur at TPBs. The electrochemically active site is also limited at the TPBs. However, in some cases, oxygen incorporation and exchange reactions are influenced by oxide materials, water vapor existence and applied potential etc. [44-46]. In this section, such influence will be described and discussed in connection with different characters of these CaF_2-type high temperature oxide-ion conductors in the oxygen incorporation and exchange reactions determined experimentally by SIMS analysis with oxygen isotope labeling method [47] for a model gold (Au) metal electrode / YSZ or GDC electrolyte. (This is because the Pt electrode has too high electro catalytic activity to trace the difference of oxygen motion between YSZ and GDC.)

Fig. 14 shows results of distribution analysis for ^{16}O and ^{18}O (isotope label) around the Au / YSZ interface as a function of depth from the surface; data were obtained under an open circuit voltage (OCV; without polarization) condition at 973K [44]. In this measurement, ^{16}O is detected at the YSZ part; Au electrode is almost removed by about 280 nm etching. Isotope ^{18}O oxygen is usually detected with the concentration of relative isotopic abundance; hence, the intensity for ^{18}O becomes high at the area where ^{16}O concentration detected is high. The higher ^{18}O concentration distribution is primarily determined by the surface isotope exchange properties as well as the diffusion properties inside solid substances. The concentration becomes high in those areas that are active for oxygen exchange reaction, the concentration becoming lower when the diffusion is faster. The observed ^{18}O concentration in Fig. 14 is high along with the TPBs of the gas / Au / YSZ even though ^{16}O concentration is not enriched at the same area. This indicates that oxygen exchange reaction takes place actively at the TPB for the Au / YSZ system under annealing treatment at 973 K and it follows the same reaction steps of Eqs. (22) to (26). At a lower temperature of 773K, oxygen exchange reaction scarcely occurred even at TPB (Fig. 15(a) [44]). Under the cathodic polarization condition, the ^{18}O incorporation reaction is apparently accelerated along with the TPB even at 773K as shown in Fig. 15(b) [44]. Reaction steps of Eqs. (25) and (26) proceed to the right-hand side under the cathodic polarization condition for the Au / YSZ system.

For the Au / GDC system, the circumstance is quite different from that for the Au / YSZ system, as shown in Fig. 16. Interestingly, the ^{18}O concentration uniformly increased on the GDC entire surface with increasing temperature. Its ^{18}O concentration remarkably increased also by applying potential of –0.3 V and was again uniform over an entire surface as shown in Fig. 17 [46]. These features indicate that, for the Au / GDC system, oxygen exchange / incorporation reactions proceed by the different mechanism from the generally-supposed sequential reaction steps of Eqs. (22) to (26) like the Au / YSZ system. Namely, the reactions occurs all over the whole GDC surface where is active for them.

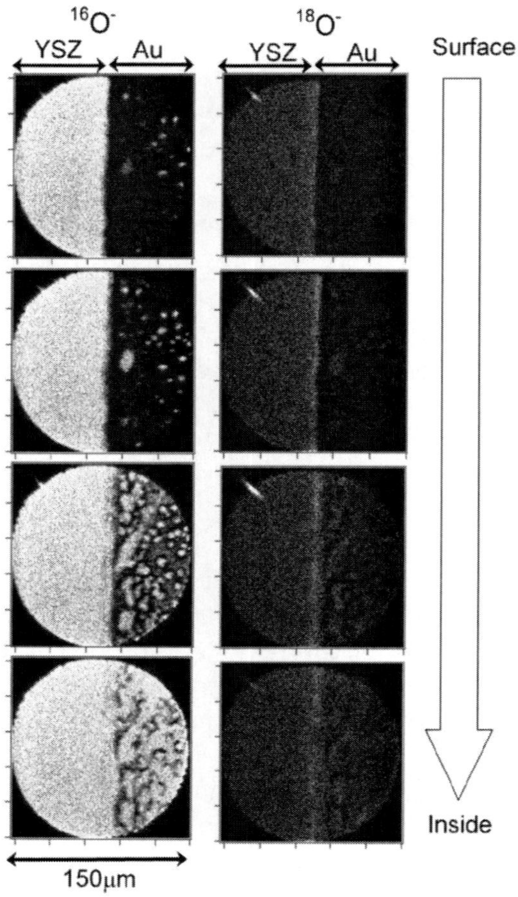

Figure 14. SIMS images of $^{16}O^-$ and $^{18}O^-$ ion on the surface around Au/YSZ interface treated at T=973 K in $p(O_2)$=7 kPa for 300 s (E=0 V). The depth of the images ranges from ca. 35 nm to 280 nm in YSZ region [44].

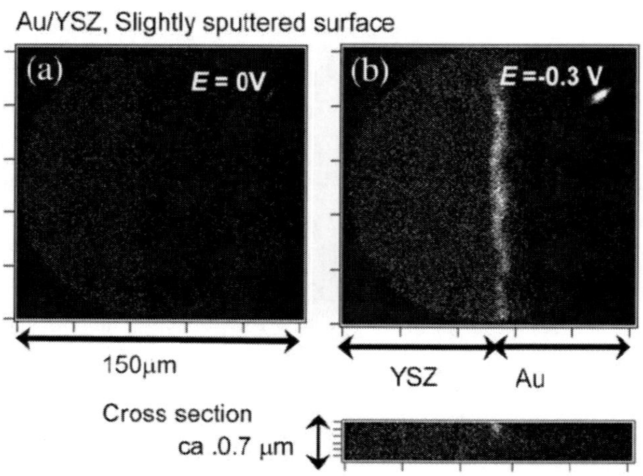

Figure 15. $^{18}O^-$ ion image of surfaces and cross sections around Au/YSZ interfaces treated at different cathodic polarization at T=773 K in $p(O_2)$=7 kPa for 300 s: (a) E=0 V, (b) E=−0.3 V. The depth of the images ranges from ca. 150 nm in YSZ region [44].

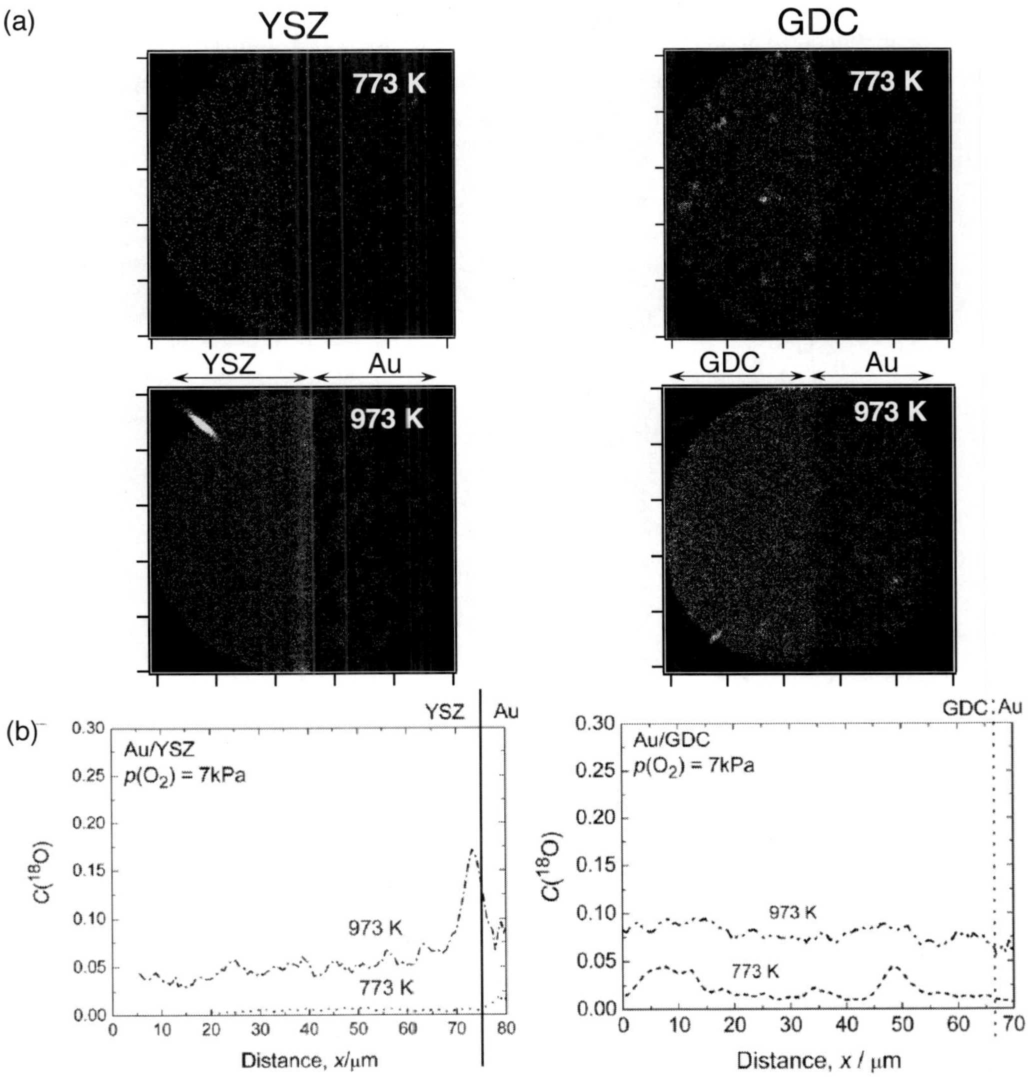

Figure 16. (a)$^{16}O^-$ and $^{18}O^-$ images around Au/GDC and Au/YSZ surfaces and (b) line profiles of ^{18}O concentration on Au/GDC or Au/YSZ treated at T=773K or 973K in dry $p(^{18}O_2)$=7 kPa for 300 s[45].

Oxygen exchange reaction is drastically enhanced by water vapor as shown in Fig. 13. Such a water vapor effect on the oxygen exchange reaction around the Au / oxides is clearly demonstrated in Fig. 18 [45]. The observed ^{18}O concentration under wet condition ($p(H_2O)$ = 2 kPa) was one order-of-magnitude higher than that under dry condition for both the Au / YSZ and the Au / GDC systems. As shown in Figs. 14-17, concentrations of ^{16}O and ^{18}O are quite small on the Au electrode surface; it should be noted that the reliability of calculated $C(^{18}O)$ (=$^{18}O/(^{16}O+^{18}O)$) is lower at the Au surface area. Generally, oxygen exchange reaction rate on the oxide surface decreased with decreasing the temperature under the same atmosphere as shown in Fig. 13, however, water vapor enhances that reaction at a temperature as low as 773 K. The oxygen exchange reaction on the oxides surface is activated by the existence of water vapor, but this does not seem to be the case for the Au surface.

These phenomena correspond to the fact that oxygen surface exchange reaction for both YSZ and GDC is enhanced to about the same level at a low temperature such as 773 K by reactions of Eqs. (18) to (21). As shown in Fig. 13, the surface oxygen exchange rates at 973 K under dry condition is essentially the same as those at 773 K for the YSZ and 873 K for the GDC, respectively, under wet condition. The water vapor effect is more significant for the YSZ case.

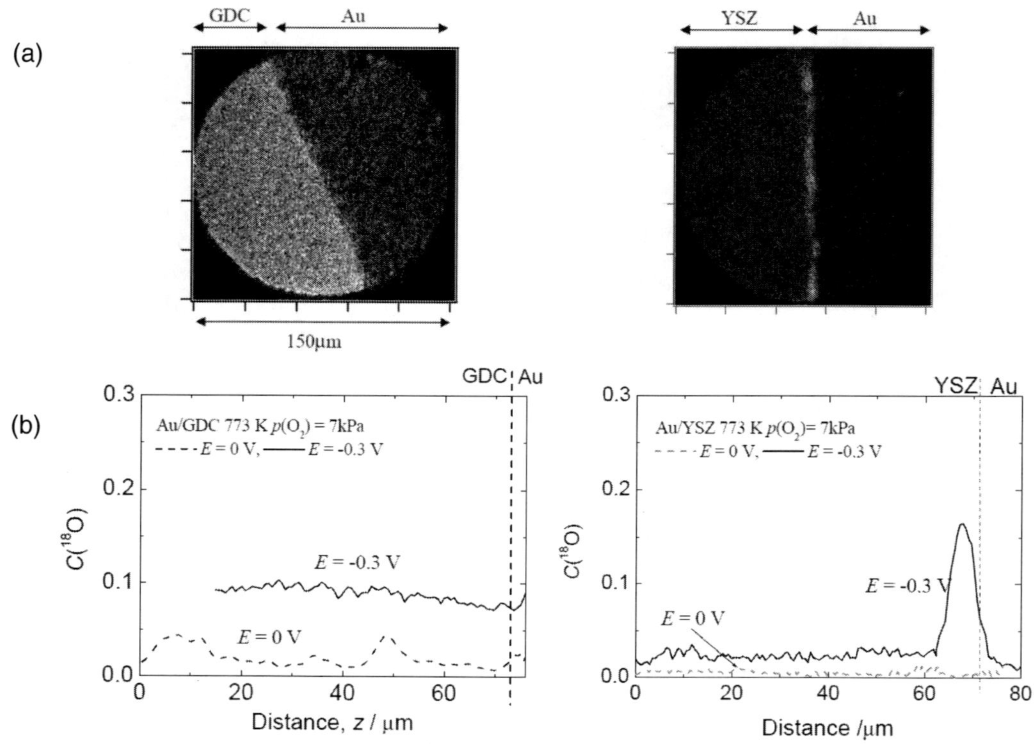

Figure 17. (a) $^{18}O^-$ images around Au/GDC and Au/YSZ surfaces at E =-0.3V and (b) line profiles of ^{18}O concentration on Au/GDC or Au/YSZ treated at T=773K in dry $p(^{18}O_2)$=7 kPa for 300 s at different cathodic polarization of E=0 and E=-0.3 V [46].

The reaction activation processes / phenomena for oxygen exchange and incorporation reactions around the Au / oxides interface are schematically summarized in Fig. 19. In the case of the Au / YSZ system, the TPB is thermally and electrochemically activated for oxygen exchange / incorporation reactions and water vapor activates the YSZ surface for oxygen exchange reaction at low temperatures. On the other hand, the GDC surface is activated equally for thermal, water vapor and electrochemical processes. These differences are discussed based on the reaction model in which different properties of YSZ and GDC, as mentioned above, are considered. The crucial difference is that all the oxide ion conductivity, the hole conductivity, the proton solubility and the surface oxygen exchange reaction constant of GDC are higher than those of YSZ; in particular, the hole conductivity and the proton solubility of GDC are two order-of-magnitude higher than those of YSZ under an oxidative condition. These differences in kinetic properties lead to the mass transport model around the TPB gas / electrode / oxide area especially for the electrochemical processes, as schematically depicted in Fig. 20 [46].

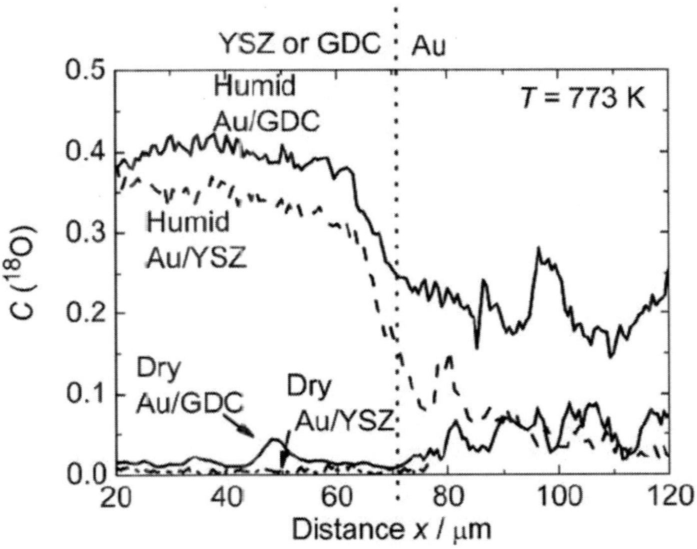

Figure 18. Line profiles of ^{18}O concentration on Au/GDC or Au/YSZ surface treatedin dry ($p(^{18}O_2)$=7 kPa) or humid ($p(^{18}O_2)$=7 kPa, $p(H_2O)$=2 kPa) atmosphereat T=773 K [45].

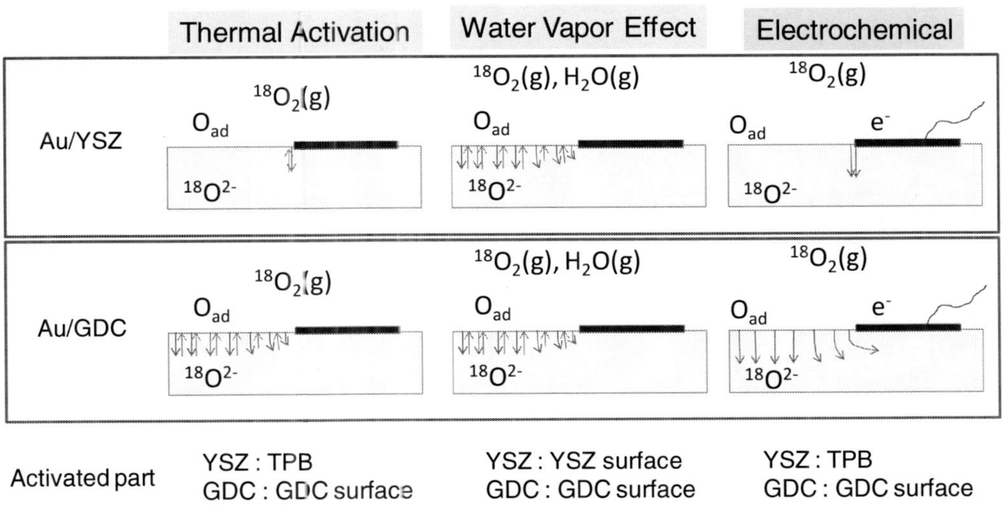

Figure 19. Schematic drawing of the activation phenomena for oxygen exchange and incorporation reactions around Au / Oxides interfaces.

In the case of YSZ, only TPB is activated in both thermal and electrochemical processes. These phenomena can be explained by the general reaction model with metal electrode expressed in Eqs. (22)-(26). In thermally activated process, oxygen dissociation reaction, Eq. (23), surface diffusion, Eq. (24), and oxygen exchange reactions, Eqs. (25) and (26), the all proceed to the right hand side. In electrochemical process, oxygen ionization reaction (Eq. (25)) and oxygen incorporation reaction (Eq. (26)) are enhanced. How does the water vapor affect on those processes at low temperatures? The catalytic reaction on the metal surface is not expected to be affected especially for the Au electrode, whereas the oxygen exchange reaction on the YSZ surface should be considerably enhanced, as seen in Fig. 13.The water adsorption mechanism on YSZ surface was intensively discussed by Raz et al. [48].A water molecule interacts with an oxide ion in the lattice, and they form two adsorbed hydroxides bridged by cations:

$$H_2O\ (g) + O_O^x\ \text{(oxide ion in YSZ lattice)} = 2OH'(ad) + V_O^{\cdot\cdot} \quad (27)$$

The adsorbed hydroxide is negatively charge and is chemically coupled with cations (that is like a chemical adsorption). As a result, the oxygen exchange between ambient water vapor and the oxide ion in the lattice occurs smoothly. In addition to surface hydroxide formation as represented by Eq. (27), we can consider the roles of oxygen vacancies and protons in the lattice as represented by Eq. (9).Here, the protons tend to migrate to the oxide ions adjacent to the dopants (that is Y ions in YSZ), whereas the oxygen vacancies are located adjacent to the host ions (that is Zr ions in YSZ) [49,50]. The water desorption reaction from YSZ lattice will take place at the dopant sites which are different from the adsorption site, where reaction (9) occurs. This proton-mediated reaction cycle enhances the isotope exchange during the adsorption-desorption process.

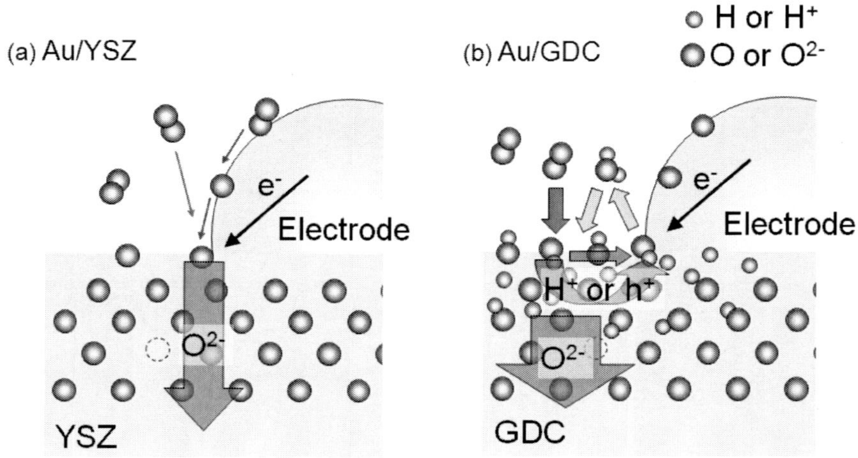

Figure 20. Schematic drawing of reaction mechanisms (a) for Au/YSZ system and (b) for Au/GDC system under cathodic polarization condition [46].

For the case of GDC, we have to suppose the common reaction mechanism for both thermal and electrochemical processes, as shown in Fig. 20; we can here expect the participation of hole and/or proton in the lattice both in oxygen exchange and oxygen incorporation reactions. As seen in Fig. 13, the surface exchange reaction for GDC is re-summarized as follows;

(1) Oxygen surface exchange reaction rate is about one order of magnitude higher than that in YSZ under dry condition.
(2) Oxygen surface exchange reaction rate is about the same as that in YSZ under wet condition.
(3) At a high temperature as 973K, oxygen surface exchange rate under dry condition is close to that under wet condition

From these facts, the GDC surface under dry condition at a high temperature as 973K is as active as both YSZ and GDC surface under wet condition. This suggests that even under "*dry*" condition, there must be some activation effect, such as water vapor, only in the GDC case. This can be ascribed reasonably to the fact that the proton (water) solubility is much higher in GDC. Even under the dry condition, dissolved protons in the GDC lattice may have "water vapor" effect. The oxygen vacancies in GDC are formed adjacent to the dopant in contrast to YSZ [51], and also protons tend to be located near the dopant. This means that the water incorporation reaction shown by Eqs. (9) or (11)can more easily take place than YSZ. This tendency provides higher proton solubility to GDC, and protons in GDC lattice should be more stable than in YSZ.

For the electrochemical process at a lower temperature as 773K, however, we need to consider the charge transfer process simultaneously with the enhanced oxygen incorporation reaction. Since the electrochemical process enhances the surface reaction much higher than that in "dry" case at 773K, we can also expect that some activation effect, such as water vapor, on the GDC surface under the polarization condition. The effect must not be limited at TPBs as shown in Fig. 17(a) (in the case of YSZ, electrochemical effect is limited at the TPBs as shown in Fig. 17(b)).Charge transfer takes place not only at TPBs but also on GDC surface; So, for this purpose, we have to consider that the charge is carried by holes and/or protons, whereby both as carrier can extend the electrochemical active sites in the order of several μm. For the hole case;

$$O_2 + 2V_O^{\cdot\cdot} = 2O_O^{x} + 4\,h^{\cdot}\text{(in electrolyte)on the GDC surface} \quad (28)$$

$$4h^{\cdot}\text{ (in electrolyte, TPB)} + 4e'\text{ (in electrode, TPB)} = \text{null} \quad (29)$$

For the proton case;

$$H_2O + V_O^{\cdot\cdot} = O_O^{x} + 2H^{\cdot}\text{(in electrolyte)on the GDC surface} \quad (30)$$

$$2H^{\cdot}\text{ (in electrolyte, TPB)} + O_O^{x} + 2e'\text{ (in electrode, TPB)} = H_2O(g) \quad (31)$$

For the proton case, emitted water vapors will cover entire surface of GDC, leading to strong enhancement of oxygen isotope exchange reaction as well as the water vapor incorporation as oxide ions and protons. Those protons will come back to the TPBs to complete the electrochemical process, since the major carrier in electrolyte is the oxide ions.

Here, we come to the quite interesting feature of this investigation which was initiated in order to identify the oxide ion flow associated with the electrochemical process; for the GDC case, however, what the isotope exchange-SIMS analysis shows us is the oxygen isotope exchange reaction enhanced by the electrochemical process and this effect is so strong that we can hardly detect the ionic flow directly associated with the electrochemical process.

According to the proton-mediated mechanism, we need the dissociated oxygen atom as well as protons in the TPB vicinity. It is reasonably assumed that there is some concentration of oxide ion vacancies in the same place. This implies that from the concentration point of view, the oxygen incorporation reaction can proceed in a similar manner to the Au / YSZ system. This suggests that the gold electrode has low activity for the oxygen incorporation but relatively high activity for the water emission as the cathodic process. This situation may

depend on electrode materials. Ceria electrolyte or interlayer are frequently used for the active cathodes such as $(La,Sr)(Co,Fe)O_3$. In such a case, the oxygen incorporation reaction may be the main process at the TPBs. Even so, the present results suggest that some contribution comes from the protons in ceria. That is, water vapors may be emitted from the electrochemically active sites in such cathode materials. In the electrochemical performance, this effect can be negligible because catalytic activity of electrode materials for electrochemical oxygen incorporation reaction is so high. On the other hand, for the degradation phenomena such as Cr poisoning or SO_2 poisoning [52-56], this may affect significantly because these degradation phenomena occur at such electrochemically active site and these reactions are strongly influenced by water vapor existence. For example, in Cr poisoning the degradation and recovery reactions are as follows;
Cr poisoning:

$$2CrO_3 + 6e' = Cr_2O_3(\text{at active site (TPB)}) + 3O^{2-} \quad (32)$$

Recovery process from Cr poisoning by water vapor:

$$Cr_2O_3(\text{at active site (TPB)}) + 2H_2O + 3/2O_2 = 2CrO_2(OH)_2(\text{gas}) \quad (33)$$

In the practical SOFC stack, Cr deposition at TPB area is suppressed when doped ceria interlayer is inserted between LSM cathode / YSZ electrolyte. Improved cells with doped ceria interlayer exhibit excellent long term stability and high tolerance against the Cr poisoning [57].

5. Carbon Deposition Behavior Around Nickel /Oxide Interface

One of the great advantages of SOFCs is fuel flexibility. Because anode is made of nickel (Ni) which has high catalytic activity and operation temperature of SOFC is high enough for reforming of hydrocarbon fuels, it is, in principle, possible to use hydrocarbon fuels by internal reforming. Reforming reaction of hydrocarbon fuels is expressed as follows:

$$C_nH_m + nH_2O = nCO + (m/2+n)H_2 \quad (34)$$

In the case where steam to carbon ratio (S/C) is low, water vapor shortage causes carbon deposition on the anode instead of Eq. (34). This is related to the thermal decomposition reaction of hydrocarbons and/or to the Boudouard reaction described as follows:

$$CO + H_2 = C + H_2O,\ 2CO = CO_2 + C \text{ and } CH_4 = C + 2H_2 \quad (35\text{-}37)$$

Anode degradation behavior related to carbon deposition has been reported in some literatures [58,59]. For the Ni / oxide composite-type so called *cermet* anode, the carbon deposition behavior, the fuel reforming activity and electrode reaction process are influenced by the oxide component in the cermet anode [60-75]. This section focuses on such carbon deposition behavior, more specifically, its difference between the Ni / YSZ system and the Ni / GDC system.

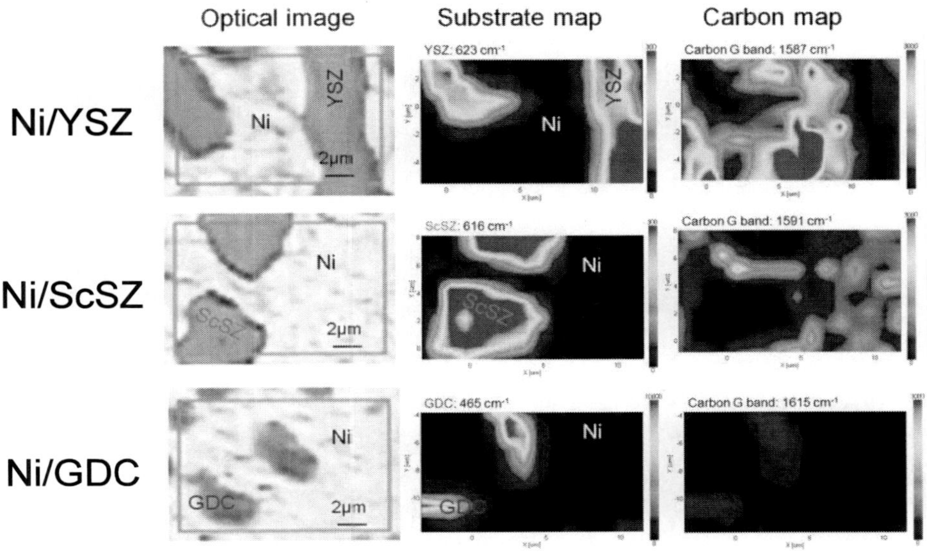

Figure 21. Optical images, and micro-Raman peak intensity maps of carbon deposition on oxide substrates and those of carbon G-band for Ni/YSZ, Ni/ScSZ and Ni/GDC system. Samples are annealed under dry Ar - 5%CH_4 at 600°C for 20min [76].

Yoshinaga et al. analyzed effects of the oxide component on the carbon deposition behavior around Ni / oxide interfaces by using 2-dimensinal model electrodes [76,77]. In this technique, a dewetting pattern (agglomerated microstructure) of Ni metal was formed on the oxide substrate. Therefore, it is possible to distinguish the Ni surface, oxide surface and TPBs. Carbon deposition behavior, such as deposited area of carbon and carbon species, was analyzed by micro-Raman spectroscopy which has very high sensitivity for carbon and high spatial resolution. Fig. 21 shows the results of mapping analysis by micro-Raman spectroscopy for the Ni / YSZ, the Ni / ScSZ and the Ni / GDC systems after carbon deposition treatment [76]. Carbon was found to be deposited on the whole Ni surface for the Ni / YSZ system. A similar carbon deposition map was observed for the Ni / ScSZ system but with a much less amount of carbon than the former. Interestingly, carbon deposition on the Ni surface was suppressed for the Ni / GDC system; slight carbon deposition seems to occur on the GDC surface. Obviously, carbon deposition on the Ni surface is largely influenced by the oxide substrate. In addition, carbon deposition behavior was found to be greatly influenced by Ni particle size (Fig. 22 [77]). Even for the Ni / GDC system, carbon deposition was observed for the small Ni particles with about several ten nm size. Oxide substrate effect on the Ni surface condition, i.e., electronic state or adsorption behavior of the surface, seems to be diminished for such smaller Ni particles. Horita et al. visually analyzed the element distribution for the Ni patterned electrode on the oxide substrate under methane reforming condition by SIMS [78,79]. Their results for the ^{12}C and ^{16}O distribution for the Ni / YSZ and the Ni / SDC systems are shown in Fig. 23 [79]. For the Ni / YSZ system, deposited carbon almost covered the Ni surface, a part of Ni surface where carbon did not deposit being covered by oxygen. On the other hand, for the Ni / SDC system, the Ni surface was completely covered with oxygen, and carbon was hardly detected. These SIMS images were obtained after slight sputtering of the surface; detected oxygen atoms were therefore originally present inside the Ni. The surface of the Ni metal was nearly oxide (i.e., oxygen dissolved Ni) for the Ni / SDC system, as confirmed by the XAFS analysis [79].

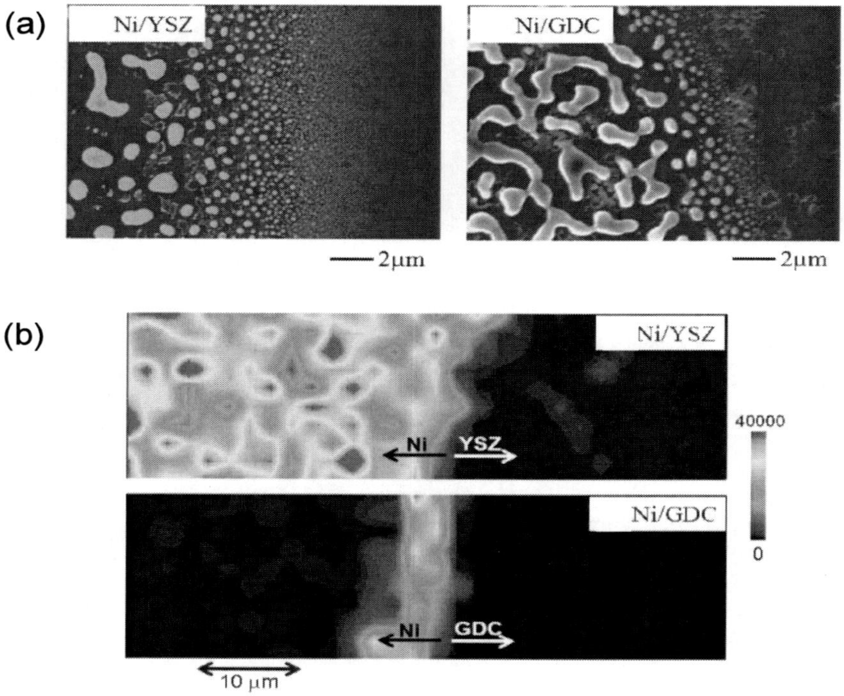

Figure 22. (a) Microstructure and (b) deposited carbon distribution obtained by micro-Raman analysis for Ni/YSZ and Ni/GDC systems. Samples are annealed under dry Ar - 5%CH_4 at 600°C for 10min and at 700°C for 10min [77].

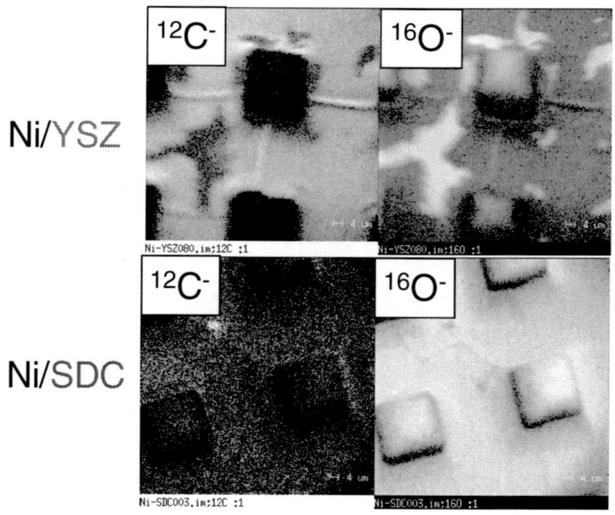

Figure 23. $^{12}C^-$ and $^{16}O^-$ images at the Ni-mesh/YSZ and Ni-mesh/SDC systems [79].

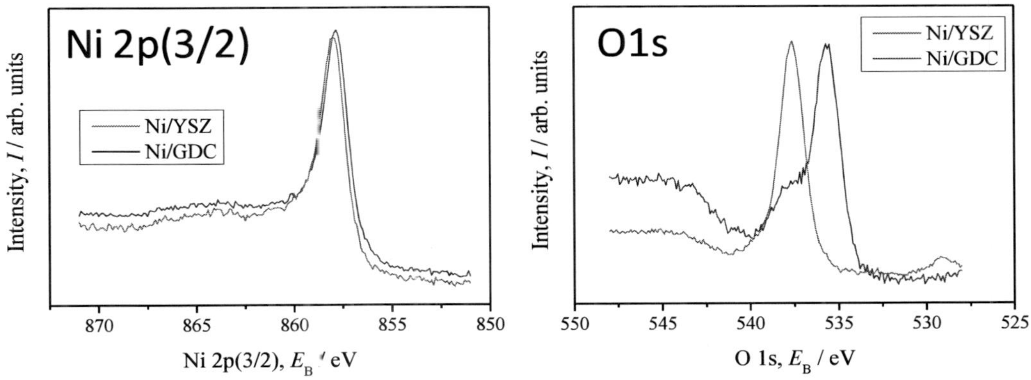

Figure 24. XPS spectra for C1s, Ni2p(3/2) and O1s for the Ni surface area of model electrodes of the Ni/YSZ system and the Ni/GDC system. The samples were annealed under 3 % H_2O - H_2 at 500°C for 10 h [75].The data are not corrected for the influence of charge-up effect; about7~8 eV both for the Ni / YSZ and the Ni / GDC systems.

Next, the above-observed difference of carbon deposition behavior on Ni surface is discussed based on the reaction model taking properly into account the difference of physicochemical properties of oxide substrates. The reason why carbon deposition does not take place on the Ni surface for the Ni / doped ceria systems should be that a much more number of oxygen atoms are strongly adsorbed on the Ni surface than for the Ni / YSZ system. Fig. 24 shows the XPS spectra for the surface of Nithin film prepared on the YSZ and the GDC substrate [75]. The samples were annealed under 3 % H_2O - H_2 at 500°C for 10h. Peak positions had to be corrected from the C1s spectra; shift was observed in the order of about 8 eV and 7 eV for the respective Ni / YSZ and the Ni / GDC systems from the reference of Ni metal; this was because of charge-up effects. From the Ni 2p(2/3) spectra, the Ni surface state was essentially the same between the Ni / YSZ and the Ni / GDC systems. On the other hand, from the O1s spectra, there was a clear difference between the Ni / YSZ and the Ni / GDC systems. Only one peak was observed at 538eV for the Ni / YSZ system. For the Ni / GDC system, the peak at 529 eV was observed in addition to a small peak at 531 eV. The adsorbed oxygen state on Ni surface is different between both systems. These two binding energies, E_b, of O1s may correspond to those for OH (ad) (at 531 eV, after correction of the charge-up effect) and O(ad) (at 529 eV, after correction of the charge-up effect) [80]. Where does such a difference come from? The key ingredient here is judged to be the different interaction mechanism with water vapor. The mass transport model around the TPB (gas / electrode / oxide) in this scheme is shown in Fig. 25 [75]. In both the Ni / YSZ and the Ni / GDC systems, adsorption of water vapor on the Ni surface occurs at first:

$$H_2O(g) = H_2O(ad) \tag{38}$$

A part of hydrogen atom of adsorbed water vapor can dissolve into Ni metal up to the solubility limit, changing the adsorbed water to hydroxyl group at the same time.

$$H_2O(ad) = OH(ad) + H(in\ Ni) \tag{39}$$

Further, the adsorbed hydroxide is dissociated into adsorbed oxygen atom.

$$OH(ad) = O(ad) + H(in\ Ni) \tag{40}$$

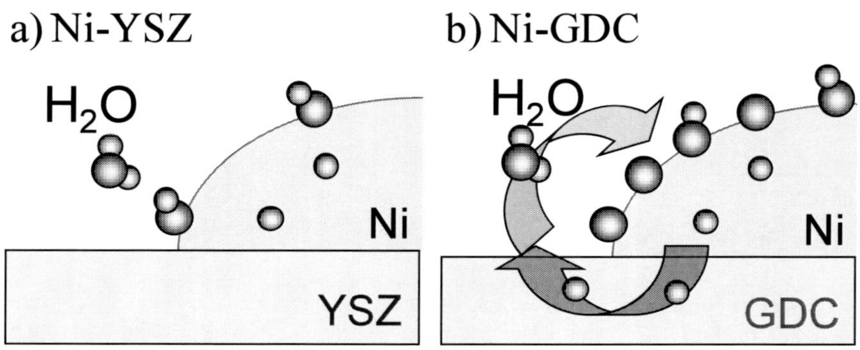

Figure 25. Schematic drawing of proposed reaction model for the Ni/YSZ and the Ni/GDC systems under OCV condition [75].

Hydrogen atom in Ni can dissolve into doped ceria as proton if the oxide has the proton solubility.

$$H(in\ Ni) = H^{\bullet}\ (in\ oxide) + e'\ (in\ oxide) \tag{41}$$

Formed electron can be used for oxygen incorporation reaction on the electrolyte surface.

$$1/2\ O_2(g) + V_O^{\cdot\cdot} + 2e' = O_O^{x} \tag{42}$$

It is controlled by the equilibrium with gas phase how these reactions proceed.

As mentioned in previous section (Figs. 8 and 9), electronic conductivity and proton solubility of GDC is six and two order-of-magnitude higher than those of YSZ, respectively. Therefore, for the Ni / GDC system, reaction (41) proceeds to a more significant level than for the Ni / YSZ system, and reactions (38) to (40) proceed in conjunction with reaction (41). In addition, oxygen nonstoichiometry (concentration of $V_O^{\cdot\cdot}$) becomes large under reducing condition and reaction (42) is also enhanced. As a result, the coverage not only of OH(ad) but also of O(ad) on Ni surface becomes higher for the Ni / GDC system. The interaction between Ni metal and O(ad) should be stronger than that for OH(ad). Strong interaction between Ni surface and O(ad) makes the surface state like an oxide. Under such condition, the penetration effect of electron on the Ni surface decreases and work function of Ni surface also decreases [81]. Carbon deposition is suppressed by this "inactivation" effect for the Ni / GDC system. Adsorption effect, which is influenced by the oxide component, on the Ni surface becomes negligible at high temperature. As a result, carbon deposition behavior is different between at873K and at 1273K. At high temperature, the predominant factor for carbon deposition behavior should be the Ni - oxide microstructure in the anode [71,72].

About Eq. (41), there can be another interpretation. This is a kind of charge transfer process. If electrons are available and supplied to Ni, the reaction can be rewritten as:

$$H(\text{in Ni}) = H^{\bullet}(\text{in oxide}) + e'(\text{in Ni}) \quad (43)$$

This process should be coupled with another charge transfer process to complete the electrochemical reaction under the OCV condition. This should be the following water vapor formation (emission) reaction:

$$2H^{\bullet}(\text{in oxide}) + 2e'(\text{in Ni}) + O(ad) = H_2O(g) \quad (44)$$

As a result of circulated reactions, oxygen atoms in the TPB vicinity are transported to surface of Ni far from TPBs. When Ni is formed as electron conductive network, those two sites can be far distant. When Ni is in particle shape, such circulated reactions have to take place within a limited area. This can be one of possible reasons why carbon is deposited on small nickel particles.

Another possible charge transfer process is the hydrogen evolution:

$$2H^{\bullet}(\text{in oxide}) + 2e'(\text{in Ni}) = H_2(g) \quad (45)$$

As a result of circulated reactions, water vapor is decomposed into O(ad) on Ni and hydrogen gases.

CONCLUSION

Cubic stabilized zirconia and rare-earth doped ceria which have the CaF_2-type cubic lattice are critical materials as high temperature oxide ion conductor in SOFCs application.

Cubic stabilized zirconia and rare-earth doped ceria are mainly oxide ion conductive at high temperatures under oxidizing condition. But their electronic conductivity, proton solubility and oxygen surface exchange reactivity are quite different from each other. Electronic conductivity of rare-earth doped ceria is much higher than that of cubic stabilized zirconia. Interaction ability with water vapor of rare-earth doped ceria is also higher than that of cubic stabilized zirconia.

The different behavior of all these charge carriers such as oxide ions, electrons, holes and protons between cubic stabilized zirconia and rare-earth doped ceria has been extensively described in this chapter, mainly based on our research results, and is shown to affect greatly the surface and interfacial electrode and mass transport behavior around the gas / electrode / electrolyte triple phase boundary (TPB). It is expected that by better controlling these phenomena we can improve not only the performance but also the reliability and durability of SOFCs.

ACKNOWLEDGMENT

We would like to express our sincere thanks to Dr. T. Horita, Dr. N. Sakai and Dr. K. Yamaji for their valuable discussions on the present topics.

REFERENCES

[1] T. Kawada and J. Mizusaki, in *Handbook of Fuel Cells – Fundamentals, Technology and Applications*, W. Vielstich, H.A. Gasteiger and A. Lamm, eds, *John Wiley & Sons,* 2003, vol.4, 987.

[2] M. Mori, T. Abe, H. Itoh, O. Yamamoto, Y. Takeda and T. Kawahara, *Solid State Ionics* (1994) 74, 157.

[3] Y. Arachi, T. Asai, O. Yamamoto, Y. Takeda, N. Imanishi, K. Kawate and C. Tamakoshi, *J. Electrochem. Soc.* (2001)148, A520.

[4] T. Ishihara, H. Furutani, M. Honda, T. Yamada, T. Shibayama, T. Akbay, N. Sakai, H. Yokokawa and Y. Takita, *Chem. Mater.*(1999) 11, 2081.

[5] A. Mineshige, Y. Ohnishi, R. Sakamoto, Y. Daiko, M. Kobune, T. Yazawa, H. Yoshioka, T. Nakao, T. Fukutsuka and Y. Uchimoto, *Solid State Ionics*(2011) 192, 195.

[6] V.V. Kharton, F.M. Figueiredo, L. Navarro, E.N. Naumovich, A.V. Kovalevsky, A.A. Yaremchenko, A.P. Viskup, A. Carneiro, F.M.B. Marques and J.R. Frade, *J. Mater. Sci.* (2001) 36, 1105.

[7] M.J. Verkerk, K. Keizer and A.J. Burggraaf, *J. Appl. Electrochem.* (1980) 10, 81.

[8] O. Yamamoto, in *Handbook of Fuel Cells – Fundamentals, Technology and Applications*, W. Vielstich, H.A. Gasteiger and A. Lamm, eds, *John Wiley & Sons,* vol.4, 1002 (2003).

[9] M. Godickmeier and L.J. Gauckler, *J. Electrochem. Soc.* (1998) 145, 414.

[10] I.K. Naik and T.Y. Tien, *J. Phys. Chem. Solids* (1978) 39, 311.

[11] H.L. Tuller and A.S. Nowick, *J. Phys. Chem. Solids*(1977) 38, 859.

[12] W. Weppner, *Z. Naturforsch.* (1976) 31a, 1336.

[13] W. Weppner, *J. Solid State Chem.* (1977)20, 305.

[14] J.H. Park, R.N. Blumenthal, *J. Electrochem. Soc.* (1989)136, 2867.

[15] T. Kawada, N. Sakai, H. Yokokawa, M. Dokiya, *Solid State Ionics*(1992) 53–56, 418.

[16] S. Lübke, H.-D. Wiemhöfer, *Ber. Bunsenges. Phys. Chem.* (1998)102, 642.

[17] S. Lübke, H.-D. Wiemhöfer, *Solid State Ionics* (1999) 117, 229.

[18] H.-D. Wiemhöfer, M. Dogan, S. Lübke, V. Muenster, New Results for Electron Transport, Chemical Diffusion and Stability of Solid Oxygen Ion Conductors, *MRS Symp. Proc.* (2003)756, 455.

[19] Y.-P. Xiong, K. Yamaji, N. Sakai, H. Negishi, T. Horita, H. Yokokawa, *J. Electrochem. Soc.* (2001)148, E489.

[20] Y.-P. Xiong, K. Yamaji, T. Horita, N. Sakai, H. Yokokawa, *J. Electrochem. Soc.* (2002)149, E450.

[21] Y.-P. Xiong, K. Yamaji, T. Horita, N. Sakai, H. Yokokawa, *J. Electrochem. Soc.* (2004) 151, A407.

[22] Y. Xiong, K. Yamaji, H. Kishimoto, M.E. Brito, T. Horita and H. Yokokawa, *J. Electrochem. Soc.* (2008)155, B1300.

[23] H. Yokokawa, T. Horita, N. Sakai, K. Yamaji, M.E. Brito, Y.-P. Xiong and H. Kishimoto, *Solid State ionics* (2004) 174, 205-221.

[24] H. Iwahara, *Solid State Ionics* (1996) 86-88, 9.

[25] S.-J. Song, E.D. Wachsman, S.E. Dorris, U. Balachandran, *Solid State Ionics* (2002) 149, 1.

[26] N. Fukatsu and N. Kurita, *Ionics* (2005) 11, 54.

[27] E. Fabbri, D. Pergolesi and E. Traversa, *Chem. Soc. Rev.*(2010)39, 4355.
[28] X. Guo, *J. Phys. Chem. Solids*(1999)60, 539.
[29] X. Guo, *J. Mater. Sci.*(1999)36, 3737.
[30] M. Hirano and E. Kato, *J. Mater. Sci.* (1999)34, 1399.
[31] N. Sakai, K. Yamaji, T. Horita, H. Yokokawa, Y. Hirata, S. Sameshima, Y. Nigara and J. Mizusaki, *Solid State Ionics*(1999)125, 325.
[32] N. Sakai, K. Yamaji, Y.P. Xiong, H. Kishimoto, T. Horita and H. Yokokawa, *J. Electroceram.*(2004) 13, 677.
[33] H. Yokokawa, T. Horita, N. Sakai, Y. Yamaji, M.E. Brito, Y.-P Xiong and H. Kishimoto, *Solid State Ionics*(2006) 177, 1705.
[34] C. Wagner, *Ber. Bunserges. Phy. Chem.* (1968) 72, 778.
[35] S. Kim, and R. Merkle, J. Maier, *Solid State Ionics* (2003) 161, 113.
[36] D. Pérez-Coll and G.C. Mather, *Solid State Ionics* (2010) 181, 20.
[37] H. Takamura and N. Takahashi, *Solid State Ionics* (2010) 181, 100.
[38] M. Shirpour, G. Gregori, R. Merkle and J. Maier, *Phys. Chem. Chem. Phys.* (2011)13, 937.
[39] S. Inami, S.Yamaguchi, H. Yokokawa, N. Sakai, T. Horita, K. Yamaji, "Water Solubility in Ceria and Zirconia Doped with Rare Earth Oxides" in *Solid State Ionics: Trends in the New Millennium*, B.V.R. Chowdari, S.R.S. Prabaharan, M. Yahaya and I.A. Talib, eds., *World Scientific Publishing Co.*(2002)pp.529-536, and unpublished data (2003).
[40] P.S. Manning, J.D. Sirman, R.A De Souza, J.A. Kilner, *Solid State Ionics* (1997)100, 1.
[41] P.S. Manning, J.D. Sirman, J.A. Kilner, *Solid State Ionics* (1997)93, 125.
[42] N. Sakai, K. Yamaji, T. Horita, H. Kishimoto, Y.P. Xiong and H. Yokokawa, *Phys. Chem. Chem. Phys.*(2003)5, 2253.
[43] N. Sakai, K. Yamaji, T. Horita, Y.P. Xiong, H. Kishimoto, and H. Yokokawa, *J. Electrochem. Soc.* (2003)150, A689.
[44] H. Kishimoto, N. Sakai K. Yamaji, T. Horita, M.E. Brito, H. Yokokawa, K. Amezawa and Y. Uchimoto, *Solid State Ionics*(2008) 179, 347.
[45] H. Kishimoto, N. Sakai K. Yamaji, T. Horita, M.E. Brito, H. Yokokawa, K. Amezawa and Y. Uchimoto, *Solid State Ionics*(2008) 179, 1343.
[46] H. Kishimoto, N. Sakai, K. Yamaji, T. Horita, M.E. Brito, and H. Yokokawa, *ECS Trans.*(2008) 13, 105.
[47] N. Sakai, K. Yamaji, T. Horita, H. Kishimoto, M.E. Brito, H. Yokokawa and Y. Uchimoto, *Appl. Surf. Sci.* (2006) 252, 7045.
[48] S. Raz, K. Sasaki, J. Maier, and I. Riess, *Solid State Ionics*(2001) 143, 181
[49] M.O. Zacate, L. Minervini, D.J. Bradfield, R.W. Grimes and K.E. Sickafus, *Solid State Ionics* (2000) 128, 243.
[50] K. Kawata, H. Maekawa, T. Nemoto, T. Yamamura, *Solid State Ionics* (2006) 177, 1687.
[51] L. Minervini, M.O. Zacate and R.W. Grimes, *Solid State Ionics* (1999) 116, 339.
[52] T. Horita, Y. Xiong, H. Kishimoto, K. Yamaji, M.E. Brito, and H. Yokokawa, *J. Electrochem. Soc.* (2010) 157, B614.
[53] T. Horita, D.H. Cho, F. Wang, T. Shimonosono, H. Kishimoto, K. Yamaji, M.E. Brito, andH. Yokokawa, *Solid State Ionics* (2012) 225, 151.

[54] S.P. Jiang, J.P. Zhang, L. Apateanu, and K. Foger, *J. Electrochem. Soc.* (2001) 148,C447.
[55] Y. Xiong, K. Yamaji, T. Horita, H. Yokokawa, J. Akikusa, H. Eto, and T. Inagaki, *J. Electrochem. Soc.* (2009) 156, B588.
[56] F. Wang, K. Yamaji, D.H. Cho, T. Shimonosono, H. Kishimoto, M.E. Brito, T. Horita, and H. Yokokawa, *J. Electrochem. Soc.* (2011) 158, B1391.
[57] H. Yokokawa, T. Horita, K. Yamaji, H. Kishimoto,.T. Yamamoto, M. Yoshikawa, Y. Mugikura,.T. Kabata and K. Tomida, *Fuel Cells - From Fundamentals to Systems*(2013) 13, 526.
[58] T. Takeguchi, Y. Kani, T. Yano, R. Kikuchi, K. Eguchi, K. Tsujimoto, Y. Uchida, A. Ueno, K. Omoshiki and M. Aizawa, *J. Power Source*(2002)112, 588.
[59] R.J. Gorte, H. Kim and J.M. Vohs, *J. Power Source* (2002)106, 10.
[60] K. Eguchi, in *Handbook of Fuel Cells – Fundamentals, Technology and Applications*, W. Vielstich, H.A. Gasteiger and A. Lamm,eds,*John Wiley & Sons,*vol.4(2003) 1057.
[61] B.C.H. Steele, P.H. Middleton and R.A. Rudkin, *Solid State Ionics* (1990)40/41, 388.
[62] H. Uchida, M. Yoshida and M. Watanabe, *J. Phys. Chem.* (1995)99, 3282.
[63] M. Watanabe, H. Uchida and M. Yoshida, *J. Electrochem. Soc.*(1997)144, 1739.
[64] C.M. Finnerty, N.J. Coe, R.H. Cunningham and R.M. Ormerod, *Catalysis Today* (1998)46, 137.
[65] E.P. Murray, T. Tsai and S.A. Barnett, *Nature* (1999)400, 649.
[66] C. Xia, F. Chen and M. Liu, *Electrochem. Solid-State Lett.* (2001)4, A52.
[67] K. Ukai, Y. Mizutani and Y. Kume, in *Proc. Solid Oxide Fuel Cells VII*, H. Yokokawa and S.C. Singhal, Eds., PV 2001-16, p. 375, The Electrochemical Society Proceedings Series, Pennington, NJ(2001).
[68] J.B. Wang, J.-C. Jang and T.-J. Huang, *J. Power Sources* (2003)122,122.
[69] S. Wang, M. Ando, T. Ishihara and Y. Takita, *Solid State Ionics* (2004)174,49.
[70] H. Sumi, K. Ukai, Y. Mizutani, H. Mori, C-J. Wen, H. Takahashi and O. Yamamoto, *Solid State Ionics* (2004)174, 151.
[71] T. Matsui, T. Iida, M. Kawano, T. Inagaki, R. Kikuchi and K. Eguchi, *ECS Trans. (Proc. SOFC X)* (2007) 7, 1437.
[72] T. Iida, M. Kawano, T. Matsui, R. Kikuchi and K. Eguchi, *J. Electrochem. Soc.* (2007) 154, B234.
[73] M. Kawano, T. Matsui, R. Kikuchi, H. Yoshida, T. Inagaki and K. Eguchi, *J. Electrochem. Soc.* (2007)154, B460.
[74] H. Kishimoto, K. Yamaji, T. Horita, Y.-P. Xiong, N. Sakai, M.E. Brito and H. Yokokawa, *IEEJ Trans. Jpn.* (2008)128, 333.
[75] H. Kishimoto, T. Shimonosono, K. Yamaji, M.E. Brito, T. Horita and H. Yokokawa, *Electrochemistry*(2013)81, 255.
[76] M. Yoshinaga, H. Kishimoto, K. Yamaji, Y.-P. Xiong, M.E. Brito, T. Horita and H. Yokokawa, *Solid State Ionics*(2011) 192, 571.
[77] M. Yoshinaga, H. Kishimoto, M.E. Brito,K. Yamaji, T. Horita and H. Yokokawa, *J. Ceram. Soc. Jpn.* (2011) 119, 307.
[78] T. Horita, K. Yamaji, T. Kato, N. Sakai, H. Yokokawa, *J. Power Sources* (2004) 131, 299.
[79] T. Horita, H. Kishimoto, K. Yamaji, N. Sakai, Y. Xiong, M.E. Brito, H. Yokokawa, M. Rai, K. Amezawa and Y. Uchimoto, *Solid State Ionics*(2006) 177, 3179.

[80] F.J. Rodrigues Nieto, E. Fachini, C.R. Cabrera, A.J. Arvia, *Thin Solid Films* (2009) 517, 1534.
[81] T.D. Pope, S.J. Bushby. K. Griffiths and P.R. Norton, *Surf. Sci.*(1991)258, 101.

In: New Research Trends of Fluorite-Based Oxide Materials ISBN: 978-1-63117-350-9
Editors: Akio Nakamura and Junichiro Mizusaki

Chapter 6

MECHANICAL PROPERTIES OF FLUORITE-STRUCTURED OXIDES

Alan Atkinson*
Department of Materials, Imperial College, London, UK

ABSTRACT

This chapter considers the mechanical properties of fluorite-structured oxides with particular emphasis given to cubic stabilized zirconia and doped ceria, both of which have potential application in electrochemical devices. They differ in that zirconia requires doping to stabilize the fluorite structure (ceria does not) and ceria has oxygen content depending on oxygen activity and temperature (zirconia does not). The mechanical properties considered are elasticity, plasticity, creep, and fracture. It is shown that all fluorite-structured oxides have similar mechanical properties at room temperature. However, there are interesting differences in behaviour between zirconia and doped ceria as the temperature is raised or doping levels and oxygen content (independently in the case of ceria) are changed. These differences are not understood, but are probably related to subtle differences in the way in which defect clusters respond to the applied stress in the two materials.

1. INTRODUCTION

Binary oxides of tetravalent large cations tend to adopt the fluorite structure: e.g., ZrO_2, HfO_2, CeO_2, ThO_2, UO_2 and PuO_2. They all have considerable technological uses, but the last three are specifically of interest in nuclear applications. In this chapter the emphasis is on ZrO_2 and CeO_2 because of their application as fast ion conductors. They differ in that Zr^{4+} is difficult to reduce and so the oxygen content of ZrO_2 is constant under typical environmental conditions of temperature and oxygen activity. On the other hand Ce^{4+} can be reduced

* E-mail address: alan.atkinson@imperial.ac.uk.

partially to Ce^{3+} at moderately high temperatures and in relatively mild reducing conditions. As a result CeO_2 is an example in which the oxygen content is variable.

This chapter is organized in the following way. First, some general comments are made about different mechanical properties and their interpretation, and then these properties are discussed for zirconia and ceria and comparisons made between them. The data presented are not exhaustive and apologies are given in advance for inevitable omissions from such a wide field.

In the literature, certain shorthand representations have developed to indicate the composition of doped oxides. These can be a source of considerable confusion because they differ for different oxides. For example, the composition $8Y_2O_3.92ZrO_2$ is often referred to as 8YSZ, whereas the composition $10GdO_{1.5}.90CeO_2$ is referred to as 10CGO (or 10GDC). To avoid confusion we will use the doped formulae, e.g., $Y_xZr_{1-x}O_{2-\delta}$ or $Gd_xCe_{1-x}O_{2-\delta}$, with the composition being specified by the value of x. Thus 10CGO has $x = 0.1$ and 8YSZ has $x = 0.15$.

2. Mechanical Properties

2.1. Elasticity

The fluorite structure is cubic and therefore the elastic response of a single crystal is characterized by a stiffness tensor with only three independent stiffness constants: C_{11}, C_{12} and C_{44}. These constants reflect the rigidity of the bonding between the atoms and are therefore expected to increase as the binding energy of the crystal increases. Conversely, as temperature increases the interatomic bond lengths, then the elastic constants tend to decrease with increasing temperature.

For an isotropic polycrystalline ceramic, only any two of: E (Young's modulus), ν (Poisson's ratio) or μ (shear modulus), are sufficient. The main microstructural parameter that influences the elastic properties is the porosity; with even small quantities of residual porosity causing a relatively large drop in elastic modulus.

In view of some issues that have appeared in the literature, and will be discussed later, it is necessary to make some observations regarding methods of measuring elastic modulus. One commonly used method is the impulse excitation technique (IET) in which a resonant vibrational mode is set up in a specimen. The resonant frequencies are usually in the kHz range and therefore the timescales are sufficiently short that point defect configurations are fixed. The other common experimental method is load-displacement, in which a static load is applied to a specimen and its deformation is recorded. The timescales for the specimen to respond to the load can vary from seconds to days. In such experiments the point defects might have sufficient time to reconfigure in response to the applied stress, depending on temperature and the details of the reconfiguration. This can lead to an extra contribution to the strain in the specimen and an apparent elastic modulus that differs from that measured by IET. It is therefore important to know which measurement method has been used.

2.2. Plastic Yield and Hardness

Plastic yield is caused by the motion of dislocations under applied stress. At room temperature and typical engineering stress levels, dislocations are not usually mobile in ceramic oxides. However, at high temperatures, or under the high stresses generated by indentation, the dislocations become mobile and the oxides can undergo plastic yield. In the fluorite structure dislocations have Burgers vector of $a/2$ <110> and can glide on {001}, {110} and {111} planes, with {001} being the easiest [1]. In a hardness indentation test, the locally high stresses are sufficient to cause plastic yield even at room temperature and the hardness (e.g., H_V in a Vickers indentation test) is generally regarded as being an approximation to the yield stress (e.g., $H_V \approx 3\sigma_y$). In ceramics the grain boundaries tend to be barriers to dislocation motion and inhibit plastic yield. This is often described by the Hall-Petch relationship between yield stress, σ_y, and grain size, d:

$$\sigma_y = \sigma_0 + \frac{A}{\sqrt{d}} \tag{1}$$

(In this chapter the symbol A is used to denote a constant in various equations. This does not imply that the constant is the same in every equation in which it appears.)

2.3. Strength and Toughness

Ceramic oxides usually fail by brittle fracture initiated at defects in the specimen. Although strength in ceramics is a useful engineering parameter, it is not a fundamental property of the material because it depends on the distribution of the critical defects that initiate the failure. As a result, strength has a rather broad statistical distribution, with a mean value that depends on the size of the specimen and the details of the applied stresses. This statistical variability is usually handled by fitting the experimental data to a Weibull distribution described by a Weibull characteristic stress and Weibull modulus. The higher the value of the Weibull modulus, the narrower is the distribution of strength and the more "reliable" the material.

The fracture properties of the material are described at a more fundamental level by the fracture toughness (or critical stress intensity), K_c, or the work of fracture, G_c. For single crystals fracture is by cleavage between particular crystallographic planes and, providing there are no irreversible losses such as plastic deformation or friction, $G_c = 2\gamma_s$, where γ_s is the surface energy of the cleaved surfaces. In a polycrystalline ceramic the same principles apply, but the fracture path can be by cleavage (trans-granular), or along grain boundaries (inter-granular).

The fracture toughness describes a critical condition for the local stress distribution around a crack tip and depends on how the crack is being loaded by the far field stress. The most common loading is opening mode (mode I) in which case the fracture toughness is denoted K_{Ic}. The toughness and work of fracture are related by

$$K_{Ic} = \sqrt{EG_c} \tag{2}$$

Since K_c depends on the mode of loading, then so does the work of fracture. When, as is usually the case, the fracture involves energy dissipation then $G_c > 2\gamma_s$, particularly at high temperatures when there is significant plastic yield to absorb energy. In some materials G_c increases as the crack extends (e.g., due to friction between crack faces).

The strength thus depends on both the toughness and the size and shape of flaws that are being stressed and act as initiation sites for fracture. The relationship takes the general form

$$\sigma_f = A\frac{K_{Ic}}{c^{\frac{1}{2}}} \tag{3}$$

where A is a dimensionless factor that depends on the type and shape of the flaw and c is its characteristic size.

The bonds at the loaded crack tip are severely distorted even when the loads are insufficient to cause crack propagation. In this condition they can be susceptible to additional chemical attack from the environment (often by water vapour) and slowly become unable to support the stresses at the crack tip. This gives rise to the phenomenon known as sub-critical crack growth (or slow crack growth, or delayed fracture, or static fatigue). The crack grows slowly at a velocity, v, determined by the stress intensity

$$v = AK^n \tag{4}$$

until the crack length reaches a value for which K becomes critical and the crack then propagates rapidly.

2.4. Creep

In creep the strain is time dependent at constant stress (or stress is time dependent at constant strain) and several processes can dominate the overall response. Creep occurs by diffusion processes and therefore is thermally activated and controlled by the atom in the structure having the lowest diffusion coefficient. In single crystals creep is dominated by the climb of dislocations away from obstacles (such as other dislocations) and is often referred to as recovery creep. The general constitutive equation for power law diffusion-controlled creep is [1]:

$$\dot{\varepsilon} = A\frac{\mu b}{kT}\left(\frac{\sigma}{\mu}\right)^n \left(\frac{b}{d}\right)^m D_0 \exp\left(-\frac{Q}{kT}\right) \tag{5}$$

where b is the Burgers vector of a dislocation. For single crystals the grain size term is absent (i.e., m=0, but then the factor "A" will be different and dependent on dislocation density).

In polycrystalline ceramics, creep is usually dominated by the diffusion of material from parts of each grain that are under compression to parts that are under tension. If the temperature is high enough, this can occur by diffusion through the crystal lattice of each grain and is known as Nabarro-Herring creep. In such a case $n = 1$ and $m = 2$. More usually however, it is diffusion along grain boundaries that dominates and the process is known as Coble creep, for which $n = 1$ and $m = 3$. In both cases there is a linear relation between the strain rate and the stress so that an analogue of Hooke's law is appropriate e.g., $\sigma = E_p\dot{\varepsilon}$ where E_p is the creep viscosity. There is also the corresponding viscous Poisson's ratio ν_p. In some materials grain sliding accompanies diffusion (via grain boundary or lattice) in a process known as Ashby-Verrall creep [1].

3. Zirconia

3.1. Elasticity

Pure zirconia exists in the cubic fluorite structure only at temperatures above 2750°C, but can be stabilized in the cubic structure by partial substitution of Zr by lower valency cations. These substitutions also result in the creation of oxygen vacancies to compensate for the lower positive charge. Since oxygen vacancies are highly mobile in the fluorite lattice, these materials have been studied for applications as oxygen-conducting solid electrolytes. It has been established that the lower valency cations giving the highest ionic conductivities are Y^{3+} (YSZ) and Sc^{3+} (ScSZ), with the composition containing 8 mol% Y_2O_3 ($x = 0.15$) being the most widely studied. Consequently, most mechanical property data are for this composition. (Zirconia with lower substitution levels than are required to stabilize the cubic phase at room temperature has some remarkable and technologically useful mechanical properties, such as high strength, high work of fracture (toughness) and super-plastic creep. These materials have been reported on extensively (e.g., [1]) and are not considered here.)

The elastic constants of cubic zirconia are shown in Table 1 and compared with those of some other fluorite oxides. It can be seen that all the oxides have very similar elastic parameters at room temperature with a Young's modulus of approximately 200 GPa and Poisson's ratio close to 0.3.

Table 1. Room temperature elastic constants of fluorite oxides (measured using IET method unless indicated otherwise)

	YSZ(x= 0.15) [7]	HfO_2	CeO_2 [22]	ThO_2 [23]	UO_2 [24]	PuO_2 [25]
C_{11}	402	547*	403	367	390	
C_{12}	95	98*	105	106	116	
C_{44}	56	81*	60	79	62	
E (polycrystal)	233	245 ($x = 0.33$, Y stabilized) [26]	249	256	205	268
ν	0.31	0.27	0.30	0.28	0.31	0.28

* Atomistic simulation [27], extrapolated to zero dopant

The temperature dependence of Young's modulus of yttria-stabilised cubic zirconia (YSZ) is shown in Figure 1 together with that of $GdO_{1.5}$-doped ceria discussed later. Two different researchers [5 and 6] report very similar results using the acoustic resonance experimental method (IET). There is a strong decrease in Young's modulus with increasing temperature to a minimum value at approximately 500°C, followed by a slow increase as the temperature rises to 1000°C. The minimum in E was observed to coincide with a maximum in the internal friction losses (out-of phase component of the elastic response). It was suggested that this behaviour is due to a transition from strongly bound defect clusters such as $\left(Y'_{Zr}\ddot{V}_O Y'_{Zr}\right)^{\times}$ (in Kröger-Vink notation) at low temperatures, to unassociated single defects at higher temperatures. The loss process would correspond to cluster defects changing orientation in the stress field due to limited defect mobility at intermediate temperatures.

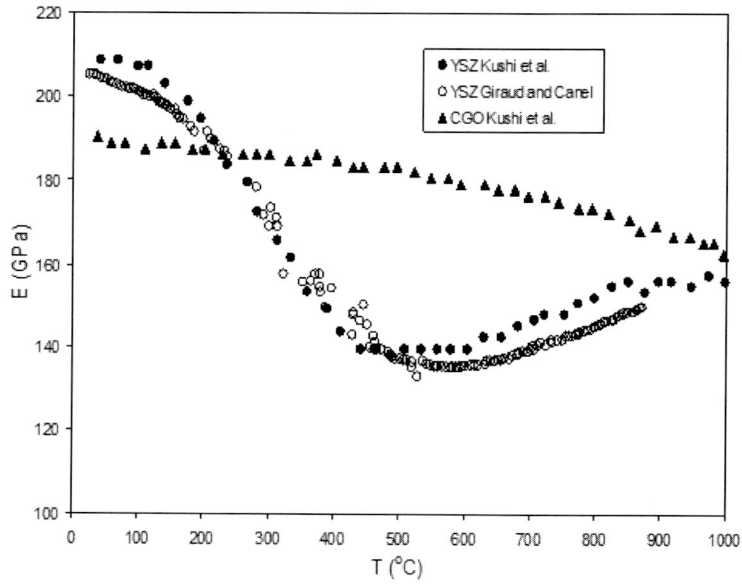

Figure 1. Young's modulus as a function of temperature measured by IET for YSZ (x =0.14) [6], YSZ (x = 0.15) [5] and CGO (x = 0.10) [6].

Kandil et al. [7] measured the bulk modulus of YSZ as a function of yttria content. They report that the bulk modulus increased as the yttria content increased. This is a rather surprising result as it might be expected that the increased concentration of oxygen vacancies resulting from the increased yttrium substitution would tend to lower the average bond strength and hence lower the bulk modulus.

3.2. Fracture and Plasticity

The plastic yield stress is reflected in the hardness values given in Table 2. The data obtained on single crystals (last row) show a very large reduction in hardness as the temperature is raised from room temperature to 600°C. This is because it is much easier for dislocations to move and multiply as the temperature is raised. Nevertheless, the implied yield stress is still high (approximately 1 GPa) as a result of the difficulty in dislocation motion even at this elevated temperature. (This is because the hardness (3 GPa) is approximately 3 times the yield stress (in section 2.2) which makes the yield stress 1 GPa.)

Some measurements of fracture toughness (also expressed as work of fracture) are summarised in Table 2. In general measurements made by observing cracks induced by indentation are less reliable than other methods, because the short crack length is difficult to measure accurately and the cracks often do not propagate far enough to sample the microstructure. The most thorough investigation was carried out by Kumar and Sorensen [8] on bulk specimens of YSZ ($x = 0.15$) using the double cantilever beam method which enabled them to identify three critical energy release rates (fracture energies). These were for crack initiation (3.5 – 4 J m^{-2}); for crack arrest (2.8 J m^{-2}); and for fast fracture (6 – 10 J m^{-2}).

Table 2. Toughness and hardness of yttria-stabilised zirconia ($x = 0.15$)

Material	Method	K_{Ic} (MPa $m^{0.5}$)	Hardness (GPa)	Work of fracture (J m^{-2})
Tape cast sheet	Pre-crack extension[28]	1.3		
Electrolyte film	Indentation [28]	1.25		
Sintered ceramic	Double cantilever beam [29]	1.3±0.3		8±2
Tape cast sheet	Double torsion [9]	1.61±0.12 at 25°C 1.02±0.05 at 900°C		12±1 at 25°C 6.7±0.3 at 900°C
Tape cast sheet	Indentation [9]	1.75±0.18	14.3±0.8	
8YSZ	Nano indentation [30]	1.79±0.11	14.2±0.7	
Single crystal	Indentation [31]		15±0.7 at 25°C 3.0±0.7 at 600°C	

Selcuk and Atkinson [9] used the double torsion method on thin sheets of the same composition and measured a fracture energy of 12 J m^{-2} at room temperature. This value is slightly higher than the fast fracture value of Kumar and Sorensen, but the two are in reasonable agreement given that the materials were from different sources and that the double torsion test does not load the crack tip in pure mode I. Selcuk and Atkinson also measured the fracture energy at 900°C and found a lower value of 6.7 J m^{-2}. Thus there is no evidence that additional plastic flow contributes significantly to the toughness of this material at this temperature and that weaker bonding (as reflected in the lower value of Young's modulus) dominates. Values of toughness measured using indentation (Table 2) are more scattered and tend to be higher, reflecting the relatively poor accuracy of this technique and the tendency to underestimate the measured crack length.

Equation 3 indicates that, provided that the distribution of flaws initiating fracture is constant, then the strength should be proportional to the toughness. Thus the strength of YSZ ($x = 0.15$) at 900°C and 950°C [10] was observed to be lower than at room temperature as expected from the lower fracture toughness at 900°C [9]. The fracture was observed to be trans-granular at room temperature, but inter-granular at high temperature (950°C).

YSZ ($x = 0.15$) has been observed to exhibit slow crack growth at room temperature and in laboratory air. The experimental results from double torsion experiments [11] and double cantilever beam experiments [8] are shown in Figure 2.

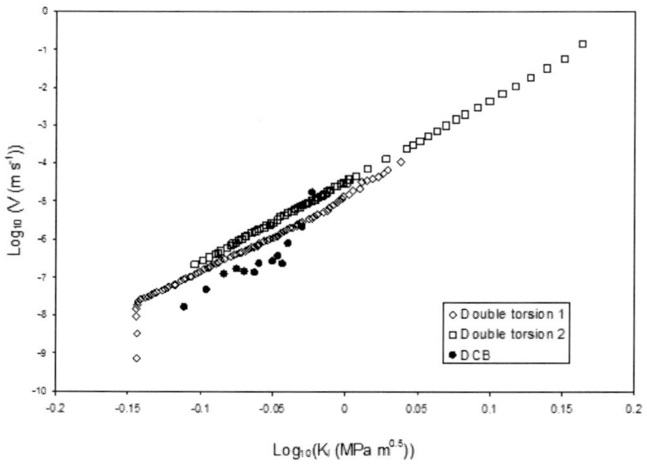

Figure 2. Sub-critical crack growth in YSZ ($x = 0.15$) at room temperature. Double torsion results are for two different specimens [9] and DCB results are by double cantilever beam method [8].

The two sets of experimental data are in good quantitative agreement with the exponent n approximately equal to 20 in equation (4). Lowrie and Rawlings [10] also found evidence of slow crack growth at room temperature by observing strength increasing with loading rate according to:

$$\sigma_f = A\dot{\sigma}^{\frac{1}{1+n}} \tag{6}$$

with $n = 11$. Although this is smaller than in the direct experiments, it is acceptable given the experimental uncertainties. Lowrie and Rawlings also found evidence of slow crack growth at 950°C, but it was not quantifiable due to creep acting simultaneously.

3.3. Creep

Creep of YSZ single crystals containing 9.4 mol% Y_2O_3 ($x = 0.17$) has been studied by Martinez-Fernandez et al. [2]. They found that after an initial transient period (strain < 5%) the creep rate at constant stress was independent of time (steady state creep). For temperatures above 1450°C the stress exponent was 4.5 and the activation energy was 6.1 eV. They concluded that creep was controlled by dislocation climb. At lower temperatures they found larger values of both n and Q, which led them to conclude that dislocation glide was also involved. They compared their single crystal behaviour with diffusion data and creep of polycrystalline YSZ at similar temperatures. In Figure 3 the creep rates of cubic zirconia single crystals and polycrystals are compared at a stress of 100 MPa based on these experiments. The polycrystalline data have $n = 1$, $m = 2$ and $Q = 5.6$ eV, indicating Nabarro-Herring creep controlled by lattice diffusion. The diffusion data give a best estimate for the lattice diffusion coefficient of the cations to be described by a pre-exponential factor $D_0 = 1\times10^{-3}$ m^2 s^{-1} and activation energy Q = 5 eV, in acceptable agreement with the polycrystalline creep data.

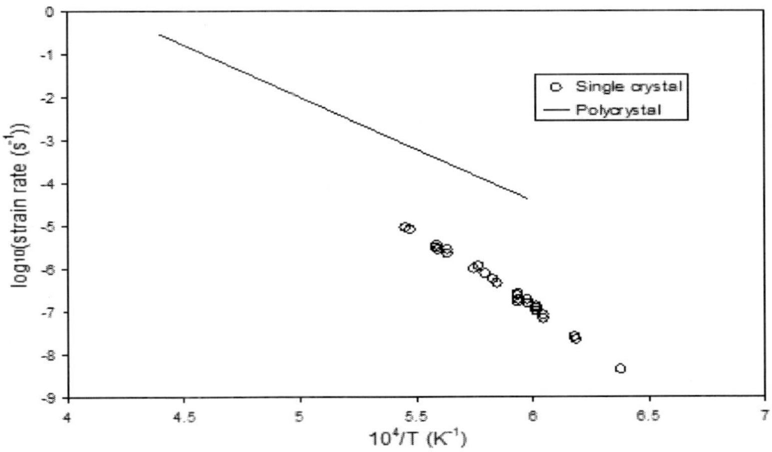

Figure 3. Comparison, at 100 MPa applied stress, between creep of single crystal cubic zirconia (YSZ, $x = 0.17$) [2] and polycrystalline cubic zirconia (YSZ, $x = 0.4$, and 5 μm grain size) [32].

Lowrie and Rawlings [10] also found linear viscous creep in polycrystalline YSZ (x = 0.15), but with a lower activation energy of 350 kJ mol^{-1} (3.3.eV). However, the temperature range was both low and narrow (925-1000°C) in their experiments and so the difference in activation energy might not be significant.

4. Ceria

Ceria differs from zirconia in two major respects. First, it exists in the cubic fluorite structure even when undoped; and second, the Ce^{4+} can be partially reduced to Ce^{3+} under appropriate conditions (high temperature and low oxygen activity). Nevertheless, when used as an oxygen ion conductor the Ce is also partially substituted by trivalent cations to generate oxygen vacancies in a similar way to zirconia. The partial reduction process, however, generates extra oxygen vacancies, and electron hopping between the two Ce valence states gives rise to electronic conductivity in addition to the oxygen ionic conductivity. As a result, the oxygen content, 2-δ, depends not only on substitution, x, but also on temperature, T, and oxygen activity, a_{O2}.

4.1. Elasticity

Kushi et al. [6] measured Young's modulus, E, as a function of temperature in air for CGO (x = 0.10) using the acoustic resonance method (IET) and their results are shown in Figure 1 compared with the results for zirconia. Although the room temperature and high temperature values are similar to zirconia, the temperature dependence is different in that the dip in Young's modulus seen for YSZ at intermediate temperatures is not seen in CGO. This is surprising because the defects in both doped oxides are expected to be very similar. If the intermediate temperature dip (and associated internal friction) in YSZ is due to defect re-configuration (re-orientation), then this should also be evident in CGO. Amezawa et al. [12] measured *E*, using the acoustic resonance method, as a function of oxygen activity for CGO (x = 0.10) and the results are shown in Figure 4, expressing *E* as a function of oxygen deficiency. They indicate that Young's modulus decreases strongly as the oxygen vacancy concentration increases due to reduction. Sato et al. [13] measured *E* in ceria, using a load-displacement method, as a function of dopant content, x, for dopants Y, Gd and Sm. The results are shown in Figure 5 and indicate similar behaviour for all three dopants, with a shallow minimum at x ≈0.2. Their values of Young's modulus are significantly lower than those reported by Amezawa et al. which could be due to errors in the displacement measurements. However, the different results could also be due to an extra contribution to strain from defect reconfiguration in the load-displacement measurements. Sato et al. [14] also measured *E* in CYO at 600°C as a function of Y content using the IET method. They found a similar dependence on Y content to that reported using load-displacement at room temperature (Figure 6). This relative insensitivity to dopant concentration, and associated oxygen vacancy concentration, was supported by results of atomistic simulation. This does not seem consistent with the sensitivity of *E* to oxygen vacancy concentration introduced by reduction. Therefore, there is apparently conflicting evidence regarding the effect of oxygen deficiency on Young's modulus of ceria. Lubomirsky and co-workers have reported many unusual phenomena related to the apparently elastic properties of doped ceria and these are summarised in Wachtel and Lubomirsky [15]. They propose that the unit cell strains, as a result of applied stress, have additional contributions due to re-configuration of point defects in the strain field which are time dependent. Hence the stress-strain response would be different for different measurement methods as discussed earlier. If this is the case for ceria, then similar behaviour would be expected for zirconia, but this does not appear to be the case. Thus it must be concluded that there is much still to be understood concerning the elastic behaviour of these oxides.

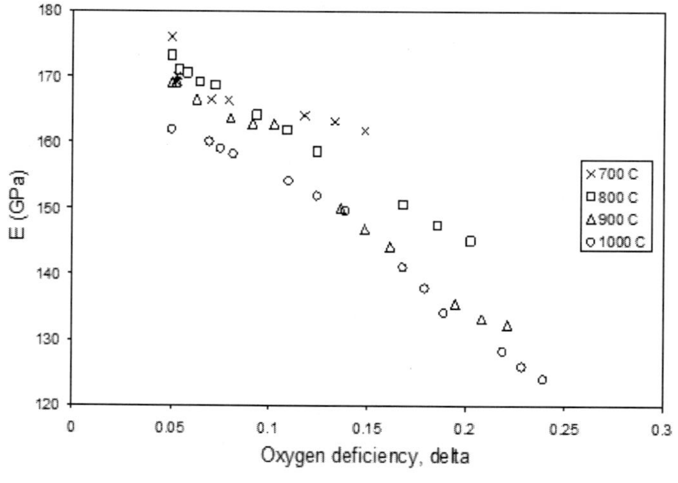

Figure 4. Young's modulus of $Ce_{0.9}Gd_{0.1}O_{2-\delta}$ measured as a function of the oxygen nonstoichiometry using the IET method [12].

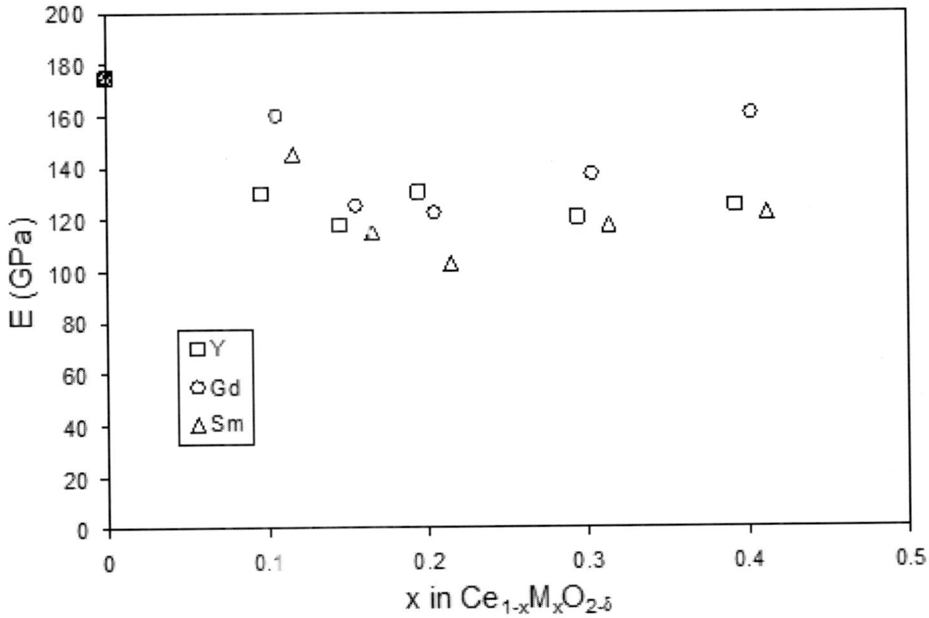

Figure 5. Young's modulus of doped ceria (M = Y, Gd, Sm) measured as a function of dopant concentration at room temperature using the load-displacement method [14].

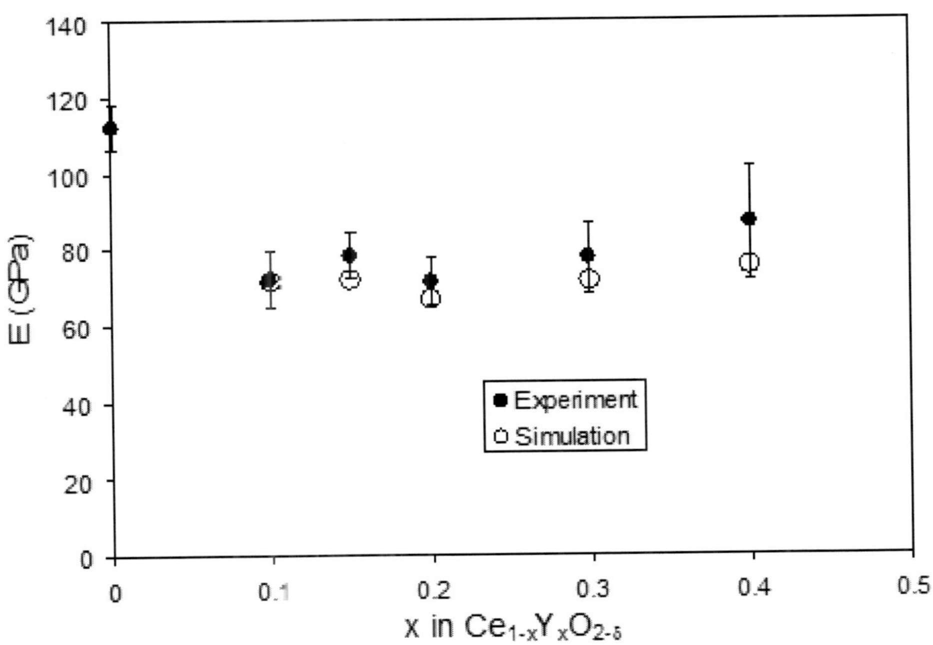

Figure 6. Effect of Y substitution level on the Young's modulus of ceria at 600°C measured using the IET method [14] and compared with atomistic simulations. The simulation results have been "normalised" to the experimental result at $x = 0.1$.

4.2. Fracture and Plasticity

Experimental data for K_{Ic} and hardness for doped ceria are summarised in Table 3. Some of the specimens were also doped with Co (typically 2 cation %) which is a commonly used sintering aid for ceria. All the toughness measurements in Table 3 were made by the indentation method and therefore the cautionary comments mentioned above for zirconia also apply here. Within these uncertainties, the fracture toughness of ceria is seen to be very similar to that of cubic zirconia. The value reported by Shemilt et al. [16] on SDC (Sc-doped ceria) is particularly high, perhaps due to underestimation of the crack length in this case.

Table 3. Toughness and hardness of doped ceria

Material	Method	K_{Ic} (MPa m$^{0.5}$)	Hardness (GPa)
10CGO+2Co	Nano indentation [30]	1.47±0.12	11.8±0.5
15CGO+2Co	Nano indentation [30]	1.39±0.11	11.1±0.3
20CGO+2Co	Nano indentation [30]	1.27±0.13	10.5±0.4
20CGO	Indentation [17]	1.53±0.2	9.0±0.7
10SDCand 20 SDC	Indentation [16]	2.4±0.3	8.7±0.3

Zhang et al. [17] report that the toughness has no dependence on grain size and that the fracture is trans-granular (as observed for cubic zirconia at room temperature). Recently Sun et al. [18] have reported molecular dynamics simulation of crack propagation in undoped and Gd-doped ceria (10% Gd_2O_3 corresponding to $x = 0.18$) crystals of different orientation. At a simulated temperature of 10 K the fracture toughness of the undoped ceria was calculated to be approximately 1.3 MPa m$^{0.5}$ (depending on crystal orientation), which is of similar magnitude to the room temperature experimental values measured for doped ceria discussed above. The initial crack was observed to propagate by a mixture of cleavage and slip, depending on crystal orientation, but not on temperature. Interestingly, the Gd doping suppressed slip and cleavage was dominant in all orientations. Literature hardness values are summarised in Table 3 and are somewhat lower than for YSZ thus indicating a lower yield stress in ceria.

4.3. Creep

Routbort et al. [19] measured creep in fine-grained CGO ($x = 0.10$) and found a stress exponent $n = 1$ and an activation energyof 480 kJ mol^{-1} (5.0 eV). These values are consistent with diffusion control (e.g., Coble or Nabarro-Herring creep) and similar to those discussed earlier for YSZ. They found no dependence on oxygen activity in the range $1 - 10^{-4}$, which is to be expected as this is not low enough to generate additional oxygen vacancies and thereby influence cation point defect concentrations which will control the creep rate. Wolfenstine [20] has compared the creep data for doped ceria and zirconia and shown that they are not only similar in magnitude, but also have a similar dependence on stress and temperature. In both materials creep is likely to be controlled by diffusion of the quadri-valent cation.

More recently Iguchi et al. [21] studied creep of Y-doped ceria ceramics in the temperature range 1050 – 1125°C and stresses 20 – 50 MPa. The stress exponent was $n = 1$, in agreement with other researchers, and the grain size exponent $m = 2$ (Equation 5) which they claim indicates control by lattice diffusion (Nabarro-Herring creep). However they report a low activation energy of 264 kJ mol^{-1} (2.7 eV) which is significantly lower than reported by others and too low for lattice diffusion. The creep rate decreased as the yttrium content increased and, for the composition with $x = 0.20$ at 1100°C, shows a maximum at an oxygen activity of 10^{-14} and then decreases on further reduction. This complex behaviour is difficult to understand at the present time.

CONCLUSION

All the fluorite-structured oxides have similar elastic properties at room temperature. However, there are interesting differences in elastic behaviour between zirconia and doped ceria as the temperature is raised. These effects are not understood, but are probably related to subtle differences in the way in which defect clusters respond to the applied stress in the two materials. In ceria, this also results in different measurement methods leading to different apparent elastic modulus values, depending on the timescale of the stress-strain response. Similarly, the effects of doping and oxygen non-stoichiometry on the elastic behaviour of these oxides are not fully understood as yet.

The hardness of cubic zirconia is slightly higher than that of doped ceria, but their fracture toughness (work of fracture) is similar. Both decrease significantly as the temperature increases. Cubic zirconia exhibits sub-critical crack growth at room temperature, but there is no reliable evidence for this at elevated temperature or in ceria.

Creep is controlled by diffusion of the cations (probably through the bulk lattice at the temperatures studied) with activation energy of approximately 5 eV. However, there are still significant discrepancies in reported data that require further clarification.

REFERENCES

[1] Dominguez-Rodriguez, A., Cheong, D.-S., Heuer, A.H., *Phil. Mag. A, Condensed Matter* (1991) *64*, 923-929.

[2] Martinez-Fernandez, J., Jimenez-Melendo, M., Dominguez-Rodriguez, A., Heuer, A.H., *J. Am. Ceram. Soc.* (1990) *73*, 2452-2456.

[3] Ashby, M.F., Verrall, R.A., *Acta Metallurgica* (1973) 21, 149-163.

[4] Basu, B., *Int. Mater. Rev.* (2005) 50, 239-256.

[5] Giraud, S., Canel, J.R.M., *J. Eur. Ceram. Soc.* (2008) 28, 77-83.

[6] Kushi, T., Sato, K., Unemoto, A., Hashimoto, S., Amezawa, K., Kawada, T., *J. Power Sources* (2011) 196, 7989-7993.

[7] Kandil, H.M., Greiner, J.D., Smith, J.F., *J. Am. Ceram. Soc.* (1984) 67, 341-346.

[8] Kumar, A.N., Sorensen, B.F., *J. Am. Ceram. Soc.* (2000) 83, 1199-1206.

[9] Selçuk, A., Atkinson, A., *J. Am. Ceram. Soc.* (2000) 83, 2029-2035.

[10] Lowrie, F.L., Rawlings, R.D., *J. Eur. Ceram. Soc.* (2000) 20, 751-760.

[11] Atkinson, A., Selçuk, A., *Solid State Ionics* (2000) 134, 59-66.

[12] Amezawa, K., Kushi, T., Sato, K., Unemoto, A., Hashimoto, S.-i., Kawada, T., *Solid State Ionics* (2011) 198, 32-38.

[13] Sato, K., Yugami, H., Hashida, T., *J. Mater. Sci.* (2004) 39, 5765 – 5770.

[14] Sato, K., Suzuki, K., Yashiro, K., Kawada, T., Yugami, H., Hashida, T., Atkinson, A., Mizusaki, J., *Solid State Ionics* (2009) 180, 1220-1225.

[15] Wachtel, E., Lubomirsky, I., *Scripta Materialia* (2011) 65, 112-117.

[16] Shemilt, J.E., Williams, H.M., Edirisinghe, M.J., Evans, J.R.G., Ralph, B.B.U., *Scr. Mater.* (1997) 36, 929-934.

[17] Zhang, T.S., Ma, J., Kong, L.B., Hing, P., Kilner, J.A., *Solid State Ionics* (2004) 167, 191-196.

[18] Sun, Y., Chen, Y., Liu, Y., Kang, G., *Computational Materials Science* (2012) 51, 181-193.

[19] Routbort, J.L., Goretta, K.C., de Arellano-López, A.R., Wolfenstine, J., *Scr. Mater.* (1997) 38, 315-320.

[20] Wolfenstine, J., *J. Power Sources* (2002) 111, 173-175.

[21] Iguchi, F., Endo, Y., Ishida, T., Yokobori, T., Yugami, H., Otake, T., Kawada, T., Mizusaki, J., *Solid State Ionics* (2005) 176, 641-644.

[22] Nakajima, A., Yoshihara, A., Ishigame, M., *Phys. Rev. B* (1994) 50, 13297-13307.

[23] Macedo, P.M., Capps, W., Wachtman, J.O., *J. Am. Ceram. Soc.* (1964) 47, 651.

[24] Landolt-Borstein Numerical Data and Functional Relationships in Science and Technology, Group III: Crystal and Solid State Physics, Vol. 11, Ed. K.-H. Hellwege, Springer-Verlag, Berlin 1979.

[25] Nakamura, H., Machida, M., Kato, M., *Phys. Rev. B* (2010) 82, 155131-155136.

[26] Dole, S.L., Hunter, O., Calderwood, F.W., *J. Am. Ceram. Soc.* (1980) 63, 136-139.

[27] Terki, R., Feraoun, H., Bertrand, G., Aourag, H., *Computational Materials Science (Proceedings of the E-MRS 2004 Spring Meeting; Symposium H: Atomic Materials Design: Modelling and Characterization)* (2005) 33, 44-52.

[28] Steinbrech, R.W., in *Surface Engineering and Functional Materials Proc. EUROMAT '97,* Vo. l 3; Netherlands Society for Materials Science, Zwijndrecht (1997) 495-498.

[29] Kumar, A.N., Sorensen, B.F., *Mater. Sci. Eng. A* (2002) 333, 380-389.

[30] Morales, M., Roa, J.J., Capdevila, X.G., Segarra, M., Piñol, S., *Acta Mater.* (2010) 58, 2504-2509 (and corrigendum: *Acta Mater.* (2011) 59, 1318).

[31] Morscher, G.N., Pirouz, P., Heuer, A.H., *J. Am. Ceram. Soc.* (1991) 74, 491-500.

[32] Dimos, D., Kohlstedt, D.L., *J. Am. Ceram. Soc.* (1987) 70, 531-536.

In: New Research Trends of Fluorite-Based Oxide Materials ISBN:978-1-63117-350-9
Editors: Akio Nakamura and Junichiro Mizusaki

Chapter 7

GRAIN-BOUNDARY STRUCTURES OF STABILIZED ZIRCONIA

Naoya Shibata and Yuichi Ikuhara*
Institute of Engineering Innovation, School of Engineering,
The University of Tokyo,Bunkyo, Tokyo, Japan

ABSTRACT

Systematic grain-boundary study in yttria-stabilized cubic zirconia using bicrystals is reviewed. It was found that the grain-boundary properties such as structures, energies and solute segregation behavior are strongly related to the grain-boundary geometries and grain-boundary planes. Microscopically, these relations can be explained by the local atomic-scale structures at grain boundaries. It was shown that the coordination-deficient cation sites in the grain-boundary core regions are the key to understand the grain-boundary properties. Thus, it is essential to understand the atomic-scale structures in order to understand the grain-boundary properties in zirconia ceramics.

1. INTRODUCTION

Zirconia (ZrO_2) ceramics exhibit several unique mechanical and electrical properties which can be utilized in various applications such as high strength materials, oxygen sensors, solid electrolyte for fuel cells, thermal barrier coating materials, and so on [1-4]. Zirconia ceramics are frequently used in the form of polycrystals which contain a number of grain boundaries as well as other structural and chemical imperfections. It has been reported that the presence of these defects, especially the grain boundaries, influences the macroscopic properties of zirconia ceramics [5-17]. For understanding the effects of these lattice defects, atomic-scale analysis is of great importance.

In pure ZrO_2, a cubic fluorite structure (c-ZrO_2) is only stable at high temperature above about 2650K [18]. For the practical use of zirconia, high-temperature cubic and/or tetragonal

* E-mail address: Yuichi Ikuhara [ikuhara@sigma.t.u-tokyo.ac.jp].

phases need to be stabilized to room temperatures because of their high performances in the materials properties. To stabilize such high-temperature phases, certain amount of stabilizer oxides such as Y_2O_3, MgO, CeO_2 *etc.* are generally added to zirconia. Therefore, zirconia is always used in a multi-component system, and the presence of such solute ions is one intrinsic feature of zirconia ceramics. Therefore, the effect of grain-boundary segregations, as well as grain-boundary structures, must be taken into account when we discuss the properties of grain boundaries in zirconia ceramics. In addition to the solute ion segregation, impurity segregation is often pointed out to be crucial for the functional properties such as oxygen ion conductivity [19,20]. In the actual materials, both types of grain-boundary segregations are supposed to simultaneously occur. Although many experimental studies have been reported as for grain-boundary segregation in zirconia ceramics, most of them have mainly focussed on the surfaces of intergranular fractured materials [21-23], which will average out the character of many grain boundaries involved there. The analyses using transmission electron microscopy (TEM) or scanning transmission electron microscopy (STEM) have been performed for the polycrystalline samples [17, 24]. These analyses themselves are very powerful to clarify the relationship between the atomic structure and segregation behaviors, however, the limitation arises from their polycrystalline nature containing many kinds of different boundaries; only the boundaries which satisfy the edge-on condition can be analyzed by high-resolution transmission electron microscopy (HRTEM).

On the other hand, theoretical approach using the space charge theory has been developed [24,25]. This theory well predicted the general features of grain-boundary segregation in zirconia, but the treatment of the grain boundaries is rather averaged and ambiguous due to the lack of knowledge on the origin of the space charge formation. One of the earliest theoretical models ascribed the origin of the space charge in zirconia grain boundary to the large interaction energies of the oxygen vacancies at the grain-boundary core regions [26]. The space charge theory may be further developed if the correlation between the grain boundary atomistic structures and space charge formations are truly understood.

Recently, detailed structure analyses of zirconia grain boundaries using bicrystals are becoming possible [27-33]. Bicrystal experiments have the possibility to shed a new light on the fundamental issues on grain boundaries in zirconia ceramics, because we can arbitrarily control the grain boundary geometries, and hence grain-boundary atomistic structures. In this chapter, we will review the systematic grain-boundary study of yttria stabilized ziroconia using bicrystals [33]. The grain-boundary properties, such as structures, energies and solute segregation behaviours are systematically investigated by controlling grain-boundary characters. Combined with the HRTEM and theoretical atomistic calculations, this chapter aims to clarify the general atomic-scale rules for the formation of grain boundaries in zirconia ceramics and the relationships between their structures and properties.

2. Zirconia Bicrystal Fabrication

The yttria stabilized cubic zirconia bicrystals were fabricated as follows. Commercially available high-purity ZrO_2-9.6mol%Y_2O_3 single crystals (Dowa Mining Co., Ltd.), which contain certain amounts of Y_2O_3 as solute ions, were used for preparing the bicrystals. At this composition, the single crystals were well stabilized to have the cubic fluorite structure at room temperature. The schematic illustrations of the bicrystals fabricated in this study are shown in Figure 1.

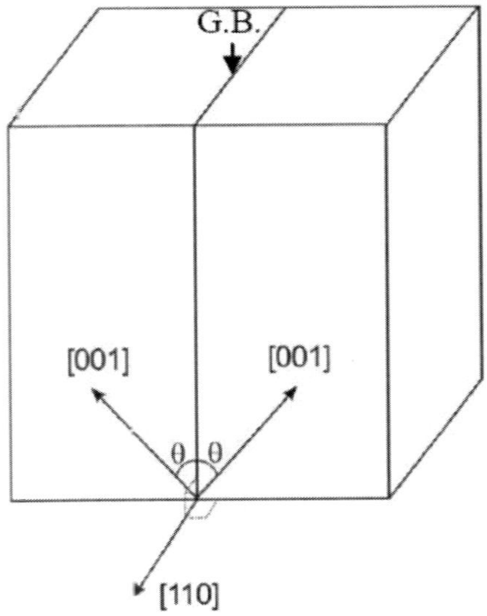

Figure 1. Schematic illustration of the bicrystal fabricated in this study. The misorientation angle, 2θ, between the [001] directions of each single crystal is systematically changed as listed in Table 1.

Symmetric tilt grain boundaries with the rotation axis of [110] were selected in this study. Here, many types of grain boundaries are covered; low-angle boundaries, high-angle random boundaries, and low Σ boundaries, according to the coincidence site lattice (CSL) theory [34]. The concept of CSL is based on the fact that certain rotations about an axis bring a lattice into partial self-coincidence. The common lattice sites (c.l.s.) then form a larger lattice known as a coincident site lattice (CSL). The CSL can also be considered as a lattice of coincidence sites in the composite lattice obtained by interpenetration of two single lattices. The points of overlap correspond to CSL points. Like any other lattice, a CSL can be defined in terms of a unit cell, the volume of which is proportional to the density of lattice sites; the higher the c.l.s. density is, the smaller the volume of the CSL unit cell becomes. Conventionally, the volume of the CSL unit cell is normalized with respect to the volume of the lattice unit cell and the result, which is necessarily an integer, n, is denoted by $\Sigma = n$. Each Σ defines a particular orientation relationship between the two lattices; thus, for two lattices in parallel orientation, $\Sigma = 1$ and the c.l.s. density is a maximum. Thus, smaller Σ indicates higher coherency in grain-boundary orientation. The single crystals were precisely oriented using X-ray diffraction (XRD), and cut so as to have the respective orientation relationship listed in Table 1. The size of each single crystal was set to be $5 \times 10 \times 10$ mm^3, so as to obtain 10 mm cube bicrystals. The flat surfaces of the single crystals were set parallel to the grain boundary planes and successively polished with 9, 3-6 and 1 μm diamond suspensions to achieve a mirror finish. The surfaces were then mechano-chemically finished using colloidal silica to obtain atomically flat surfaces. These two single crystals were subsequently joined at 1873K for 15h in air. The typical view of the bicrystals is shown in Figure 2.

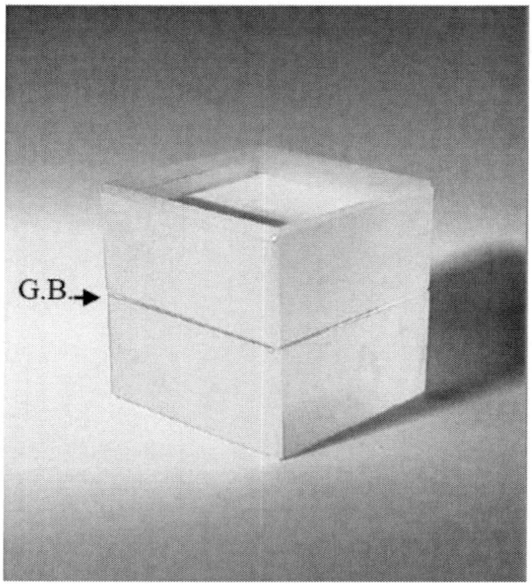

Figure 2. Typical view of a bicrystal fabricated in the present study.

Table 1. The list of tilt angles, grain-boundary types and grain-boundary planes of the bicrystals

2θ (degree)	Type	G.B. plane
5.0	Low-angle	~{110}
20.0	Random	
39.0	Σ=9	{221}
50.4	Σ=11	{332}
60.0	Random	
70.6	Σ=3	{111}
109.4	Σ=3	{112}
129.6	Σ=11	{113}
141.0	Σ=9	{114}

3. Grain-Boundary Energies

The grain-boundary energies of the bicrystals were evaluated by the thermal grooving technique [35]. The surfaces of cut plates from the bicrystals were mechanically polished with diamond suspension to achieve a mirror state. Then, the specimens were annealed at 1673K for 2h in air to groove the grain-boundary regions. The groove profiles were then measured directly from the surface images obtained by the atomic force microscopy (AFM). According to the thermal grooving theory, three forces are balanced at the grooves as schematically shown in Figure 3. The following equation expresses the force balance at a boundary groove;

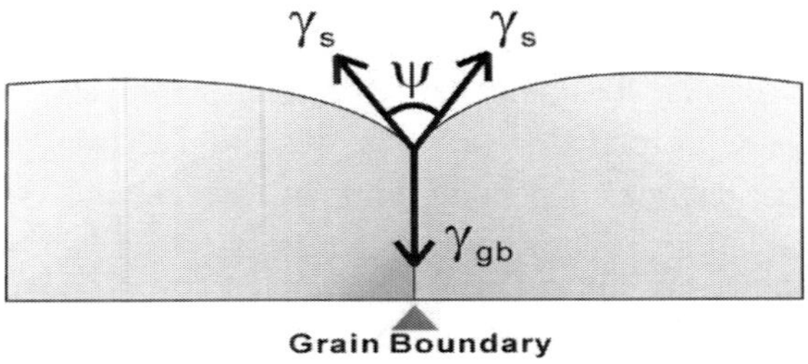

Figure 3. Schematic illustration of the ideal force balance at a grain boundary groove.

$$2\gamma_{gb} = 2\gamma_s \cos(\psi/2)$$

where γ_{gb}, γ_s and ψ are the grain-boundary energy, the surface energy and the dihedral angle of the groove, respectively. The grain boundary energies were estimated by substituting the measured dihedral angles obtained from the groove profiles in this study. The average surface energy of polycrystalline ZrO_2-8mol%Y_2O_3 (cubic) at the same temperature (1673K) was used for γ_s, i.e., ~1.21 J/m [36].

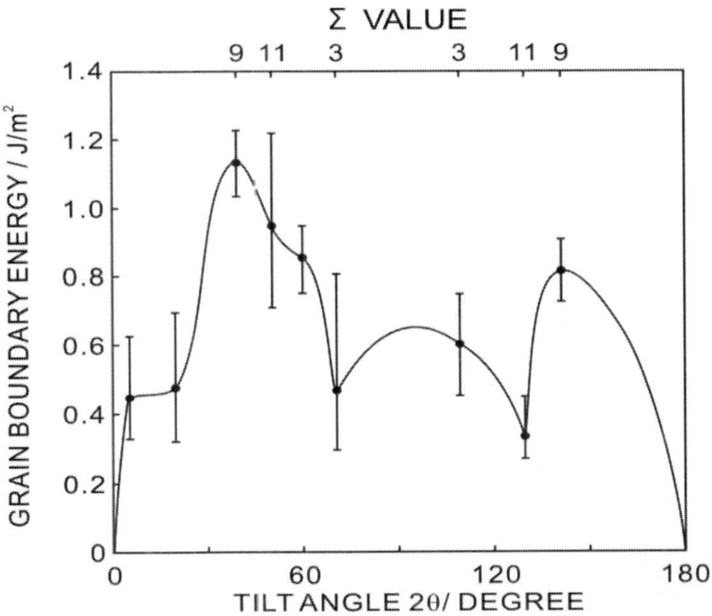

Figure 4. Experimental grain-boundary energy vs. misorientation angle plots for yttria-stabilized zirconia.

Figure 4 shows the grain-boundary energy vs. misorientation plots for the series of [110] tilt grain boundaries. The horizontal axes correspond to the misorientation angle, 2θ, and the vertical axes correspond to the grain-boundary excess energies estimated by the thermal grooving method. It is seen that the grain-boundary energies are strongly dependent on the misorientation angles, and there are two apparent energy cusps. Those two cusps are appeared at 2θ = 70.6°, Σ = 3, {111} boundary and 2θ = 129.6°, Σ = 11, {113} boundary, which showed comparable grain-boundary energies. It is also noticed that the grain boundaries with the same Σ values but different grain-boundary planes have different grain-boundary energies.

In addition, the lower angle $\Sigma = 9$ and $\Sigma = 11$ grain boundaries showed higher grain-boundary energies compared with others. It should be noted that, in polycrystalline materials, many orientation relationships are present. However, the present results suggest that individual GBs should have very different energies depending on their own characters. In theory, the system should prefer more stable, low-energy GBs, but because of the restraint of the grain orientations and depending on the process conditions, many types of grain-boundary orientations can be formed in reality.

4. Grain-Boundary Structures

Figure 5 shows a series of HRTEM images of the grain boundary regions in the bicrystals. The observations were made from the common [110] directions to set all the grain boundaries in the edge-on conditions. The misorientation angles of 2θ were systematically changed from the left hand side of the image to the right hand side. There were no secondary phases such as amorphous phases along the grain boundaries, and the both crystals were well bonded in an atomic level to form the grain boundaries. It is noted that the grain-boundary structures are strongly dependent on the grain boundary geometries.

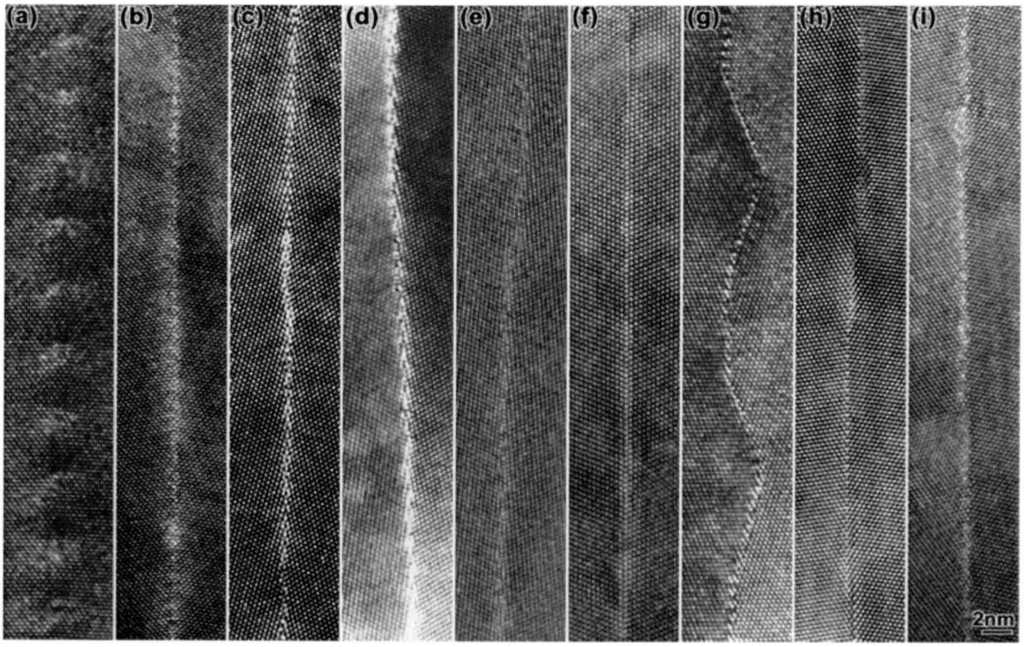

Figure 5. HRTEM images of the bicrystals fabricated in this study. The misorientation angle 2θ is systematically changed from the left hand side as follows: (a) 5.0°, (b) 20.0°, (c) 39.0° $\Sigma = 9$, {221}, (d) 50.4 °$\Sigma = 11$, {332}, (e) 60.0°, (f) 70.6° $\Sigma = 3$, {111}, (g) 109.4° $\Sigma = 3$, {112}, (h) 129.6° $\Sigma = 11$, {113}, (i) 141.0° $\Sigma = 9$, {114}.

At the low-angle boundary region of 2θ = 5.0°, the grain boundary was composed of periodical array of dislocations with the Burgers vector of b = *a*/2<110>, where *a* corresponds to the lattice parameter of the present yttria-stabilized cubic zirconia [30]. The angle deviation from the perfect $\Sigma = 1$ orientation was thus compensated by the periodical introduction of dislocations. As the misorientation angle becomes larger than ~20°, the grain boundary structure can hardly be defined as the array of dislocations. In this case, the grain-boundary structures were rather considered to be the combination and array of locally relaxed structural segments, usually called the structure units. The concept of the structure units was firstly proposed in metal grain boundaries [37], but the applicability of this concept to other type of materials is becoming noticed. In the present high-angle grain boundaries, however, situations are rather complicated because pure zirconia is composed of two ions, Zr^{4+} and O^{2-}. Strictly speaking, the definition of the structure units in pure zirconia must be regarded as the smallest periodic structure segment in the periodic grain-boundary structure, including both cation and anion sublattices. However, it is very difficult to determine the structure of anion sublattice directly from the HRTEM images because the images only correspond to the position of cation sublattices in the present conditions, which was verified by the multi-slice image simulations as described in the subsequent sections. Therefore, in this chapter, the structure units will be simply defined by the cation sublattice. The approximation will be shown to be permissible, as compared with the theoretical calculations.

It is also noticed that there are some grain boundaries which showed very unique structural morphologies. The $2\theta = 109.4°$, $\Sigma = 3$, [110] / {112} grain boundary (g) showed facetted feature along the boundary, although the $2\theta = 70.6°$, $\Sigma = 3$, [110] / {111} boundary (f) showed very coherent structure in spite of the same Σ orientations. Moreover, the $2\theta = 129.6°$, $\Sigma = 11$, {113} grain boundary (h) showed periodical stepped structures along the boundary. The steps may be related to the introduction of displacement shift complete (DSC) dislocations for compensating the slight mistilt from the perfect $\Sigma = 11$ orientation. The grain-boundary structure in between these steps were different from that of $2\theta = 50.4°$, $\Sigma = 11$, [110] / {332} grain boundary (d) with the same geometrical orientation. It is also the case in $\Sigma = 9$ grain boundaries with different grain-boundary planes, {221} and {114}. Grain-boundary atomic structures thus strongly depend on their geometries, and especially the grain boundary planes. Furthermore, CSL theory should have important meanings to zirconia, because certain low Σ grain boundaries showed very unique structure characteristics. The detailed discussion on the atomic-scale grain-boundary structures will be made in the following sections.

5. Grain-Boundary Solute Segregation

The single crystals of cubic zirconia used for bicrystal fabrication contained certain amount of yttrium ions as solute, in order to stabilize high-temperature cubic phase at room temperatures. These solutes are expected to affect the grain-boundary properties by segregation, which will complicate the understanding of grain-boundary phenomena in zirconia. However, even in the practical usage, zirconia ceramics are always used by adding certain kinds and amount of oxides, such as Y_2O_3, MgO, CeO_2 *etc.*, to stabilize high temperature cubic and/or tetragonal phases. Therefore, solute segregation is the essential phenomenon for grain boundaries in zirconia ceramics.

Figure 6 shows the amount of yttrium segregation vs. misorientaion plot obtained by the systematic TEM - energy dispersive X-ray spectroscopy (EDS) analysis. The vertical axis indicates the normalized intensity ratio, YKα / ZrKα of the EDS counts, and the horizontal axis the misorientation angles around the [110] tilt axis. The errors inthe EDS counts are at the level of about 5 %. The intensity ratio corresponds to the relative content of yttrium ions at the grain-boundary cores. The black circles indicate the data obtained exactly from the grain-boundary cores, while white circles indicate those obtained from the grain interiors of the both-side crystals. The data of grain interiors correspond to the content of yttrium ions at the points about 200 nm away from the grain boundaries, and those data are likely to be almost constant for all the bicrystals fabricated in this study. It should be noted that these segregation data correspond to the amount of yttrium ions at the "core" of the respective grain boundary, and may be slightly different from the total amount of the segregated ions with some width along the boundary. However, these data should enable us to directly relate the amount of segregation to the grain-boundary core structures.

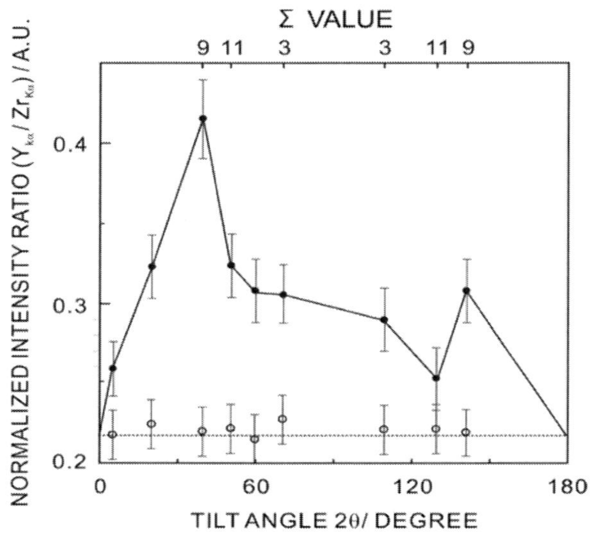

Figure 6. Experimental grain-boundary solute (Y) segregation vs. misorientation angle (2ϑ) plots for yttria-stabilized zirconia.

It is noticed that the contents of yttrium ions were always higher at the grain-boundary cores than at the grain interiors in all the grain boundaries analyzed here. Thus, the grain-boundary segregation of yttrium ions occurred in all the fabricated bicrystals. It is also pointed out that the amount of yttrium segregation drastically changes according to the misorientation angles, and the $2\theta = 39.0°$, $\Sigma = 9$, [110] / {221} grain boundary (Figure 5(c)) has the highest amount of yttrium segregation, while the small cusp of segregation is found at the $2\theta = 129.6°$, $\Sigma = 11$, [110]/{113} grain boundary. The grain-boundary segregation is also strongly dependent on the grain-boundary plane, because the grain boundaries with same Σ orientations but with different grain boundary planes showed different amount of yttrium segregation. This indicates that the grain-boundary segregation is closely related to the local atomistic structures, rather than the misorientation between the adjacent crystals in cubic zirconia. It should be noted that the amount of segregation at the $2\theta = 5.0°$ small-angle grain boundary corresponds to that of the grain-boundary dislocation cores. The nearly perfect crystal region in between the dislocations showed almost the same amount of yttrium content with that of the grain interiors. The yttrium segregation only took place at the grain boundary dislocation core in this case.

6. ATOMIC STRUCTURES OF THE SPECIAL GRAIN BOUNDARIES

The grain boundary energies and solute segregation behavior shown in the previous sections suggested that they were likely to be strongly dependent on the grain boundary atomistic structures. In this section, three high-angle grain boundaries with peculiar characteristics in their energies and segregation behavior were selected, i.e., $2\theta = 39.0°$, $\Sigma = 9$, {221} boundary, $2\theta = 70.6°$ $\Sigma = 3$, {111} boundary and $2\theta = 129.6°$, $\Sigma = 11$, {113} grain boundary (Figure 5c, f and h, respectively). The extraction of the factors responsible for the notable behaviors at these boundaries will be attempted through the systematic comparison of these three grain boundaries.

6.1. Σ = 3, {111} Grain Boundary (Figure 5f)

Figure 7 shows a magnified HRTEM image of the $\Sigma = 3$, {111} grain boundary. The incident beam direction was parallel to the [110] direction, and the boundary was set at the edge-on condition. This grain boundary showed a cusp in grain boundary energy vs. misorientation plot, as shown in the previous section (Figure 4). The inset in the HRTEM image is the diffraction pattern obtained from the grain boundary area, which confirms that the both crystals were well oriented to have a $\Sigma = 3$ relationship. The HRTEM image showed that the grain boundary has a coherent and flat structure on an atomic scale, with the boundary plane parallel to the {111} type planes of both grains. Besides the atomically coherent regions, periodical interfacial dislocations were also observed along the boundary by the preliminary conventional TEM observations. Since the grain boundary structure at the atomic level would be easily affected by very small misalignments during fabrication of the bicrystal, these additional defect structures must be caused by the accommodation of a slight deviation from the perfect $\Sigma = 3$ orientation, rather than the major intrinsic structure of the $\Sigma = 3$ grain boundary. In this section, however, we focus on the intrinsic structure of the $\Sigma = 3$, {111} boundary. According to the through-focus images combined with the HRTEM image simulations, the white dots in the HRTEM image correspond to the cation column along the [110] direction under this imaging condition. In cubic zirconia, the cation sublattice takes a face-centered cubic (fcc) structure, while oxygen sublattice adopts a simple cubic structure, and these sublattices can be separated as different columns along the [110] direction. Since contrasts associated with the oxygen columns are too weak to be visualized, the HRTEM image is related to the cation sublattice in cubic zirconia. It can be seen that the cation sites of both grains are nearly symmetrical around the grain boundary. The {111} planes maintained almost the same spacing across the grain boundary. The {111} type planes of cubic zirconia can be described as the stacking of one cation polar plane followed by two anion polar planes with different stacking sequences as schematically shown in Figure 8. From the experimental observations and electrical neutrality criterion, it is likely that such a stacking sequence continues across the grain boundary. However, the position of the oxygen sublattice cannot be confirmed solely from the HRTEM image. Therefore, combined with the above experimental results, the static lattice calculations were conducted to find the equilibrium structure of the $\Sigma = 3$, {111} grain boundary. Although the real $\Sigma = 3$, {111} grain boundary is accompanied by a certain amount of Y^{3+} segregation, the calculations were carried out for ideal pure cubic zirconia to obtain a clear picture of the atomic structure associated with the grain boundary geometry. It is generally accepted that the substitution of Zr^{4+} site by Y^{3+} causes the introduction of oxygen vacancies to preserve total charge neutrality. Y^{3+} segregation should therefore introduce oxygen vacancies around the boundary. However, the effects of oxygen vacancies on the boundary structure were not taken into account in the present calculation. Figure 9 shows the lowest energy structure for the $\Sigma = 3$, {111} boundary in pure cubic zirconia.

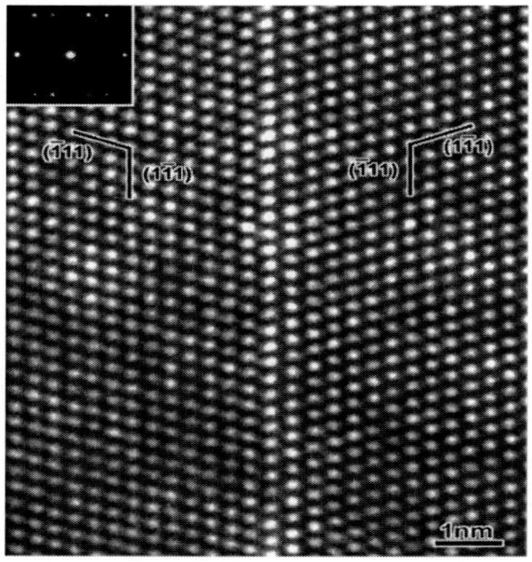

Figure 7. HRTEM image of $2\theta = 70.6\,°\Sigma = 3$, {111} symmetric tilt grain boundary viewed from the [110] direction. Inset is the diffraction pattern obtained from the boundary area.

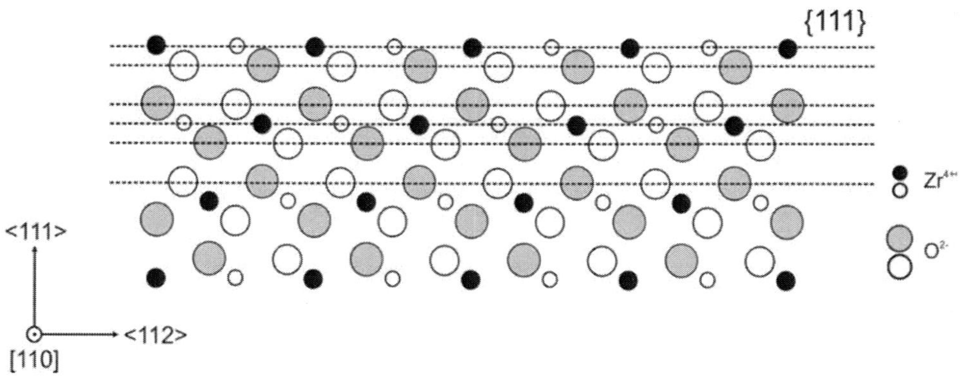

Figure 8. Schematic illustration of {111} plane stacking sequence in cubic zirconia viewed from [110] direction. Large and small circles correspond to O^{2-} and Zr^{4+} ions, and the ions denoted by open and filled circles are located on different layers along the [110] direction.

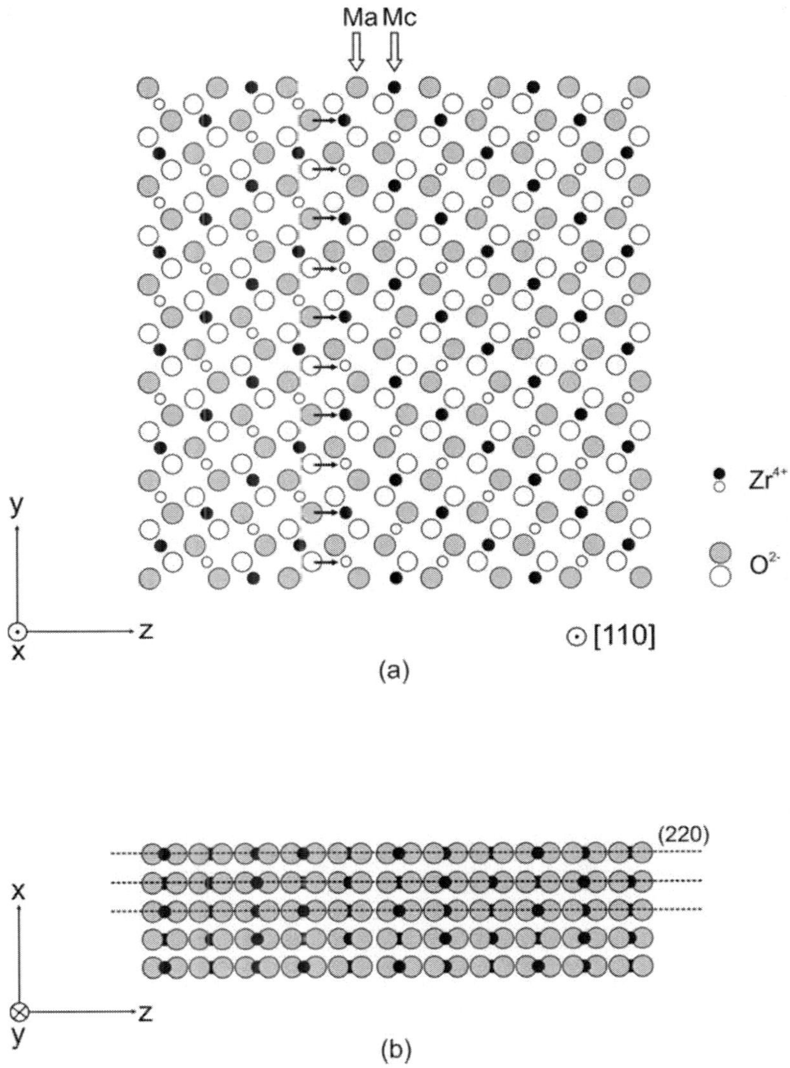

Figure 9. (a) The lowest energy structure of the $\Sigma = 3$, {111} grain boundary predicted by the systematic static lattice calculations. Mc and Ma indicate the mirror planes of cation and anion sublattices, respectively. Note that cation sites with seven-fold coordination are formed along the boundary, as indicated by small arrows. (b) The view of the structure from the *y* direction. The (220) lattices are continuously connected across the boundary.

Large and small circles correspond to oxygen and zirconium ions, respectively, and the ions denoted by open and filled circles are located on the different (220) layers parallel to [110] direction. The grain boundary excess energy was calculated to be 2.12 J/m^2. There are no other equilibrium structures that exhibit local minima in the boundary energy, although a wide variety of translation states were considered in the present calculation. This structure has no translation states parallel to the boundary plane, but a slight expansion of 0.02 nm occurs perpendicular to the boundary. Therefore, apart from the boundary expansion and the small atomic displacements, the cation sublattice is nearly mirror symmetric with respect to the cation mirror plane (Mc). The cation configuration is almost the same as that of the $\Sigma = 3$, {111} boundary structure in fcc metals [38], as already suggested from the present HRTEM image. On the other hand, the anion sublattice is also mirror symmetric with the anion mirror plane (Ma), which is located two layers apart from the Mc. This structure minimizes the oxygen-oxygen interactions across the grain boundary, as compared with a possible boundary structure with both sublattices having an identical mirror plane. This comparison is demonstrated in Figure 10. In the structure model (b) where both sublattices have an identical mirror plane, oxygen ions on the same layer are positioned very close to each other across the mirror plane. This should be energetically unfavorable, as also suggested by the present calculations. The result indicates that the oxygen-oxygen interactions may play a critical role on the final grain-boundary excess energies.

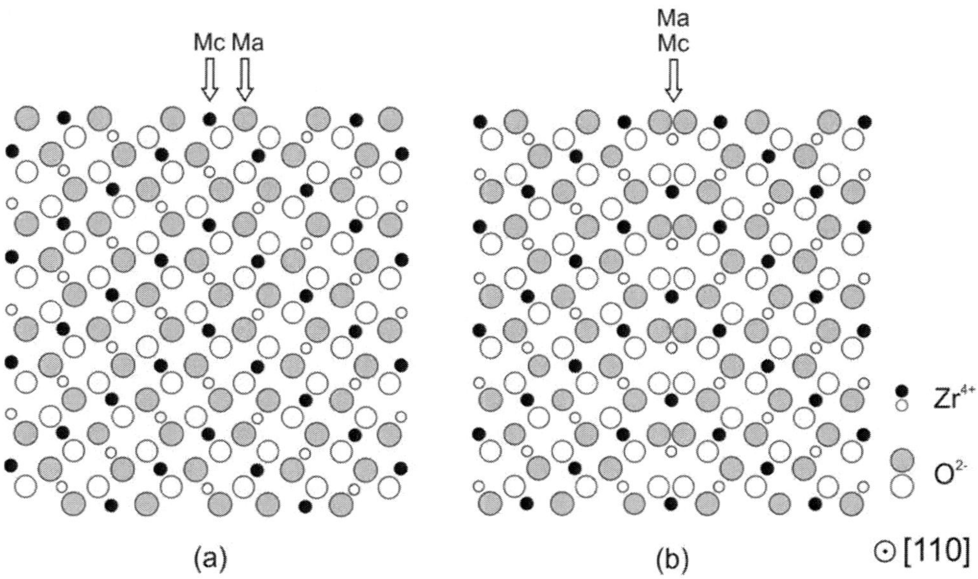

Figure 10. (a) The theoretically obtained grain-boundary structure model in which the cation and anion sublattices have the different mirror planes. (b) A possible grain-boundary model in which the cation and anion sublattices have the identical mirror plane. Note that the oxygen ions on the same layers are positioned too close to each other in the model (b).

To examine the (220) lattice continuity across the grain boundaries, the calculated model viewed from *y* direction is shown in Figure 9(b). It is obvious that the (220) lattices of the one side crystal are smoothly connected to those of the adjacent crystal without any distortions. The connection may be one of the characteristic features for the [110] tilt grain boundaries, as will be discussed later.

Figure 11 shows the comparison between the experimental HRTEM image and the simulated HRTEM image based on the boundary structure model in Figure 9. The simulated image is in good agreement with the experimental HRTEM image, indicating that the atomic structure of the yttrium-segregated boundary can be predicted using the pure cubic zirconia model approximation. The segregation of Y^{3+} ions and oxygen vacancies does not seem to cause dramatic change in the experimental image. Y^{3+} ions must segregate to the present grain boundary to substitute for the cation sites, as in the case of the formation of ZrO_2-Y_2O_3 solid solutions. It is also noticed that there are certain cation sites along the boundary at which the local coordination number is different from that of the grain interior, indicated by the small arrows in Figure 9(a). The cations occupy the whole sites in one {111} cation polar plane adjacent to the mirror plane of the cation sublattice, and are forced to have seven-fold coordination of oxygen ions owing to the restraint of the boundary atomic configuration, although the cations in the cubic fluorite structure have eight-fold coordination. It is noted that the coordination deficient cation sites still keep the cubic hexahedron coordination polyhedra as in the perfect cubic fluorite structure. The abrupt change in local coordination state at the boundary core may be related to the peculiar energy and segregation characteristics of the Σ = 3, {111} boundary in zirconia ceramics. It is quite possible that the local changes in coordination introduce the local strain and/or electrical charge around the boundary, and should enhance the boundary segregation. The effect of such sites will be discussed more quantitatively through the comparison with other grain boundaries.

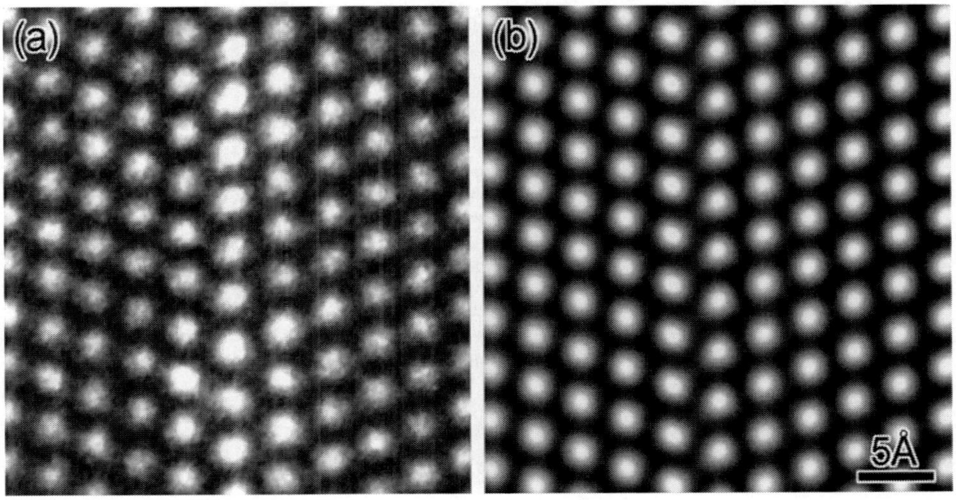

Figure 11. Comparison between (a) the experimental HRTEM image and (b) the simulated HRTEM image using the lowest-energy structure model shown in Figure 9. The image simulation was performed on the condition that a defocus value is -50nm and a film thickness is 4nm.

6.2. Σ = 11, {113} Grain Boundary (Figure 5(h))

Figure 12 shows a HRTEM image of the Σ = 11, {113} grain boundary. This image corresponds to the boundary regions in between the boundary steps observed in the Figure 5(h). Regarding these steps as secondary structures for the perfect Σ = 11, {113} grain boundary, the intrinsic structure of the boundary must be represented in the image. It is clear that the grain boundary structure can be described by the periodical array of structure units as superimposed by the solid quadrilateral prisms. Since the white dots of the HRTEM image correspond to the cation sublattice as in the Σ = 3, {111} boundary, the cation frame structure of the present boundary should, at least, have the mirror symmetric structure in the [110] projection.

To theoretically predict the stable atomic structure of the Σ = 11, {113} grain boundary including oxygen sublattice, systematic static lattice calculations were also performed here. Figure 13 is the lowest energy grain-boundary structure predicted by the static lattice calculations. The calculated grain boundary excess energy was 1.78 J/m^2, which is lower than that of the Σ = 3, {111} grain boundary. Large and small circles correspond to oxygen and zirconium ions, respectively, and the ions denoted by open and filled circles are located on the different (220) layers perpendicular to [110] direction. The structure of cation sublattice is almost mirror symmetric across the grain boundary plane.

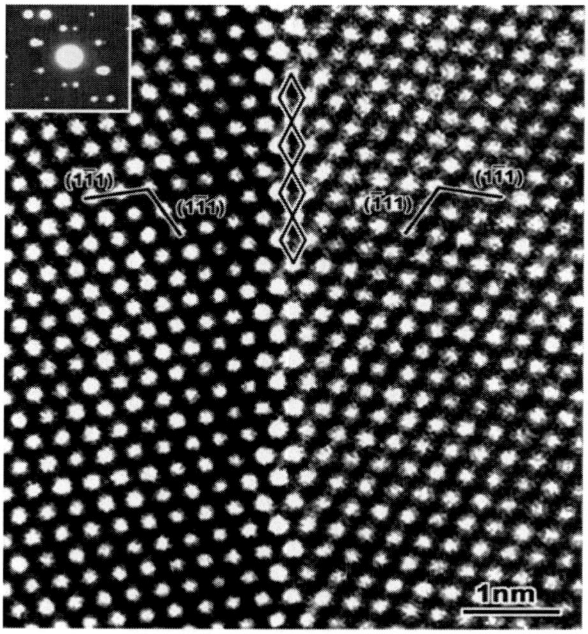

Figure 12. HRTEM image of the $\Sigma = 11$, {113} grain boundary. Inset is the diffraction pattern obtained from the boundary area. Solid lines indicate the periodical array of structure units along the grain boundary.

This structure has no translation states parallel to the boundary plane as in the $\Sigma = 3$, {111} grain boundary, but showed slight expansion of 0.04 nm perpendicular to the boundary plane. The calculated structure also showed the periodical array of quadrilateral prism structure units in cation sublattice along the grain boundary, as superimposed by the solid lines. Similar to the $\Sigma = 3$, {111} grain boundary, it is obvious that the (220) lattices of the one side crystal are smoothly connected to those of the adjacent crystal without any distortion in the lowest energy structure as shown in (b). The (220) lattice continuity may be one of the characteristic features of the low energy structures in the [110] axis tilt boundaries.

The comparison between the experimental and simulated HRTEM images based on the predicted structure model is shown in Figure 14. The calculated model well simulates the experimental image, indicating the validity of the theoretically obtained model. Therefore, it may be concluded that the atomic structure of the present yttrium segregated $\Sigma = 11$, {113} boundary can also be predicted using the pure cubic zirconia model approximation, and the segregation of Y^{3+} ions and oxygen vacancies does not seem to largely affect the grain boundary structure, such as creating new atomic columns and/or causing rearrangement of grain-boundary atomic structure. Therefore, the segregation of Y^{3+} ions to the present boundary probably takes place in the substituting manner for the cation sites around the grain boundary.

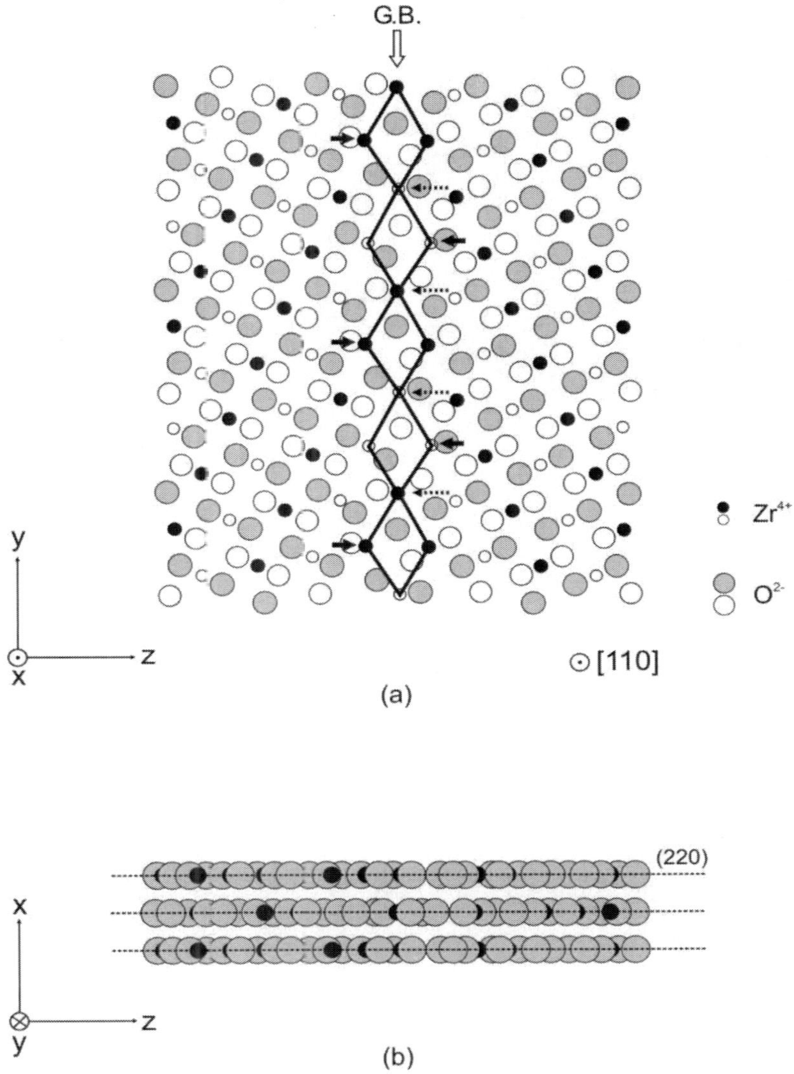

Figure 13. (a) The lowest energy structure of the Σ = 11, {113} grain boundary obtained by the systematic static lattice calculaticns. Note :hat cation sites with six-fold coordination are formed along the boundary, as indicated by the small solid arrows. There are also another peculiar cation sites with eight-fold coordination, but with different :oordination polyhedra, as indicated by the small dashed arrows. (b) The view of the structure from the *y* direction. The (220) lattices are continuously connected across the boundary.

As indicated by the solid arrows in Figure 13(a), there are also certain cation sites at which the local coordination number is different from that of the grain interior. These cation sites have six-fold coordination of oxygen ions, almost keeping cubic hexahedron coordination polyhedra. It is :nteresting to point out that the formation of such coordination deficient cation sites is simi:ar to the Σ = 3, {111} boundary, although the coordination numbers were different in ea:ch case. It should be pointed out that there were also another peculiar cation sites with al:nost eight-fold coordination, but with different coordination polyhedra, as indicted by the dashed arrows.

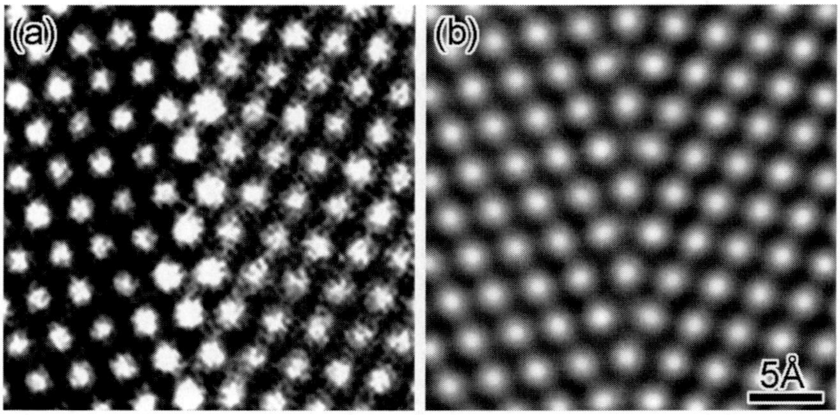

Figure 14. Comparison between (a) the experimental HRTEM image and (b) the simulated HRTEM image using the lowest-energy structure model shown in Figure 13. The image simulation was performed on the condition that the defocus value is -65nm and a film thickness is 5nm.

6.3. Σ =9, {221} Grain Boundary (Figure 5c)

Figure 15 shows the HRTEM image of the Σ = 9, {221} grain boundary. The inset in the HRTEM image is the diffraction pattern obtained from the grain-boundary area including both adjacent crystals. The pattern clearly shows that the both crystals were well oriented to have a Σ = 9 orientation relationship. It is noted that the grain-boundary structure appears to be very narrow in width, and consist of periodical array of structure units. There is no widely "disordered" or "rearranged" region along the boundary, and the peculiar characteristics in the energy and segregation (both very high) should be related to the grain boundary core structures.

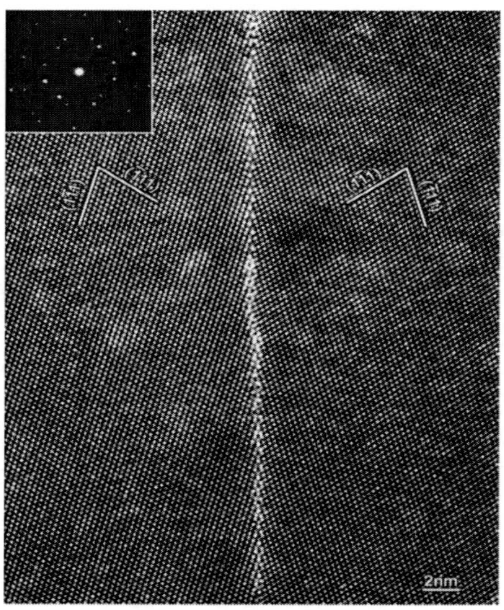

Figure 15. HRTEM image of Σ = 9, [110] / {221} symmetric tilt grain boundary viewed from the [110] direction. Inset is the diffraction pattern obtained from the boundary area.

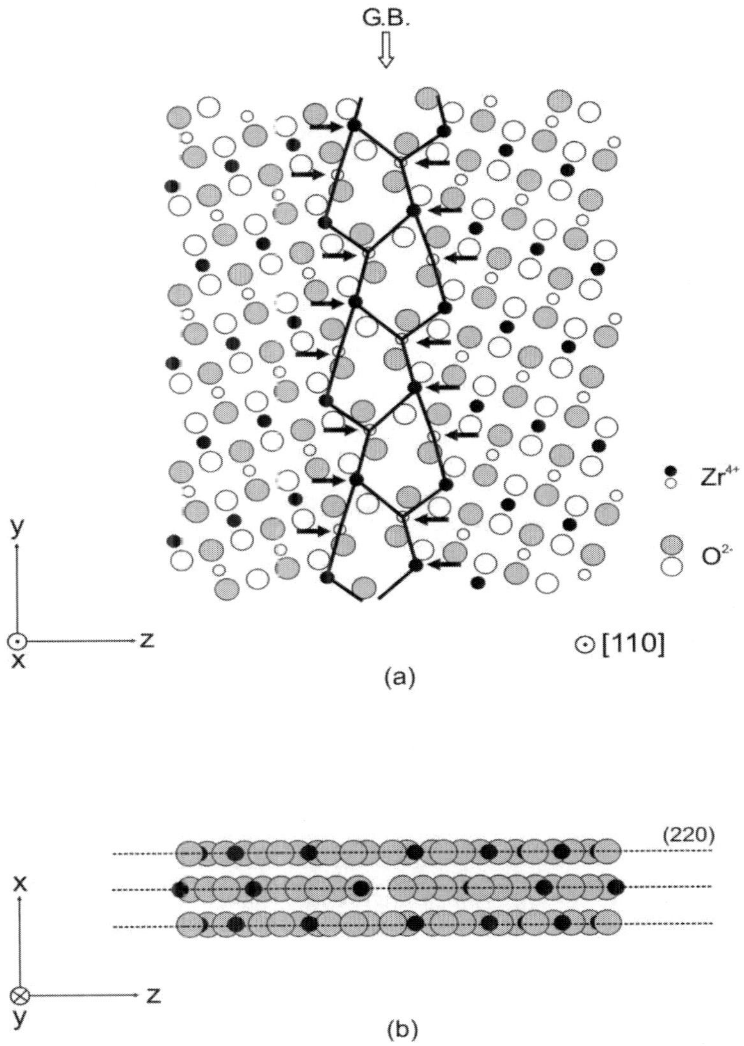

Figure 16. (a) The lowest energy structure of the Σ = 9, {221} grain boundary obtained by the systematic static lattice calculations. Note that cation sites with seven-fold coordination are densely formed along the boundary, as indicated by the small solid arrows. (b) The view of the structure from the *y* direction. The (220) lattices are continuously connected even in this highly relaxed boundary.

To theoretically predict the stable atomic structure, we again performed the systematic lattice statics calculations based on the charge neutrality assumption. Figure 16 shows the lowest energy grain-boundary structure predicted by the static lattice calculations. The grain-boundary excess energy was calculated to be 3.01 J/m^2, which is much higher than the previous two grain boundaries. It was found that there were several structures with local energy minimum in the course of the systematic translation. However, all of these boundary structures had similar cation sublattice structures to the lowest energy structure, accompanied with slight difference in oxygen sublattice structure. The lowest energy grain-boundary structure had a rather large boundary expansion of ~0.1 nm, resulting in somehow free volumes at the core of the boundary. It is also obvious in this boundary that the (220) lattices of the one side crystal are smoothly connected to those of the adjacent crystal without any distortion as shown in (b).

The open-spaced structures at the grain-boundary core may arise channeling contrast in the HRTEM image where there are no atomic columns. The imaging of the HRTEM image was made on the condition that the cation sites are imaged by white dot contrast, according to the image simulation combined with the through focus imaging. However, different from the direct imaging systems such as high-angle annular dark-field scanning transmission electron microscopy (HAADF-STEM) [39], it becomes rather ambiguous to determine the atomic position of the grain boundary directly from the HRTEM images if non-atomic columns arise channeling contrasts in addition to the real atomic columns. To avoid such ambiguity, the atomic-resolution high-voltage electron microscopy (ARHVEM) with a sufficiently thin sample was used to directly determine the atomic column positions of cation sublattice at the present grain boundary.

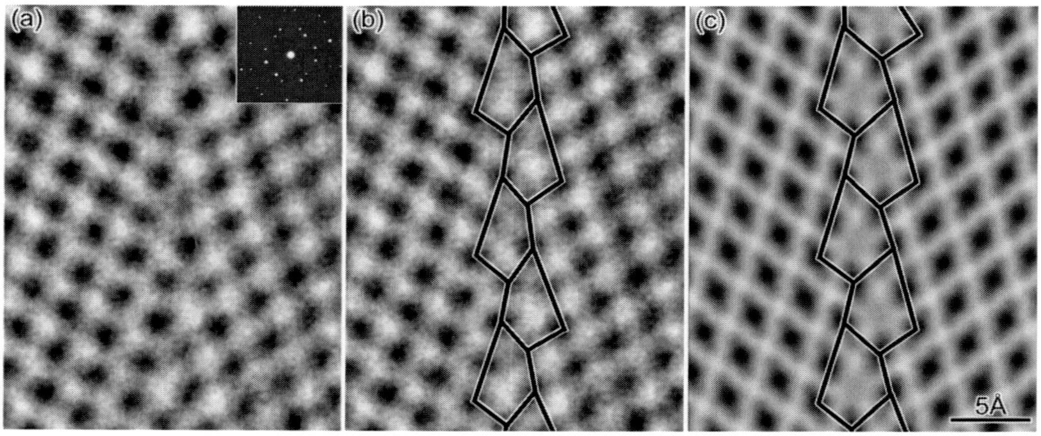

Figure 17. (a) The experimental ARHVEM image of the $\Sigma = 9$, {221} grain boundary. Inset is the diffraction pattern obtained from the boundary area. In this condition, the black dots in the image correspond to the position of the cation column along the [110] direction. (b) A periodic array of asymmetric structure units can be found along the boundary, indicated by the solid lines. (c) The simulated ARHVEM image using the lowest-energy structure model shown in Figure 16. The image simulation was performed on the condition that the defocus value is -38nm and a film thickness is 4nm.

Figure 17(a) shows the ARHVEM image of the $\Sigma = 9$, {221} grain boundary. In this case, the image was taken under near Scherzer defocus of about -38nm, resulting in the atomic columns imaged as black dots reflecting their potentials. The black dots were imaged as somehow elongated in the [001] directions because of the presence of anion sublattice very close to the cation sites in this direction. The cation sublattice shows the periodical array of the asymmetrical structure units at the grain boundary, as indicated by the solid lines in the image in (b). The open spaced cation sublattice structure is directly observed in this micrograph. Figure 17(c) shows the calculated ARHVEM image based on the predicted model as shown in Figure 16. The simulated image roughly agreed with the experimental image as for the periodical array of asymmetrical structure units. The yttrium ions should segregate to the boundary in the substituting manner, as is the case in the above-mentioned two boundaries. However, the calculated image shows relatively expanded feature at each structure unit than the experimental image. The difference may be derived from the slight difference in the boundary expansion. In the real grain boundary, the intense segregation of yttrium ions is present, and it should cause the introduction of oxygen vacancies to compensate the charge differences associated with the substitution of Y^{3+} at the Zr^{4+} site. The accumulation of oxygen vacancies to the grain boundary core region accompanied by the segregation of yttrium ions has been experimentally suggested in other type of grain boundary in yttria-stabilized cubic zirconia [40]. The introduction of oxygen vacancies may relax the interactions between the neighboring oxygen ions inside the structure units. If the repulsion between the neighboring ions in the structure units is effectively reduced, it may be possible that the structure units slightly shrink with the cation frame structures being preserved.

Similar to the previous two low-energy grain boundaries, the calculated grain-boundary structure possesses the cations sites with different coordination state along the grain boundary, indicated by the solid arrows in Figure 16(a). These sites have seven-fold coordination of oxygen ions, and almost keep the cubic hexahedron polyhedra. The density of the coordination deficient cation sites appears to be much higher than that of the other two grain boundaries.

7. Validity of the Pure Cubic Zirconia Approximation

The present results showed that the grain boundary stable atomistic structures in yttria-stabilized cubic zirconia can be theoretically predicted by using empirical potentials assuming the pure cubic zirconia at 0K. The coincidence between the experiments and calculations may indicate that the framework grain-boundary structures in stabilized cubic zirconia are mainly dominated by the crystal nature of the cubic fluorite symmetry, and not so much affected by the solute segregation and/or vacancy formation. The solute segregation of Y^{3+} ions to the grain boundary should just increase the Y^{3+} ions content of the cation columns around the boundary, but not to create new boundary atomic columns which were absent in the pure model. However, the slight change in the boundary expansion seemed to take place because of the difference in ionic radius and/or in effective charge from the host cation accompanying the introduction of oxygen vacancies. Basically, the grain boundary solute segregation in zirconia is considered to occur in the substitutional manner as in the bulk lattice.

It has been reported that the further addition of Y^{3+} ions to the cubic stabilized zirconia forms the ordered defected fluorite structures [41]. The report suggests that further substitution of Y^{3+} ions to the cation sites should not destroy the cubic fluorite symmetry to a large extent. On the other hand, if the amount of Y^{3+} ions decreases to less than ~8mol% Y_2O_3, the cubic fluorite structure will transform into a tetragonally distorted form. In the present study, the segregation always took place in the direction to increase the amount of Y^{3+} ions around the boundary, and this may be one of the reasons why the cation frame structures of the grain boundary did not seem to be so much affected by the Y^{3+} segregations. The validity of the pure cubic zirconia approximation is thus considered to be related to the full stabilization of cubic phase at this composition, and structural stability of cation sublattice toward the increase in Y^{3+} content.

The validity of the present approximation enables a correlation to be made between the atomic structure and other factors of grain boundaries in zirconia ceramics. In real alloyed materials, the structure formation and solute segregation at grain boundaries are very difficult to separate, and they are treated as concurrent phenomena. We can only experimentally obtain the grain-boundary features for the specimens after the Y^{3+} segregation. However, using the present approximation, structural heterogeneity in a boundary of a given misorientation can be extracted from the computer simulation study. Such structural heterogeneity must also be significant features of grain boundaries in the yttria-stabilized cubic zirconia as long as the present approximation holds. The following sections will show the strong correlation between the structural heterogeneity in the pure zirconia model and the experimentally obtained properties of yttria-stabilized cubic zirconia grain boundaries.

8. Formation of the Coordination-Deficient Cation Sites

The atomistic calculations clearly showed that the relaxed grain-boundary structures tend to possess cation sites with coordination deficiencies. This is in completely different manner from the grain boundaries in covalent materials, as reported in Si, Ge and SiC [42]. The tendency of creating miscoordination site at the core of the boundaries is rather similar to fcc metal grain boundaries with metallic bonds [43]. In the case of covalent material grain boundaries, structure relaxation occurs in the manner not to form coordination deficient sites with dangling bonds. This is related to high excess energies of the dangling bonds in covalently bonded materials [44]. So, the difference in structural relaxation mode must be closely related to the difference in the nature of chemical bond; ionic bond is mainly based on the electrostatic Coulombic force which is a kind of central forces. It is interesting to point out that the coordination deficient sites still keep the cubic coordination polyhedra after the effective atomistic structural relaxations. This tendency may indicate that grain-boundary structures are relaxed so as to retain the bond length between the respective ions similar to those in grain interiors.

It should be mentioned that these coordination deficient sites are different from the sites which have point defects (vacancies) in their nearest-neighbor coordination sites. In the atomistic simulation, the calculation cells were constructed assuming the total charge neutrality inside the cell, and the presence of point defects were not taken into account in this case. Nevertheless, the grain-boundary structures showed the relaxation to create such sites, depending on the restraint of initial boundary atomistic configuration. Since the calculated structure models were seen from the [110] direction, these miscoordinated sites were periodically aligned along the direction, forming rather ordered features. Basically, these sites are structurally peculiar sites and the grain-boundary core frame structures can be characterized by the accumulation of these coordination deficient cation sites.

To evaluate the effect of the coordination deficient cation sites, the densities of the sites are estimated from the calculated grain boundary models. We assume that each grain boundary can be defined as one atomic plane without any extension to the neighboring planes, and therefore, the densities are simply estimated by dividing the number of the coordination deficient cation sites around the boundary by the boundary area. Thus, we calculate the densities as if the coordination deficient sites were projected on to one grain-boundary atomic plane. In fact, every coordination deficient site of the respective boundary is positioned within very narrow region less than 1nm perpendicular to the boundary planes. The widths are also effectively narrow compared with the resolution of the present experimental microanalyses. In this estimation, the bond distortions, and the difference in coordination numbers are not taken into account for the first approximation. The allowance of the bond lengths between Zr^{4+} ion sites and coordinating O^{2-}ion sites are assumed within the deviation of 25 % from those in grain interior. It has been reported theoretically that the apparent bond break occurred at the point of about 20% deviation from the bulk bond length in SiC grain boundary [45] (Kohyama 1999). Although the bonding nature is different, we refer to this report, and the 25% criterion is adopted to cover all the critical bond distances around the 20%. The 25% bonding length is effectively shorter than the bond length between the Zr^{4+} site and the second nearest neighbor O^{2-} sites in the grain interior (90% deviation). Table 2 shows the estimated densities of the miscoordinated cation sites per unit boundary area for the respective grain boundaries. The densities are not simply as a function of tilt angles or Σ values. It is noted that the density is much higher in the Σ = 9, {221} boundary than the other two grain boundaries. In the following, we will compare the densities of the miscoordinated cation sites with grain-boundary properties.

Table 2. Densities of the coordination deficient sites for the three high-angle grain boundaries

G. B. type	Density (Å^{-2})
Σ = 3, {111}	0.089
Σ = 11, {113}	0.046
Σ = 9, {221}	0.179

9. The Coordination-Deficient Cation Sites vs. Grain-Boundary Energy

The experimentally and theoretically obtained grain-boundary energies are listed in Table 3. The calculated energies are overestimated than the experimental values. The overestimation of the energy may be partly attributed to the neglection of entropy effect for the present high-temperature grooving condition. In addition, the presence of yttrium solutes and oxygen vacancies can affect the boundary energy in real materials. However, the calculated grain-boundary energies can be used for relatively evaluating the energy difference between different grain boundaries, since the calculations were made with the same potentials and almost the same calculation cell sizes. It is interesting to point out that the calculation results show the same tendency to the experimental results as for the relative orders in grain-boundary energies, although the experimental energies included the effect of grain boundary solute segregation.

This may suggest that the internal energy has the dominant effect on the total grain boundary energy, which must be the strong function of the grain-boundary atomistic configuration. Fig, 18(a) shows the relationship between the calculated grain-boundary energies and the density of coordination deficient cation sites at the respective grain boundaries. There is a clear relation between those values; i.e., the grain-boundary energies increase as the densities of the coordination-deficientcation sites increase. Similarly, the relationship between the experimental excess energies and the density of the coordination deficient sites at respective grain boundaries is plotted in Figure 18(b). Even in the experimental data, similar correlation can be found. This result indicates that the grain-boundary energies are closely related to the coordination-deficient cation sites at the grain-boundary cores.

Table 3. Comparison between experimental and theoretically-calculated grain boundary energy

G. B. type	Av. G.B. energy (exp.) [J/m^2]	G.B. energy (calc.) [J/m^2]
Σ = 3, {111}	0.47	2.12
Σ = 11, {113}	0.33	1.78
Σ = 9, {221}	1.13	3.00

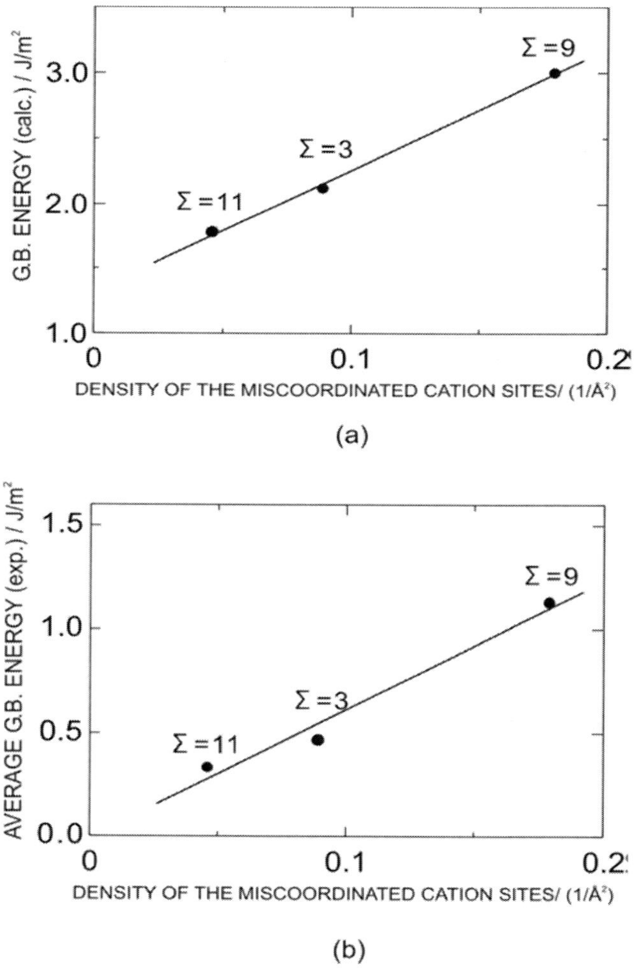

Figure 18. The relationships between the densities of the coordination deficient cation sites and (a) theoretically and (b) experimentally obtained grain-boundary energies.

It may be puzzling that if the coordination deficient cation sites are the major origin of the grain-boundary excess energy, why the grain-boundary relaxation took place in the manner to create such sites? After the detailed analysis of the grain-boundary structures, it was found that the grain boundaries seem to relax in order to retain the interatomic distances similar to the grain interiors in spite of creating the coordination deficient cation sites. This means that the creation of the coordination deficient cation sites less increases the boundary energy than the structural relaxation to preserve the coordination state as in covalent materials, although certain amount of energy increase would be unavoidable.

Here, one may recall that similar correlation between grain-boundary energies and local atomistic bonding environments at the boundary cores has been suggested in grain boundaries of fcc metals. The model, so-called 'broken-bond model', proposed by Wolf [43], showed close correlation between miscoordinations at grain boundaries and their energies mostly at high-angle grain boundary regions. In this model, the number of broken nearest-neighbor bonds per unit area was extracted as the important parameter for determining the grain-boundary energy. However, in this stage, the applicability seems to be limited to the mono-atom metal systems which have only one species of atomic bond. In the case of ionic crystals such as zirconia, there are more than one bond species, i.e., Zr-Zr, Zr-O and O-O, and there are long-range ionic interactions. The broken-bond model could then become much more complex, and may need several assumptions. The present model of extracting the coordination-deficient cation sites at the boundary core region seems to be, as the first approximation, successful for relating the structures and energies in zirconia ceramics. In the present study, we ignored the difference in coordination numbers of the miscoordinated sites in the estimation of the site densities. Although the density of the coordination deficient cation sites appeared to have more significant effect than the coordination number of each miscoordinated sites in this case, the inclusion of such effect must be needed for the further precise quantifications as in the broken-bond model.

10. The Coordination-Deficient Cation Sites vs. Grain-Boundary Solute Segregation

Figure 19 shows the relationship between the amount of yttrium segregation (experimental) and the density of the coordination-deficient cation sites at the respective grain boundaries. It is noticed that there is also a clear correlation between these values; i.e., the grain-boundary segregation increases as the density of coordination deficient cation sites increases.

Y^{3+} segregation at the grain-boundary core in yttria-stabilized zirconias has been ascribed to the grain-boundary excess free volumes [46]. This suggestion seems to be mainly based on the strain relaxation argument that the yttrium ions have the larger ionic radius than the zirconium ions. However, this explanation seems not to simply apply to the present cases, especially in the case of the $\Sigma = 3$, {111} grain boundary. This boundary has very coherent and closely packed structure without having apparent excess free volume. The amount of grain-boundary expansion is the smallest in the $\Sigma = 3$, {111} grain boundary, although the amount of yttrium segregation was higher than the $\Sigma = 11$, {113} grain boundary. Thus, the coherent $\Sigma = 3$, {111} grain boundary exhibited exceptional behavior in the context of free volume arguments. The coordination deficient site criterion seems to explain the present grain-boundary segregation phenomena more successfully. However, it must be noted that in general, the density of the coordination-deficient cation sites should have some correlation to the grain-boundary excess free volumes.

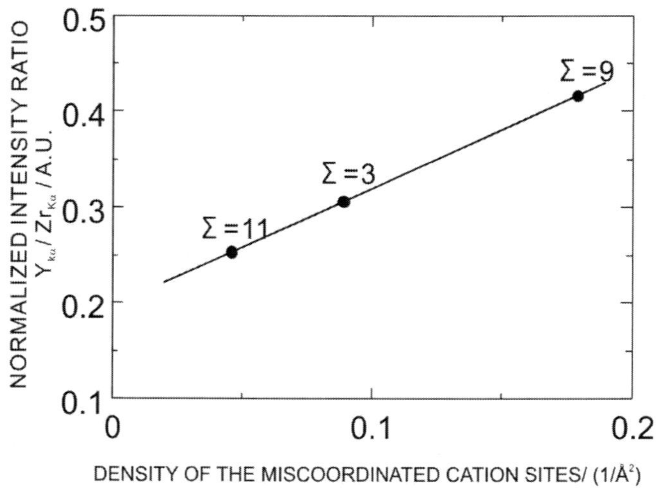

Figure 19. The relationship between the densities of the coordination deficient cation sites and the amount of yttrium segregations (exp.). The normalized intensity ratio of YKα/ZrKα was used as a measure of Y^{3+} segregation to the grain boundary. There is a good correlation between density and intensity ratio.

The present correlation indicates that the coordination-deficient cation sites offer the possibility of having lower substitution energy for Y^{3+} ions than the sites in the grain interior in terms of the strain relaxation effects. Moreover, the accumulation of such sites at the grain-boundary core may attract Y^{3+} ions also electrically to the boundary. In terms of the space charge theory, Y^{3+} segregation is supposed to be enhanced by the presence of positive charges at the grain-boundary core [47] because of its negative effective charge in the Zr^{4+} sites. Normally, this positive charge is thought to be originated from the interstitial Zr and/or Y sites, or O vacancies at the boundary core. The coordination-deficient sites at the boundary core may be regarded as being a locally nonstoichiometric region even though the overall charge neutrality has been preserved. There is a possibility that the coordination-deficient cation sites at the grain-boundary core enhance the Y^{3+} segregation in either or both of the elastic and electric ways.

11. Comparison with FCC Metals and Other Fluorite-Type Oxides

In spite of their similar tendency, there are certain grain boundaries which show apparently different tendencies in grain-boundary energies and solute segregation behavior between yttria-stabilized cubic zirconia and fcc metals. The differences are summarized as follows,

a) The energies of the $\Sigma = 3$, {111} and $\Sigma = 11$, {113} boundaries are comparable in zirconia, while the $\Sigma = 3$, {111} boundary has much lower energy (almost zero) in fcc metals.
b) The solute segregation actually occurred in the $\Sigma = 3$, {111} boundary of zirconia, while there is no indication of solute segregation to the same boundary in fcc metals.

These differences can be explained in terms of the local atomic structures. In the case of fcc metals, the local coordination number of each atom is preserved at twelve across the $\Sigma = 3$, {111} twin boundary. This means that the $\Sigma = 3$, {111} boundary orientation enables highly coherent boundary structure, and that must be why the $\Sigma = 3$, {111} boundary in fcc metals has no tendency to be wetted by solute atoms, and possesses very low excess energy. On the other hand, in the case of zirconia, the difference in local coordination occurs even at such very coherent orientations due to the presence of anion sublattice. Moreover, the presence of anion sublattice resulted in the $\Sigma = 11$, {113} boundary to have the lower density of such sites and in turn lower excess energy, which is completely different from the fcc metal case. Thus, the different behavior in cubic zirconia is considered to arise from the difference in the local coordination states because of the presence of oxygen sublattice.

Recently, atomic-scale analysis on grain boundaries in different fluorite-type materials have been reported [48-50]. In CeO_2, it has been shown that the valence state of Ce atoms at $\Sigma = 5$, {210} grain-boundary cores are reduced from that of grain interiors [48]. This indicates that the grain-boundary cores are tend to be nonstoichiometric, and possibly oxygen vacancies are preferentially formed. Recent report also showed that this tendency may be dependent on the grain-boundary characters [49].

In CeO_2 and UO_2, the atomic structures of $\Sigma = 3$, {111} boundary have been predicted by the first-principles calculations [49, 50]. The predicted most stable structures are very similar to that determined for yttria-stabilized zirconia in this study. The similarity suggests that the framework grain-boundary structures should be dominated by the crystal nature of the cubic fluorite symmetry as in zirconia, but further studies are necessary to understand the fundamental relationships between grain-boundary structures and properties in these fluorite-type materials.

CONCLUSION

Systematic grain-boundary study in yttria-stabilized cubic zirconia using bicrystals was reviewed. It was found that the grain-boundary properties such as structures, energies and solute segregation behavior are strongly related to the grain-boundary geometries and grain-boundary planes. Microscopically, these relations can be explained by the local atomic-scale structures at grain boundaries. It was shown that, as the first approximation, the coordination-deficient cation sites in the grain-boundary core regions are the key to understand the grain-boundary properties. Thus, to fundamentally understand the grain-boundary phenomena in zirconia ceramics, it is essential to understand the atomic-scale structures of the grain boundaries.

REFERENCES

[1] T.H. Etsell, and S.N. Flengas, *Chem. Rev.* (1970) 70, 339.

[2] Y. Arachi, H. Sakai, O. Yamamoto, Y. Takeda and N. Imanishi, *Solid State Ionics* (1999) 121, 133.

[3] R.H. Hannink, P.M. Kelly and B.C. Muddle, *J. Am. Ceram. Soc.* (2000) 83, 461.

[4] P.K. Wright and A.G. Evans, *Current Opinion in Solid State and Materials Science* (1999) 4, 255.

[5] A.I. Ioffe, M.V. Inozemtsev, A.S. Lipilin, M.V. Perfilev and S.V. Karpachov, *Phys. Stat. Sol.* (1975) 30, 87.

[6] J.-H. Lee, T. Mori, J.-G. Li, T. Ikegami, M. Komatsu and H. Haneda, *J. Electrochem. Soc.*, 147, 2822 (2000).

[7] X. Guo and R.-Z. Yuan, *J. Mater. Sci. Lett.* (1995) 14, 499.

[8] X. Guo, *Solid State Ionics* (1997) 96, 247.

[9] X. Guo and R.-Z. Yuan, *Solid State Ionic s* (1995) 80, 159.

[10] S.P.S. Badwal and J. Drennan, *J. Mater. Sci.* (1987) 22, 3231.

[11] S.P.S. Badwal and A.E. Hughes, *J. Euro. Ceram. Soc.* (1992) 10, 115.

[12] E.P. Butler and J. Drennan, *J. Am. Ceram. Soc.* (1982) 65, 474.

[13] A.P. Santos, R.Z. Domingues and M. Kleitz, *J. Euro. Ceram. Soc.* (1998) 18, 1571.

[14] K.E. Adham and A. Hammou, *Solid State Ionics* (1983) 9-10, 905.

[15] S. Stemmer, J. Vleugels and O.V.D. Biest, *J. Euro. Ceram. Soc.* (1998) 18, 1565.

[16] P. Mondal, A. Klein, W. Jaegermann and H. Hahn, *Solid State Ionics* (1999) 118, 331.

[17] Y. Ikuhara, P. Thavorniti and T. Sakuma, *Acta mater.* (1997) 45, 5275.

[18] R.J. Ackermann, E.D. Rauh and C.A. Alexander, *High temp.* (1975) 7, 304.

[19] B.J. Verkerk, B.J. Middelhuis and A.J. Burggraaf, *Solid State Ionics* (1982) 6, 158.

[20] P.S. Badwal and S. Rajendran, *Solid State Ionics* (1994) 70/71, 83.

[21] A.E. Hughes and B.A. Sexton, *J. Mater. Sci.* (1989) 24, 1057.

[22] G.S.A.M. Theunissen, A.J.A. Winnubst and A.J. Burggraaf, *J. Mater. Sci.* (1992) 27, 5057.

[23] M.M.R. Boutz, C.S. Chen, L. Winnubst and A.J. Burggraaf, *J. Am. Ceram. Soc.* (1994) 77, 2632.

[24] M. Aoki, Y.M. Chiang, I. Kosacki, L. Jong-Ren Lee, H. Tuller and Y. Liu, *J. Am. Ceram. Soc.* (1996) 79, 1169.

[25] X. Guo, *Solid State Ionics* (1995) 81, 235.

[26] D. Bingham, P.W. Tasker and A.N. Cormack., *Phil. Mag. A* (1989) 60, 1.

[27] Y. Lei, Y. Ito, N.D. Browning and T.J. Mazanec, *J. Am. Ceram. Soc.* (2002) 85, 2359.

[28] E.D. Dickey, X. Fan and S.J. Pennycook, *Acta mater.* (1999) 47, 4061.

[29] E.D. Dickey, X. Fan and S.J. Pennycook, *J. Am. Ceram. Soc.* (2001) 84, 1361.

[30] N. Shibata, T. Yamamoto, Y. Ikuhara and T. Sakuma, *J. Electron Microscopy* (2001) 50, 429.

[31] N. Shibata, N. Morishige, T. Yamamoto, Y. Ikuhara and T. Sakuma, *Phil. Mag. Lett.* (2002) 82, 175.

[32] N. Shibata, F. Oba, T. Yamamoto, Y. Ikuhara and T. Sakuma, *Phil. Mag. Lett.* (2002) 82, 393.

[33] N. Shibata, F. Oba, T. Yamamoto and Y. Ikuhara, *Phil. Mag. A* (2004) 84, 2381.

[34] D.G. Brandon, B. Ralph, S. Ranganathan and M.S. Wald, *Acta metall.* (1964) 12, 813.

[35] A.P. Greenough and R. King, *J. Inst. Metals* (1951) 79, 415.

[36] A. Tsoga and P. Nikolopoulos, *J. Mater. Sci.* (1996) 31, 5409.

[37] A.P. Sutton and V. Vitek, *Phil. Trans. R. Soc. London Ser. A* (1983) 309, 1.

[38] H. Ichinose and Y. Ishida, *Phil. Mag. A* (1981) 43, 1253.

[39] S.J. Pennycook, B. Rafferty and P.D. Nellist, *Microsc. Microanal.* (2000) 6, 343.

[40] Y. Lei, Y. Ito and N.D. Browning, *J. Am. Ceram. Soc.* (2002) 85, 2359.

[41] J. Kondoh, S. Kikuchi, Y. Tomii and Y. Ito, *J. Electrochem. Soc.* (1998) 145, 1536.

[42] M. Kohyama, *Modelling Simul. Mater. Sci. Eng.* (2002) 10, R31.

[43] D. Wolf, *J. Appl. Phys.* (1990) 68, 3221.

[44] M. Kohyama, R. Yamamoto and M. Doyama, *Phys. Status Solidi* (1986) 137, 11.

[45] M. Kohyama, *Phil. Mag. Lett.* (1999) 79, 659.

[46] Z. Mao, S.B. Sinnott and E.D. Dickey, *J. Am. Ceram. Soc.* (2002) 85, 1594.

[47] X. Guo, *Solid State Ionics* (1995) 81,235.

[48] H. Hojo, T. Mizoguchi, H. Ohta, S. D. Findlay, N. Shibata, T. Yamamoto and Y. Ikuhara, *Nano Lett.* (2010) 10, 4668.

[49] B. Feng, H. Hojo, T. Mizoguchi, H. Ohta, S. D. Findlay, Y. Sato, N. Shibata, T. Yamamoto and Y. Ikuhara, *Appl. Phys. Lett.* (2012) 100, 073109.

[50] P.V. Nerikar, K. Rudman, T.G. Desai, D. Byler, C. Unal, K.J. McClellan, S.R. Phillpot, S.B. Sinnott, P. Peralta, B.P. Uberuaga and C.R. Stanek, *J. Am. Ceram. Soc.* (2011) 94, 1893.

In: New Research Trends of Fluorite-Based Oxide Materials ISBN:978-1-63117-350-9
Editors: Akio Nakamura and Junichiro Mizusaki

Chapter 8

TETRAGONAL ZIRCONIA POLYCRYSTALLINE (TZP) AS A NEW DENTAL MATERIAL

Motohiro Uo *
Advanced Biomaterials Section, Graduate School of Medical and Dental Sciences, Tokyo Medical and Dental University, Bunkyo-ku, Tokyo, Japan

ABSTRACT

Zirconia is one of the most focused dental materials because of its high mechanical strength, chemical stability and good estheticity. Recent progress of computer aided machining assisted the dental application of the zirconia. In this chapter, the basics of the tetragonal zirconia polycrystalline (TZP), the process to the dental application and their future problems are reviewed and discussed.

1. INTRODUCTION

Zirconia has been one of the most interested ceramic materials as a refractory, structural and engineering ceramics because of its excellent physical and chemical properties, e.g., hardness, wear resistance, heat resistance, chemical durability and ionic conductivity. Pure ZrO_2 exhibits fluorite type cubic structure (Fm3-m) at high temperature (> 2370°C) and ambient pressure, tetragonal structure ($P4_2/nmc$) at intermediate temperature (1200 ~ 2370°C) and monoclinic structure ($P2_1/c$) at low temperature (< 950°C). Figure 1 shows the structure of the tetragonal and the monoclinic phases. The transformation from tetragonal to monoclinic phase at around 950°C, which accompanies a large volume expansion of approximately 4%, results in the crack propagation and fracture of sintered monolith. This volume change with the phase transformation should be controlled for the fabrication of zirconia.

* E-mail address: uo.abm@tmd.ac.jp.

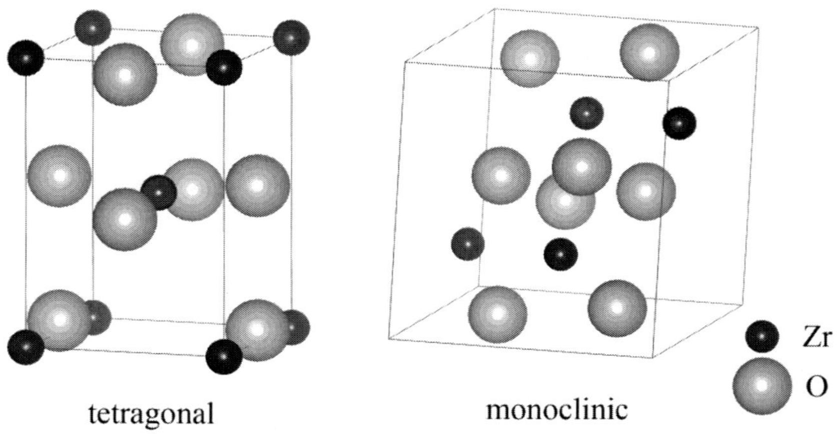

Figure 1. Crystal structure of tetragonal and monoclinic ZrO_2. The more distorted and dilated nature of the latter is apparent.

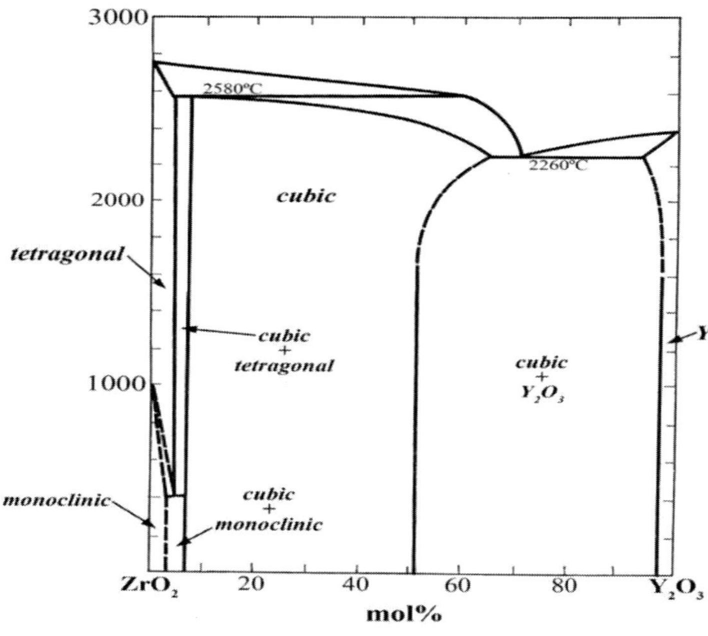

Figure 2. Phase diagram of ZrO_2-Y_2O_3 system.

Addition of other oxides (e.g., MgO. CaO, Y_2O_3 and CeO_2) to ZrO_2 stabilize the tetragonal and/or cubic phase. Figure 2 shows the phase diagram of ZrO_2-Y_2O_3 system [1]. By small addition of Y_2O_3, the region of tetragonal phase is expanded to the lower temperature, then it becomes the metastable phase at room temperature, thus "partially stabilized zirconia (PSZ)" and "tetragonal zirconia polycrystalline (TZP)" could be prepared. Garvie et al. [2,3] showed that the fracture toughness was improved by the phase transformation from tetragonal to monoclinic ($t \rightarrow m$) induced by stress. When stress applied to this metastable tetragonal phase, the stress-induced $t \rightarrow m$ transformation, which accompany with volume expansion, occurred. As shown in Figure 3, the monoclinic phase was formed at the tip of crack with the transformation. The monoclinic phase applies the compression stress caused by its own volume expansion to the tip of crack, thus efficiently inhibits crack propagation and leads to high toughness [4]. Since then, zirconia has been used for wear applications, thermal barrier coatings, optical fiber connectors, and most recently, for biomedical applications because of its improved performance [5-7].

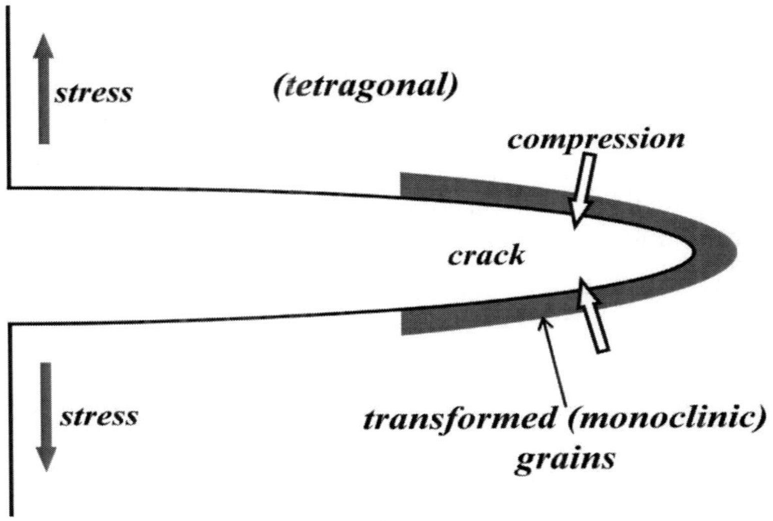

Figure 3. The schematic representation of the stress-induced transformation toughening of PSZ and TZP.

2. CERAMICS IN DENTISTRY

As the dental restorative materials, metallic materials, e.g., precious alloys and chromium based alloys and titanium, have been widely used. Metallic restorations have high mechanical toughness and good dimensional accuracy with die casting technique. However, from the viewpoint of the estheticity and prevention of the metal allergy, "metal-free restoration" had been requested. Ceramics show white and opaque appearance which is suitable as the esthetic restorations. In addition, their high chemical stability and (heavy metal free) composition is suitable for the prevention of metal allergy. However, the mechanical properties of conventional ceramics was not enough to form the dental restorations.

As the esthetic restorations, the porcelain fused metal was applied at first. Figures 4(a) and (b) show the schematic diagram of the porcelain fused metal and ceramic restoration, respectively. In the former (a), the core is formed by the dental alloys with die casting. The front side (labial side) is covered by the feldspathic porcelain. Its process and the structure is similar to the "porcelain enamel". The porcelain fused metal consists of several different porcelain layers. The inner porcelain is opaque to mask the metallic appearance and the outer porcelain is rather translucent to show the glossy appearance similar to natural tooth surface. This system has advantages from both the porcelain (estheticity) and the alloys (toughness and dimension accuracy), but, the dark line observed at the gingival line of the porcelain fused metal restoration spoil its estheticity. Also, the core alloy involves the possibility to cause metal allergy. Then, the all-ceramic restorations (b) have been developed. Current all-ceramic restorations are roughly classified into three groups as (1) silica-based glass ceramics, (2) glass-infiltrated ceramics and (3) zirconia [8-13].

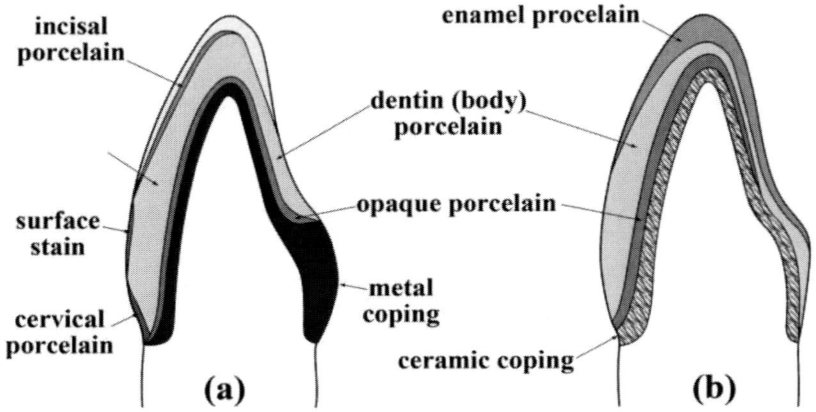

Figure 4. The schematic illustration of the cross section of the porcelain fused metal (a) and ceramic (b) restorations.

Table 1 lists up such currently available commercial (non-zirconia) ceramic restorations belonging to (1) and (2) groups with their typical mechanical properties. The silica-based glass ceramics contain the leucite ($KAlSi_2O_6$) or lithium disilicate ($Li_2Si_2O_5$) as the main crystalline phase. The glass-infiltrated ceramics consist of high volume fraction of crystalline oxides (Al_2O_3 or ZrO_2) with glassy matrix. Sintering process, which is most important in ceramic fabrication, accompanies large volume shrinkage. The dental restoration requires accurate dimensional fit to the tooth and/or oral cavity. Thus, those ceramic restorations are processed by heat press or machining to avoid the misfit caused by the sintering shrinkage. The application of those silica and alumina based restorations had been mostly limited to the low stress restorations such as anterior tooth or single tooth restoration , due to their limited mechanical strength.

Table 1. Typical commercial dental ceramic systems and their various properties

Bland name	Manu-facturer	Composition and condition of block	Flexural strength (MPa)	Fracture toughness (MPa $m^{1/2}$)
Vitablocs Mark II	Vita	Feldspartic	102 [14)], *94 [15)]	1.37 [15)]
IPS Empress	Ivoclar	Leucite	134 [16)], *137 [15)] *106 [17)]	1.74 [16)],2.18 [15)] , 1.2 [17)]
IPS Empress II	Ivoclar	Lithium disilicate	359 [18)], *306 [17)]	2.9 [17)]
IPS e.max	Ivoclar	Lithium disilicate	406 [14)] 300-420 [19)]	2.0-2.5 [19)]
ProceraIn-Ceram Alumina	Nobel Biocare	glass infiltrated Al_2O_3	352 [16)], *594 [17)]	4.49 [16)], 4.4 [17)]
Procera All-Ceram	Nobel Biocare	Al_2O_3	687 [16)]	4.48 [16)]

Flexural strength: * = 3-point bending, others = bi-axial strength.
Fracture toughness: indentation method.

3. Zirconia in Dentistry

For the highly stressed restorations, such as molar or long span restorations, further mechanical strength and fracture toughness had been required. It was well known that zirconia has the highest fracture toughness and also chemical stability in the ceramic materials, then zirconia has been applied as the dental restorative materials. Zirconia (and all other ceramic materials) shows large sintering shrinkage, thus the dimensional misfit caused by the shrinkage should be avoided. The progress in the computer aided designing and machining (CAD/CAM) enabled the preparation of the dental restorations using zirconia with high dimensional accuracy [20].

Zirconia restorations are mainly fabricated by the numerical controlled machining of sintered or pre-sintered zirconia block. Full sintered block of zirconia for hard machining is processed by hot isostatic press (HIP) which is effective to increase the density and improve the mechanical properties of ceramics [21]. Then, the restorations fabricated by the machining of full sintered block has excellent mechanical properties and good dimensional accuracy, however, that requires longer working time and higher cost. The machining of pre-sintered zirconia block (green zirconia) is usually processed by cold isostatic press (CIP) and firing at lower temperature (around 1000°C) for full sintering. The block is machined into the enlarged size of the purposed restorations considering the firing shrinkage (around 20%). The pre-sintered block is softer than the full sintered block and suitable for saving the machining time and tool consumption.

Table 2. The commercial dental zirconia systems and their typical properties

Bland name	Manufacturer	Composition and condition of block	Flexural strength (MPa)	Fracture toughness (MPa $m^{1/2}$)
In-Ceram Zirconia	Vita	Glass infiltrated Ce-TZP/Al_2O_3	592 [22], 542 [23]	5.56 [23]
In-Ceram YZ-Cubes	Vita	Pre-sintered Y- TZP	1107 [18], *793 [14]	
Everest ZS	Kavo	Pre-sinteredY- TZP	*1155 [24]	3.17 [25]
Everest ZH	Kavo	HIPed Y- TZP	*>1200 [24]	4.52 [25]
IPS e.max ZirCAD	Ivoclar	Pre-sintered Y- TZP	801 [14], >900 [19]	5.5 [19]
Procera Zirconia	Nobel Biocare	Pre-sintered Y- TZP	**1208 [26]	
Cercon	Dentsply	Pre-sintered Y- TZP	927 [18], 911 [22], 1141 [23]	6.27 [23], 6.8 [28]
LAVA	3M ESPE	Pre-sintered Y- TZP	1185 [27]	6.6 [28]
InCoris ZI	Sirona	Pre-sintered Y- TZP	1200 [29]	5.9 [29]
ZENOSTAR	Wieland	Pre-sintered Y- TZP	1200 [30]	5.0 [30]
DC-Zircon	DCS Dental	Pre-sintered Y-TZP	840 [31]	7.4 [31]
Katana	Noritake	Pre-sintered Y-TZP	1200 [32]	
Denzir	Decim	HIPed Y-TZP	1000-1300 [33]	9-10 [33]
Nano-ZR	Panasonic Healthcare	HIPed Ce-TZP/ Al_2O_3	1422 [34]	VN11.03 [35] 19.02 [36]

Flexural strength: * = 3-point bending, ** = 3-point bending, others = bi-axial strength.
Fracture toughness: VN =, V-notch beam, others = indentation method.

The mechanical properties of the commercial dental zirconia systems were shown in Table 2. Most of these commercial dental zirconia systems are fabricated by employing the machining of pre-sintered block to save the machining time and tool consumption. However, higher flexural strength and fracture toughness are observed with HIPed zirconia. Figure 5 shows the relation between the fracture toughness and the flexural strength of typical dental zirconia systems. Compared to the conventional silica based dental ceramics listed in Table 1 (and the glass infiltrated zirconia: In-Ceram Zirconia in Table 2), all zirconia products show much higher flexural strength and fracture toughness. Among them, Nano ZR, which consist of HIPed CeO_2 stabilized zirconia/alumina composite, showed the most superior mechanical properties.

In the clinical application of those all-ceramic restorations, the durability is required during long term usage in oral cavity. Table 3 shows the results of five years survival test of zirconia and other all-ceramic restorations.

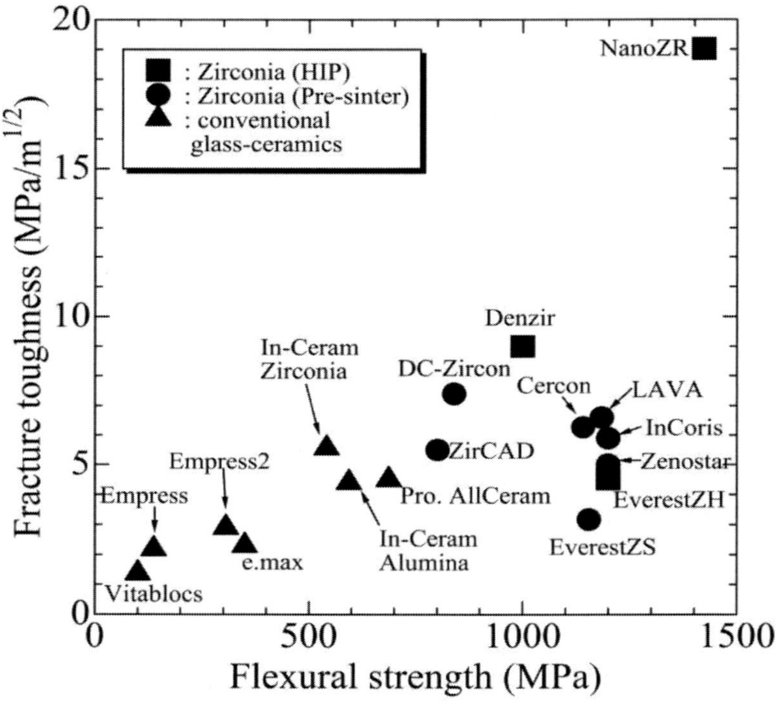

Figure 5. The relationship between the fracture toughness and the flexural strength of typical dental zirconia systems.

The survival probability of zirconia seems to be lower than the other ceramic restorations. However, the result of zirconia consists of various products fabricated by many different manufacturers and operators. Thus, the zirconia restorations would be supposed to be comparable to other all-ceramic restorations in the long term usage for clinical application. In case of conventional porcelain fused metal restoration, 5 years survival probability is reported as 98% for the high gold alloy and 84% for titanium. Hence, the all-ceramic restorations are judged to possess almost comparble long-term durability to porcelain fused metal.

Table 3. Five years survival probability in clinical application of all-ceramic dental restorations

Brand name	5 years survival probability (%)	Ref.
IPS Empress II	93	[37]
In-Ceram Alumina	88-90	
In-Ceram Zirconia	94.4	
Zirconia (Cercon, Lava, Digicon, DC-Zircon, Denzir)	59.19-98.53	[38]

4. Alumina Toughened Zirconia (ATZ) and Zirconia Based Nanocomposite (NanoZR)

Further improvement of fracture toughness of zirconia restoration is required to create the long and thin (narrow) restorations and to survive longer time intraoral usage. Zirconia/ alumina composites (zirconia toughened alumina (ZTA) and alumina toughened zirconia (ATZ)) have been developed and applied as the toughened ceramics. Figure 6 shows the composition dependence of the frexural strength and the fracture toughness of ZTA/ATZ [40]. 20wt% alumina containing zirconia shows the highest frexural strength and the fracture toughness, thus, this composition is widely used as ATZ restoration.

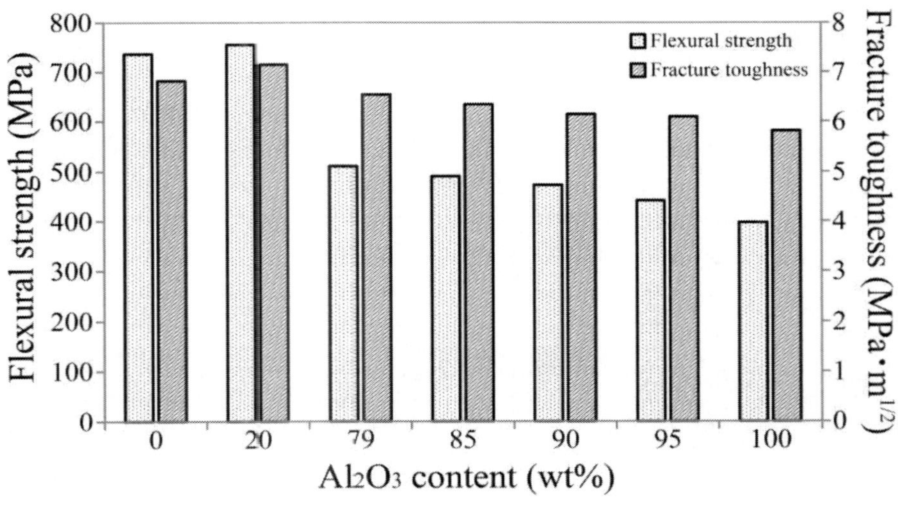

Figure 6. The composition dependence of the frexural strength and the fracture toughness of ZTA/ATZ.

Recently developed Ce-TZP/Al_2O_3 nanocomposite (NanoZR) consists of nano particle dispersion between zirconia and alumina in a geometrically reciprocal manner [41-43], as schematically depicted in Figure 7: Nano-sized Al_2O_3 and Ce-TZP particles are inter-penetrated into each other. The frexural strength and the fracture toughness of NanoZR are highest in the commercial dental zirconia materials, as shown in Figure 5. The strengthening mechanism was suggested that the intragranular nanodispersion particles with several ten nanometer size would suppress the grain growth and decrease the flaw size. In addition, the hydrothermal degradation of NanoZR was reported to be negligible [34,35]. Thus, it have been graded as one of the most favorable as the dental restorative materials.

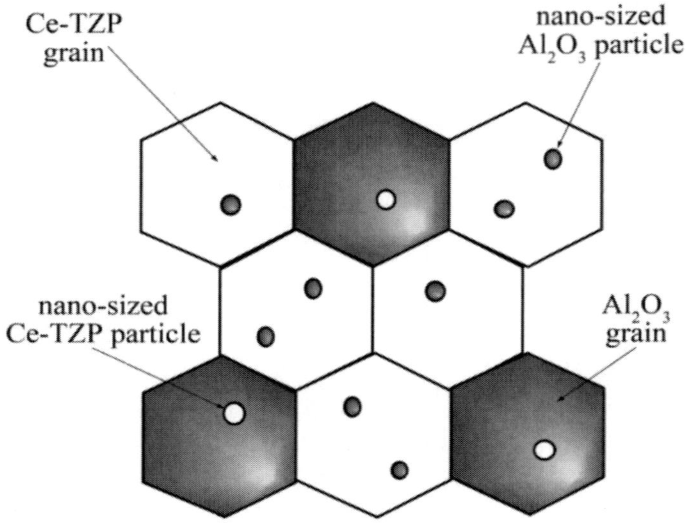

Figure 7. The schematic drawing of the micro structure of the Ce-TZP/Al_2O_3 nanocomposite (NanoZR).

5. Sintering Process and Microstructure of Zirconia

Most of the zirconia powder is processed by liquid phase synthesis (e.g., hydrolysis, spray pyrolisys or emulsion combustion). Figure 8 shows SEM images of the typical zirconia (Y-TZP) powder. The powder has a spherical shape with several ten micrometers (a) consisting of the coagulation of the primary particles with several ten nanometers in diameter (b). Those powders show good powder compression property giving a high green compact density.

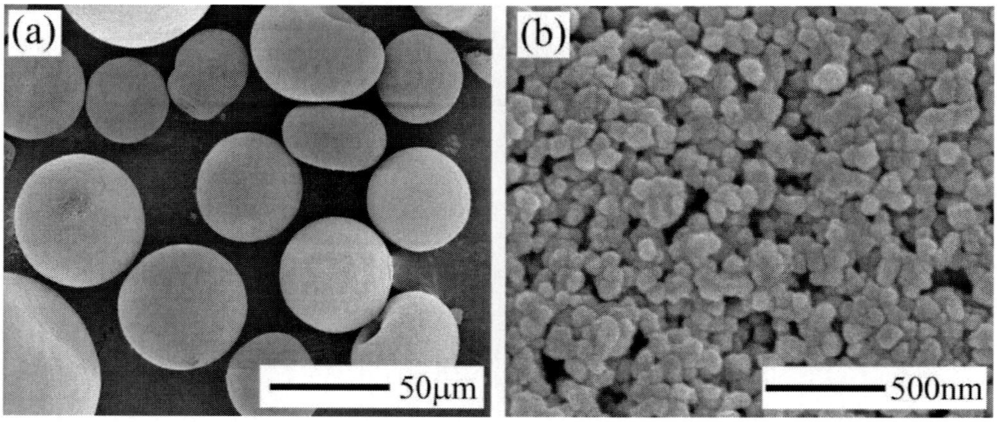

Figure 8. SEM images of typical Y-TZP powder (Daiichi Kigenso Kagaku Co., ltd.). (a) secondary particles, (b) primary particles.

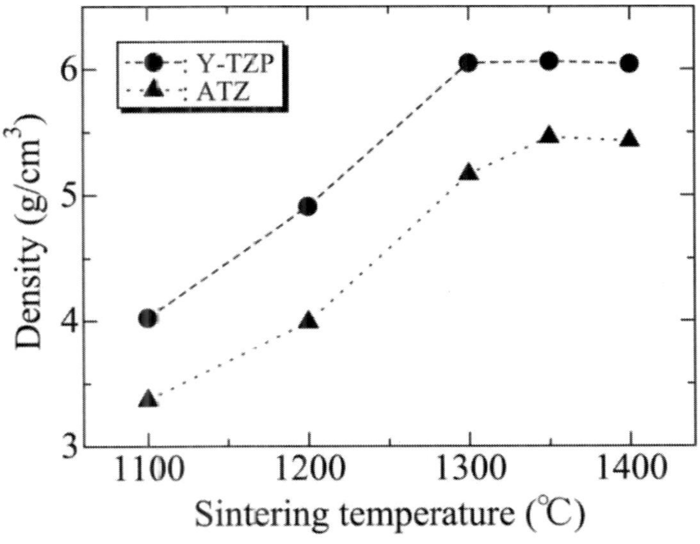

Figure 9. The sintering temperature dependence of the density of Y-TZP and ATZ.

The dependence of the density of sintered Y-TZP and ATZ block on the sintering temperature is shown in Figure 9. The density of sintered Y-TZP block achieved the theoretical density (6.05g/cm^3) at 1300°C. Also, ATZ was fully sintered at 1350°C. The grain morphology of Y-TZP and ATZ sintered at 1350°C and 1450°C, observed by TEM, is shown in Figure 10 [44].

Figure 10. TEM observation of the grain morphology and size of Y-TZP (left) and ATZ (right) sintered at 1350°C (upper) and 1450°C (lower).

Y-TZP and ATZ sintered at 1350°C showed similar grain size. Y-TZP sintered at 1450°C (Figure 10(c)) showed excess grain growth, while ATZ sintered at 1450°C (Figure 10(d)) showed less grain growth (than that of Y-TZP). Thus, grain growth was relatively suppressed in ATZ. Gupta et al. proposed that the critical grain size exists in TZP to obtain the high strength materials. Materials with a grain size > 0.32 μm contained substantial amount of the monoclinic phase, which resulted in the decrease of the mechanical properties [45]. The grain size observed in Figure 10 was almost less than 0.3 μm when sintered at 1350°C. When sintered at 1450°C, Y-TZP contained the grains larger than 0.5 μm, however, ATZ still contained the smaller grain ~0.3 μm. This grain growth suppression would be the reason of the superior mechanical strength of ATZ, which is seen in Figure 6.

This grain growth of Y-TZP with increase in the sintering temperature could be also observed in EBSP (electron backscattering pattern). Figure 11(a) shows the typical backscattered electron pattern (Kikuchi pattern) of the tetragonal zirconia. From this pattern, the crystallographic orientation of each point on the zirconia surface is represented as orientation maps in Figure 11(b) and (c) for Y-TZP surface sintered at 1350°C and 1450°C, respectively. Areas of individual crystal grains are shown in the same contrast. Y-TZP sintered at 1450°C (c) clearly shows larger grain sizes compared with that sintered at 1350°C (b), and the estimated grain sizes using EBSP were in agreement with grain sizes estimated from TEM images (Figure 10). Grain size estimation using EBSP could be carried out using the polished surface, therefore, the grain size observation could be easily made with simple specimen preparation compared to TEM observation.

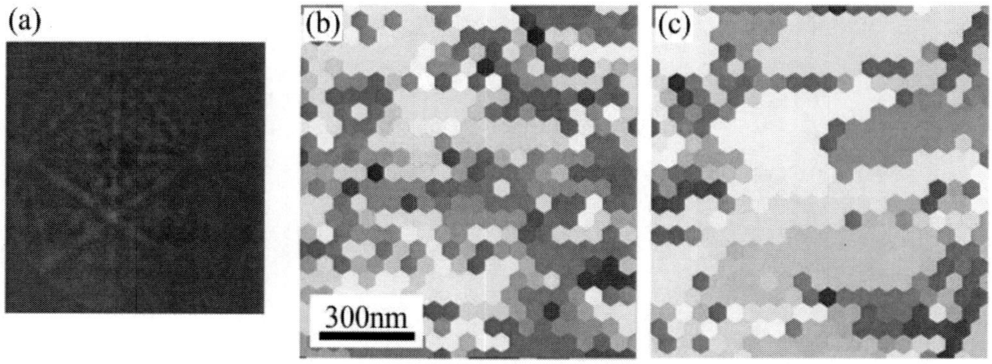

Figure 11. The electron backscattering pattern (EBSP) of the sintered Y-TZP. (a) Kikuchi pattern of backscattered electron, orientation mapping of Y-TZP sintered at 1350°C (b) and 1450°C (c).

6. Low Temperature Degradation (LTD)

One of the concerned problems of zirconia is the low-temperature degradation (LTD) in the presence of moisture [46-50]. This is a kinetic phenomenon in which the polycrystalline tetragonal material slowly transforms into monoclinic zirconia at low temperature less than approximately 400°C, depending on the stabilizer used, its concentration, and the grain size, etc. In the application of zirconia to dental restorations, LTD is a vital concern because those restorations are constantly used in a humid atmosphere at body temperature over a long period (of time) and besides it is autoclaved for sterilization.

LTD occurs on zirconia surface, thus, it could be estimated by the grazing angle X-ray diffraction (XRD) method with low incident angle. Figure 12 shows the typical XRD patterns of Y-TZP after LTD (autoclaved at 140°C in distilled water (DW) for 3 days) [44]. In the XRD pattern obtained by the 2θ–θ method, strong peaks representing the tetragonal phase were observed, while peaks assigned to the monoclinic phase were not distinct. With the grazing angle method (θ = 1° and 2°), peaks assigned for monoclinic zirconia were clearly seen. The depth probed by grazing angle XRD could be varied by changing the incident angle of the X-ray, and was estimated approximately as 1.2 and 2.3 μm for the incidence angle of 1° and 2°, respectively. Thus, the monoclinic phase formation with LTD is estimated to be limited to the surface layer with thickness of several micrometer. The alumina-toughened zirconia (ATZ) can be made to have both further-improved mechanical strength and durability for LTD [51,52].

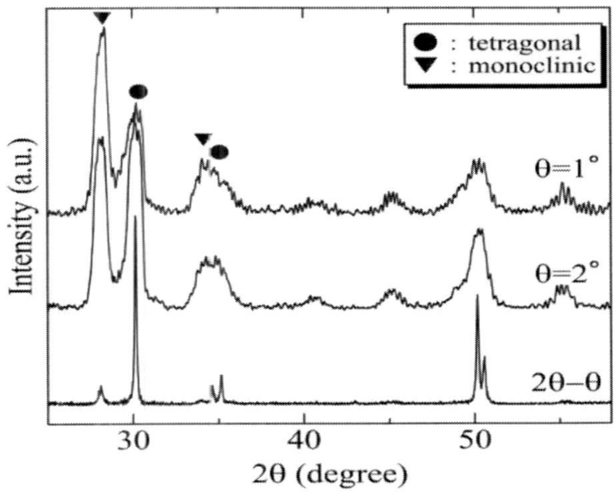

Figure 12. The typical XRD patterns of Y-TZP after LTD (autoclaved at 140°C in distilled water (DW) for 3 days).

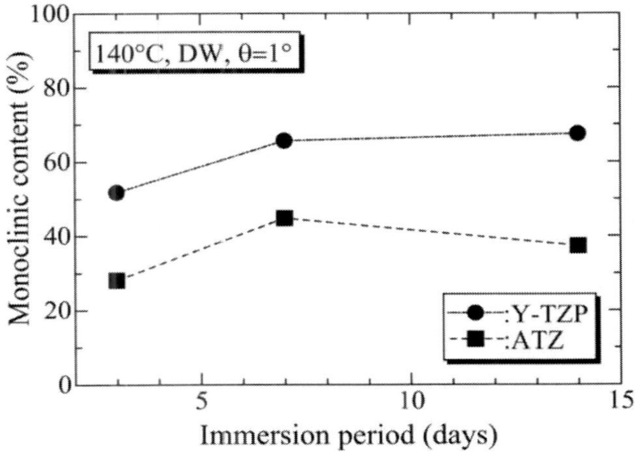

Figure 13. The difference in the monoclinic phase formation of Y-TZP and ATZ at different autoclaving times in distilled water (DW).

The difference in the monoclinic phase formation of Y-TZP and ATZ at different autoclaving times in DW are shown in Figure 13. The monoclinic zirconia content of Y-TZP was greater than that of ATZ for each autoclaving time. At a treatment time of 14 days, the monoclinic zirconia content of Y-TZP was 67 vol%, compared with 37 vol% for ATZ. These results indicate that phase transformation from tetragonal to monoclinic was suppressed in ATZ. The tetragonal to monoclinic transformation is known as the thermal activation process [52] and the autoclaving at 140°C for 7 days would correspond to more than 200 years at 37°C. Therefore, ordinary intra-oral use of Y-TZP and ATZ would result in hydrothermal degradation of just less than 1 μm below the surface, which would hardly affect the properties of zirconia (especially in ATZ).

The effect of various solution soaking and the resultant LTD on the mechanical strength (biaxial flexural strength) of Y-TZP and Ce-TZP/Al_2O_3 nanocomposite (NanoZR) was studied by Ban et al. and the results are shown in Figure 14 [34]. NanoZR was not significantly degraded after soaking. In contrast, the biaxial flexure strengths of Y-TZP was decreased from 1046 to 892 MPa (15%) with autoclaving (at 121°C for 10 days in DW). Therefore, the Ce-TZP/Al_2O_3 nanocomposite is much more durable to LTD compared to the conventional Y-TZP.

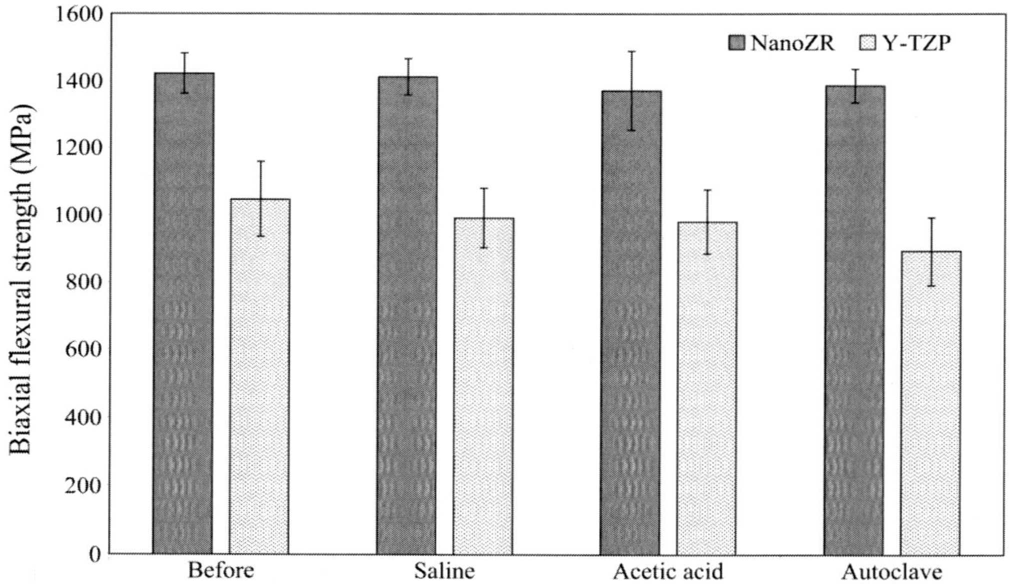

Figure 14. Change in the biaxial flexural strength of Y-TZP and Ce-TZP/Al_2O_3 nanocomposite (NanoZR) after soaking various solutions and autoclaving at 121°C in DW for 10 days.

7. Bonding Property of Dental Zirconia to Dental Porcelains or Dental Adhesives

In actual application for dental restorations, zirconia is veneered with dental porcelain in order to improve its estheticity (Figure 4b). Thus, the long-term stability of both Y-TZP framework and veneered porcelain are required. The fused porcelain is specially prepared for the zirconia core to fit its thermal expansion coefficient to that of zirconia (10.5 $\times 10^{-6}$/°C). In the clinical studies of porcelain veneered Y-TZP core ceramics, the weak performance of the veneering porcelains and low adhesion to the zirconia frame-work were reported [53,54]. Especially, chip-off fractures of the veneering porcelain due to its weak interaction with the zirconia core were problematic as the most frequent cause for failures of zirconia restorations.

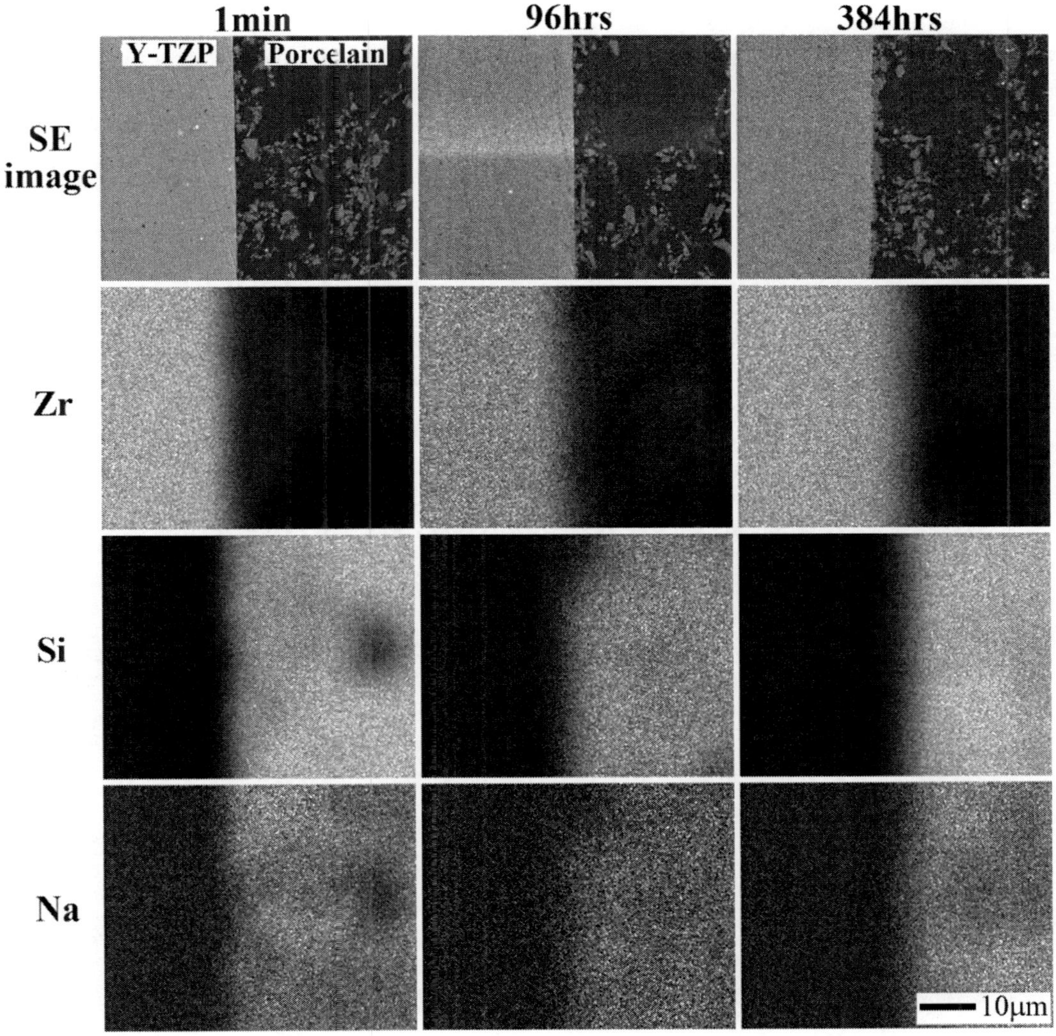

Figure 15. SEM images and the elemental distribution images of the porcelain/zirconia interface with various firing period (at $T = 940°C$).

For the strong adhesion of the veneering porcelain with the zirconia, the formation of the continuously mixed layer of the mutual components would be indispensable at their interface. Figure 15 shows SEM images and elemental distribution images (by SEM/EDS) around the porcelain/zirconia (Y-TZP) interface which was fused with the manufacturer's instruction (940°C x 1min) and with the excess fusing periods (48 hrs and 384 hrs) [55]. In the SEM images, zirconia/porcelain interface was clearly divided to each layer and no mutual elemental diffusion was observed. In the elemental distribution images, Zr diffusion from zirconia core to porcelain and transition layer (and also Si and Na diffusion from porcelain to zirconia core) could be observed on the porcelain/zirconia interface. However, the lateral resolution of SEM/EDS is supposed to be a few μm, therefore those profiles do not necessarily suggest the mutual elemental diffusion through the zirconia/porcelain interface. Zirconia is chemically quite a stable material which is applicable as a crucible for molten metal and oxides. Therefore, the fused porcelain would be merely mechanically retended (but not enough chemically bonded) to the zirconia core. Figure 16 shows the bonding strength between the various core ceramics and the veneering porcelain [56]. For the veneering, the manufacturers' recommendation products for each cores were used.

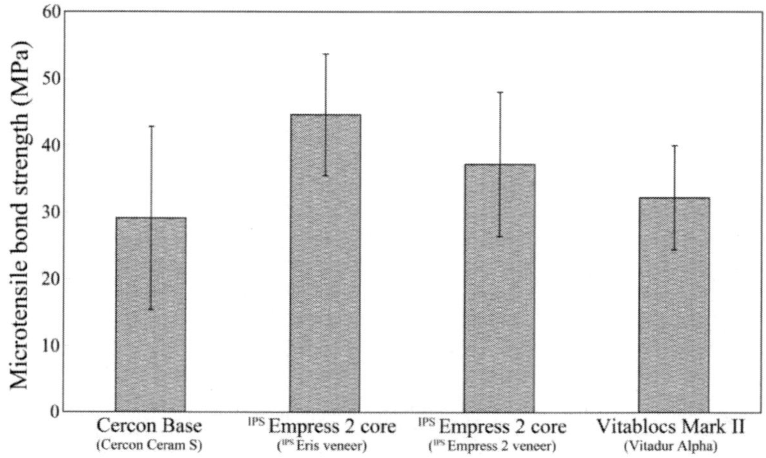

Figure 16. Micro tensile bond strength between the cores and veneering porcelains.(Core). Cercon Base: Y-TZP, IPS Empress 2 core : Glass infiltrated lithium disilicate, Vitablocs Mark II : nepheline crystallized glass matrix (Veneer), Cercon Ceram S: Feldspathic porcelain, IPS Eris veneer: Fluoroapatite feldsparthic porcelain, IPS Empress 2 veneer: Fluoroapatite feldsparthic porcelain, Vitadur Alpha: Alumina containing glass matrix (See Table 1 for each product specification).

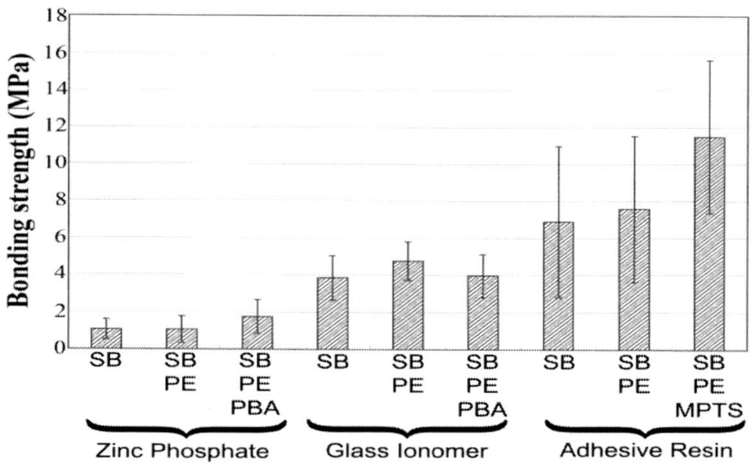

Figure 17. Bonding strength of Y-TZP with various dental cements: SB: sandblast with SiC, PE: phosphate etching (K-etchant gel, Kuraray), PBA: surface treatment with the porcelain bond activator (Kuraray), MPTS: silanization with γ-MPTS, Zinc phosphate cement (Elite Cement, GC), Glass ionomer cement (Fuji I, GC). Adhesive rein composite cement (Panavia 21, Kuraray).

In this report, Y-TZP (Cercon Base) showed lower bonding strength with the veneer compared to the conventional ceramic core materials. Guess et al. also reported that the shear bond strength between the Y-TZP core and the veneer was less than half of the combination with the conventional gold alloy and the veneer [57]. Zirconia dental restorations are usually bonded to the remaining tooth using dental adhesives (dental cements). High chemical stability of zirconia was also of great concern to optimize the bonding reaction with the dental cements. Figure 17 shows the effects on the bonding strength of surface treatments using sandblast and/or silanization with γ-methacryloxypropyl-triethoxysilane (MPTS) in combination with the various dental cements [58]. The adhesive resin cement showed high bonding strength and MPTS treatment was effective to improve the bonding strength. Sandblast with a impact of abrasives (Al_2O_3 or SiC) are widely used to increase the combined mechanical and chemical retention of various dental materials. In case of zirconia, the impact of sandblast would induce the phase transformation of the Y-TZP surface into monoclinic, then the effect to the mechanical properties have been reported [59-62]. Therefore, the sandblast should be carefully applied to the zirconia.

CONCLUSION

High estheticity, chemical stability and biocompatibility are required as the dental restorative materials. Zirconia provides both high mechanical properties (e.g., fracture toughness, bending strength and hardness) and high chemical stability. Previously, the machining cost of sintered zirconia and the dimensional change with sintering shrinkage of pre-sintered zirconia block were of problem. But, recent progress in the precision machining has enabled zirconia dental restorations to attain the sufficient dimensional accuracy with economically-feasible machining cost. According to the review in 2009, more than 20 manufacturers have commercialized their own processing systems of the dental zirconia restorations [63] and the number of manufacturers are still increasing. Recent competition among the manufacturers acts to improve the quality of dental zirconia restorations and to decrease the processing cost. In addition, the recent product of Ce-TZP/Al_2O_3 nanocomposite has been steadily improved both in mechanical property and low-temperature degradation (LTD) durability over the conventional Y-TZP product. From the patient side, the request for the metal-free restorations is increasing because of estheticity and metal allergy safety. Thus, zirconia-based restorations are now becoming one of the most promising materials for alternatively replacing the conventional metallic restorations.

REFERENCES

[1] E.M. Levin, C.R. Robbins and H.F. McMurdie, *Phase diagram for ceramists*, Figs. 354, The American Ceramic Society, Ohio (1964).
[2] R.C. Garvie and P.S. Nicholson, *J. Am. Ceram. Soc.* (1972) 55, 152.
[3] R.C. Garvie, R.H. Hannink and R.T. Pascoe, *Nature* (1975) 258, 703.
[4] R.H. Hannink, P.M. Kelly and B.C. Muddle, *J. Am. Ceram. Soc.* (2000) 83, 461.
[5] C. Piconi and G. Maccauro. *Biomater.* (1999) 20, 1.

[6] I. Denry and J.R. Kelly, *Dent. Mater.* (2008) 24, 299.
[7] B. Al-Amleh, K. Lyons and M. Swain, *J. Oral. Rehabilitaion* (2010) 37, 641.
[8] K.J. Anusavice, *J. Am. Dent. Assoc.* (1993) 124, 72.
[9] W. Höland, V. Rheinberger, E. Apel, C. van't Hoen, M. Höland, A. Dommann, M. Obrecht, C. Mauth and U. Graf-Hausner, *J. Mater. Sci.: Mater. Med.* (2006) 17, 1037.
[10] H.J. Conrad, W.J. Seong and I.J. Pesun, *J. Prosth. Dent.* (2007) 98, 389.
[11] S. Ban, *Jpn. Dent. Sci. Rev.* (2008) 44, 3.
[12] W. Höland, M. Schweiger, R. Watzke, A. Peschke and H. Kappert, *Expet. Rev. Med. Dev.* (2008) 5, 729.
[13] I. Denry and J.A. Holloway, *Materials 3* (2010), 351.
[14] L. Buso, O.B. Oliveira-Júnior, F. Hiroshi Fujiy, G.H. Leão Lombardo H. Ramalho Sarmento, F. Campos and R.O. Assunção Souza, *Minerva. Stomatol.* (2011) 60, 311.
[15] D.G. Charlton, H.W. Roberts and A. Tiba, *Quintessence Int.* (2008) 39, 573.
[16] W.C. Wagner and T.M. Chu, *J. Prosth. Dent.* (1996) 76, 140.
[17] M. Guazzato, M. Albakry, S.P. Ringer, M.V. Swain, *Dent. Mater.* (2004) 20, 441.
[18] J. Chai, F.C. Chu, T.W. Chow and B.M. Liang, *Int. J. Prosthodont.* (2007) 20, 587.
[19] C. Ritzberger, E. Apel, W. Höland, A. Peschke and V.M. Rheinberger, *Materials* (2010) 3, 3700.
[20] T. Miyazaki, Y. Hotta, S. Kuriyama and Y. Tamaki, *Dent. Mater. J.* (2009) 28, 44.
[21] K. Tsukuma and M. Shimada, *Am. Ceram. Soc. Bull.* (1985) 64, 310.
[22] Y.M. Chen, R.J. Smales, K.H.K. Yip and W.J. Sung, *Dent. Mater.* (2008) 24, 1506.
[23] H. Yilmaz , C. Aydin and B.E. Gul , *J. Prosth. Dent.* (2007) 98, 120.
[24] http://www.kavo-everest.com/EN/Downloads/Processing-instructions.aspx
[25] G. Mitov, J. Gessner, U. Lohbauer, K. Woll, F. Muecklich and P. Pospiech, *Dent Mater.* (2011) 27, 684.
[26] M.M. Aboushelib, *Materials* (2010) 3, 2975.
[27] P. Pittayachawan, A. McDonald, A. Petrie and J.C. Knowles, *Dent.Mater.* (2007) 23, 1018.
[28] S.K. Nemli, H. Yilmaz, C. Aydin, B.T. Bal, T. Tıras, *J. Biomed. Mater. Res. Part B* (2012) 100B, 416.
[29] http://td.sirona.com/pdf/6229418.pdf
[30] http://www.zenostar.de/fileadmin/assets/pdf/english/Zenostar_Zahnarztbroschuere_e.pdf
[31] M. Guazzato, M. Albakry, S.P. Ringer and M.V. Swain, *Dent. Mater.* (2004) 20, 449.
[32] http://www.noritake-dental.co.jp/katana/system/system-materials.html#pro05
[33] http://www.denzir.se/pages/default.asp.phdID=73
[34] S. Ban, H. Sato, Y. Suehiro, H. Nakanishi and M. Nawa, *J. Biomed. Mater. Res. Part B* (2008) 87B, 492.
[35] S. Ban, Y. Suehiro, H. Nakanishi and M. Nawa, *J. Ceram. Soc. Jpn.* (2010) 118, 404.
[36] M.N. Aboushelib, C.J. Kleverlaan and A.J. Feilzer, *J. Prosthodont.* (2008) 17, 538.
[37] A.J. Raigrodski, *J. Prosthet. Dent.* (2004) 92, 557.
[38] J.S. Schley, N. Heussen, S. Reich, J. Fischer, K. Haselhuhn and S. Wolfart, *Eur. J. Oral. Sci.* (2010) 118, 443.
[39] M. Walter, P.D. Reppe, K. Böning and W.B. Freesmeyer, *J. Oral. Rehab.* (1999) 26, 91.

[40] M.C. Corrêa de Sá e Benevides de Moraes, C.N. Elias, J.D. Filho and L. Guimarães de Oliveira, *Mater. Res.* (2004) 7, 643.
[41] K. Niihara, *J. Ceram. Soc. Jpn.* (1991) 99, 974.
[42] M. Nawa, S. Nalamoto, T. Sekino and K. Niihara, *Ceram. Int.* (1998) 24, 497.
[43] M. Nawa, N. Bamba, T. Sekino and K. Niihara, *J. Eur. Ceram. Soc.* (1998) 18, 209.
[44] Y. Kawai, M. Uo, Y. Wang, S. Kono, S. Ohnuki and F. Watari, *Dent. Mater. J.* (2011) 30, 286.
[45] T.K. Gupta, F.F. Lange and J.H. Bechtold, *J. Mater. Sci.* (1978) 13, 1464.
[46] E.M. Moser, B.A. Keller, P. Lienemann and P. Hug, *Fresen. J. Anal. Chem.* (1993) 346, 255.
[47] J.D. Lin and J.G. Duh, *J. Am. Ceram. Soc.* (1998) 81, 853.
[48] J. Chevalier, B. Cales and J.M. Drouin. *J. Am. Ceram. Soc.* (1999) 82, 2150.
[49] X. Guo, *J. Mater. Sci.* (2001) 36, 3737.
[50] J. Chevalier, L. Gremillard, A.V. Virkar and D.R. Clarke, *J. Am. Ceram. Soc.* (2009) 92, 1901.
[51] A. Nevarez-Rascon, A. Aguilar-Elguezabal, E. Orrantia and M.H. Bocanegra-Bernal, *Int. J. Refract. Met. H.* (2009) 27, 962.
[52] J. Schneider, S. Begand, R. Kriegel, C. Kaps, W. Glien and T. Oberbach, *J. Am. Ceram. Soc.* (2008) 91, 3613.
[53] I. Sailer, A. Feher and F. Filser, *Quintessence Int.* (2006) 37, 685.
[54] A.J. Raigrodski, G.J. Chiche and N. Potiket, *J. Prosthet. Dent.* (2006) 96, 237.
[55] Y. Kawai, M. Uo, F. Watari, *Nanobiomed.* (2010) 2, 31.
[56] M.N. Aboushelib, N. de Jager, C.J. Kleverlaan and A.J. Feilzer, *Dent. Mater.* (2005) 21, 984.
[57] P.C. Guess, A. Kulis, S. Witkowski, M. Wolkewitx, Y. Zhang and J.R. Strub, *Dent. Mater.* (2008) 24, 1556.
[58] M. Uo, G. Sjögren, A. Sundh, M. Goto, F. Watari and M. Bergman, *Dent. Mater. J.* (2006) 25, 626.
[59] T. Kosmac, C. Oblak, P. Jevnikar, N. Funduk and L. Marion, *Dent. Mater.* (1999) 15, 426.
[60] M. Guazzato, L. Quach, M. Albakry and M.V. Swain, *J. Dent.* (2005) 33, 9.
[61] R.G. Luthardt, M. Holzhu¨ter, O. Sandkuhl, V. Herold, J.D. Schnapp and E. Kuhlisch, *J. Dent. Res.* (2002) 81, 487.
[62] Y. Zhang, B.R. Lawn, E.D. Rekow and V.P. Thompson, *J. Biomed. Mater. Res. Part B* (2004) 71B, 381.
[63] S.O. Koutayas, T. Vagkopoulou, S. Pelekanos, P. Koidis and J.R. Strub, *Eur. J. Esthet. Dent.* (2009) 4, 348.

In: New Research Trends of Fluorite-Based Oxide Materials ISBN:978-1-63117-350-9
Editors: Akio Nakamura and Junichiro Mizusaki

Chapter 9

COMPUTER SIMULATION OF FLUORITE-BASED OXIDES FOR NUCLEAR APPLICATIONS

Tatsumi Arima*
Graduate School of Engineering, Kyushu University,
Fukuoka, Japan

ABSTRACT

In this chapter the methodology and the main results of our classical molecular dynamics (MD) calculations employing Born-Mayer-Huggins or Busing-Ida type interatomic potential function are presented and discussed to elucidate some basic thermo-physical properties of fluorite-type actinide- and zirconia-based oxide nuclear fuels with intrinsic or extrinsic lattice defects. These include; oxygen and metal diffusion behavior for bulk and grain boundaries of UO_2, its based UO_2-PuO_2 mixed oxide and related multi-component oxides and yttria-stabilized zirconia, *etc.*, their thermal conductivity in both equilibrium and non-equilibrium MD schemes in connection with their oxygen non-stoichiometry, and the melting phenomena of UO_2 studied by solid/liquid two-phase simulation (TPS) method.

1. INTRODUCTION

Molecular dynamics (MD) simulations have been developing for over the last 50 years [1]. Recently, the first-principle MD simulation based on the quantum mechanics has also emerged, which, however, can deal only a relatively small number of particles. Here, MD means the numerical simulation technique based on the Newtonian equations, which is often called the "classical" MD. Nowadays, the classical MD calculation can deal multi-billion particles with modern supercomputer system [2]. As for nuclear fuels, the MD study of UO_2 being widely used in the light water reactors has started over 30 years ago [3]. Up to now,

* E-mail address: arima@nucl.kyushu-u.ac.jp.

various thermo-physical and chemical properties of UO_2 have been MD simulated, and their quantitative assessments made by several researchers [4,5].

In the nuclear fuel industry, besides UO_2, the heavier actinide oxides, e.g., plutonium, neptunium and americium dioxides (PuO_2, NpO_2 and AmO_2, respectively), *etc*. are becoming increasingly more important. These actinides are radio-toxic alpha emitters which would be hazardous to human health and environment. Therefore, their experimental usage requires special radioactive facilities. As these oxide fuels are inevitably subject to nuclear-fission associated high temperature, steep temperature gradient and high radiation *etc* conditions, they exhibit the following additional and specific complex behavior: (1) actinides and oxygen re-distribution; (2) pore (or void) formation and migration; (3) fission products distribution, *etc* [6]. These and associated material-property changes could have severe effects on the crucial 'safety' issue of nuclear fuel management, *e.g*., the lowering of thermal conductivity and melting point, local burn-up and chemical reaction with Zircaloy cladding, and so on. The MD simulation is a powerful computational tool for clarifying in an atomistic scale many basic thermo-physical properties of these oxide fuels existing behind and governing such complex practical material problems in severe irradiation conditions. With this motivation we have conducted for many years a number of MD-simulations research on the basic mass- and heat-transport and melting properties of these oxide-fuel materials. In this chapter. the main results of those studies are presented and discussed, including their some prospects in the future.

2. Fluorite-Based Structures of Nuclear Fuel Oxides

Actinide dioxides used as nuclear fuel have fluorite-based structure, as depicted schematically in Figure 1(a-d). Since actinides are transition elements, they have multiple oxidation states ranging from +2 to +7 in aqueous solution. In solids the valence state changes mainly between +3 and +6, depending on the oxygen potential, *i.e*., the $O_2(g)$ partial pressure and temperature, of the system. As the fluorite structure generally has the much more rigid cation sublattice than the oxygen sublattice, such cation-valence change forms various oxygen lattice defects: Figure 1 shows their schematic structure variations in case the oxygen-to-metal (O/M) ratio is varied around the ideal (stoichiometric) value of 2: (a) the ideal fluorite structure of stoichiometric UO_2 and PuO_2, *etc*.; (b) the defect-fluorite structure of hyper-stoichiometric UO_{2+x} with Willis 2 : 2 : 2-type cluster containing two oxygen vacancies (V_Os), four oxygen interstitials (O_is) and higher valent (+5 or +6) U ions [7]; (c) the defect-fluorite structure of hypo-stoichiometric (U, Pu)O_{2-x} mixed oxide with oxygen vacancy. (Of course, hyper-stoichiometric (U, Pu)O_{2+x} mixed oxide with oxygen interstitial is also important here, for the Pu content in the usual MOX fuel is generally quite low; less than ~10 %). As discussed later, the exact defect structure of 'Willis cluster' still remains unclear even today. In the hypo-stoichiometric mixed oxides (MOX), a part of Pu ions is in 3+ state, and to keep electrical neutrality oxygen vacancy is formed. Here, most of these crystal-structure drawings from crystallographic data were made using the VESTA program [8]. To be noted here is that even the stoichiometric UO_2, PuO_2, and their UO_2 - PuO_2 mixed oxides, *etc.* in Figure 1(a) have in fact a small amount of thermally-produced intrinsic (Shottky and Frenkel) defects at high temperature which are neglected (i.e. not depicted) here, and eventually the existence of such intrinsic defects largely determines their many important high-temperature properties such as oxygen and metal diffusion and melting *etc* discussed in this chapter.

Zirconia in Figure 1(d) is not only widely used as oxygen conductor but also is expected as a prominent nuclear fuel and radioactive nuclear waste-form. Zirconia doped with actinides is one of so-called "inert matrix fuels (IMFs)", for it normally does not contain U which produces the heavier actinides through neutron absorptions and β-decays in nuclear reactor [9]. Such zirconia-based IMF has some additional advantages, such as the superior radiation tolerance, chemical and structural stability either in high-temperature irradiation condition or in disposal environment. However, IMFs are unsuitable for reprocessing because of the chemical stability. IMFs are generally multi-component oxides typically containing yttrium (Y) as a cubic-fluorite stabilizer, gadolinium (Gd) or erbium (Er) as a burnable poison and Pu and/or heavier actinides as the fuel element. Oxygen vacancy is therefore introduced with addition of such lower-valent cations than +4.

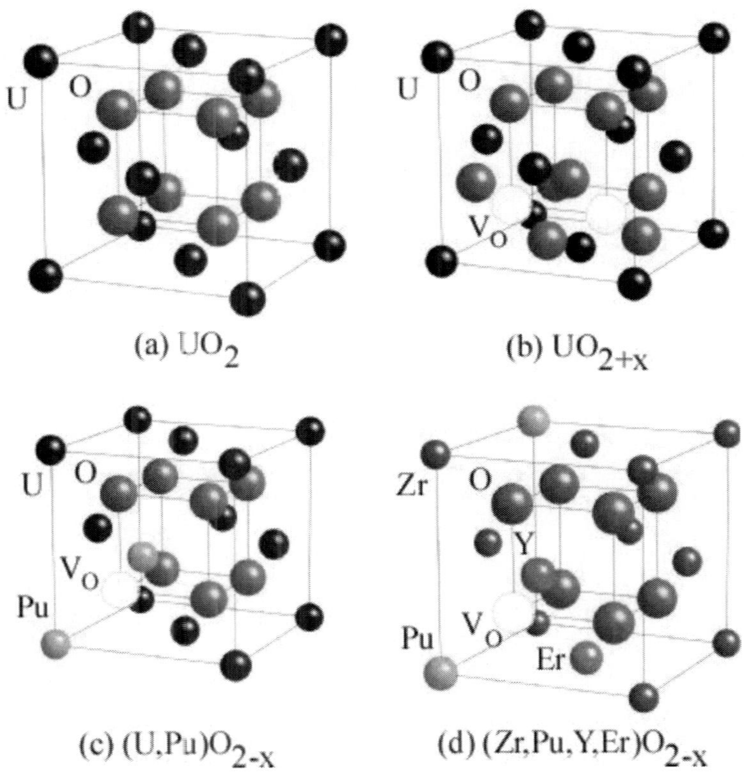

Figure 1. Schematic crystal and defect structure of fluorite-based oxide nuclear fuels; (a) stoichiometric UO_2; (b) hyper-stoichiometric UO_{2+x}; (c) hypo-stoichiometric $(U,Pu)O_{2-x}$; (d) zirconia -based inert matrix (Vo denotes oxygen vacancy, i.e., missing oxygen site).

3. Classical Molecular Dynamics Simulation

The MXDORTO program customized by ourselves was used for MD simulations [10]. The original MXDORTO developed by Kawamura was built for the equilibrium MD (EMD) simulation, and was suitable for mainly studying the orthogonal system, e.g., orthorhombic, tetragonal, cubic structured materials [11]. It is essential for the MD simulations to consider what type of interatomic potential function is used, because it governs the motion of particles in the simulated system. Thereby, several items, e.g., reproducibility, physical meaning, calculation cost *etc* are to be taken into consideration. We employed the Born-Mayer-Huggins potential with fully ionic model (FIM) [11,12] or the Busing-Ida potential with the partially ionic model (PIM) [10,11,13], depending on the specific purpose. These potential functions, respectively, are given by;

$$U_{\mathrm{FIM}}(r_{ij}) = \frac{z_i z_j e^2}{r_{ij}} + A_{ij}\exp\left(-\frac{r_{ij}}{\rho_{ij}}\right) - \frac{C_{ij}}{r_{ij}^6}, \tag{1}$$

$$U_{\mathrm{PIM}}(r_{ij}) = \frac{z_i z_j e^2}{r_{ij}} + f_0(b_i + b_j)\exp\left(\frac{a_i + a_j - r_{ij}}{b_i + b_j}\right) - \frac{c_i c_j}{r_{ij}^6}, \tag{2}$$

where f_0 in Eq. (2) is an adjustable parameter. Potential parameters A_{ij}, ρ_{ij} and C_{ij} are given to the ion pair *i-j*, and a_i, b_i and c_i, are given to the ion of type *i*. The first term of the right-hand side of both potential equations stands for the long-range Coulombic interaction, for which evaluation Ewald summation algorithm was introduced into the program [14]. Other terms stand for short-range interactions: the second one is the repulsive potential between ionic cores; the third one van der Waals interaction. Thus, both potential functions have essentially the same physical meaning, although the latter is more transferable than the former. In the FIM, z_i is the formal charge of type *i* ion, e.g., +4 for Zr^{4+} ion. On the other hand, the effective charge is given to each ion in the PIM, and the ionic bonding character of 67.5% is assumed, e.g., +2.7 for U^{4+} ion. These potential parameters are determined so as to reproduce the experimental thermal expansion coefficient and compressibility of the target system [8,12,15-17]. In addition, the new potential developed especially for UO_2 by Yakub was used to simulate the diffusion behavior of actinide and oxygen ions [18]. It is given by

$$U_{\mathrm{Yakub}}(r_{ij}) = \frac{z_i z_j e^2}{r_{ij}} + A_{ij}\exp\left(-\frac{r_{ij}}{\rho_{ij}}\right) - \frac{C_{ij}}{r_{ij}^6} + D_{ij}\left(\left[\exp\left(-\beta_{ij}(r_{ij} - r_{ij}^*)\right) - 1\right]^2 - 1\right) \tag{3}$$

In the above potential function, the forth term in the right hand side is introduced to account for the U-O covalent bonding, and 55.52 % of ionicity is given for the effective ionic charge states. Other potential parameters are given in Ref. [18].

The equations of motion are given by the (classical) Newton's second law,

$$m_i \frac{d^2\vec{r}_i}{dt^2} = F_i = \sum_{j(\neq i)} F_{ij} \tag{4}$$

where m_i is the mass of the *i*-th particle, $\vec{r}_i$ is position and F_{ij} the force acting on each particle pair. This force F_{ij} is related to to the interatomic potential function U_{ij} as

$$F_{ij} = -\nabla U_{ij}. \tag{5}$$

In the MD simulation, several hundred or thousands of ions are generally provided in a supercell, and a three-dimensional (3D) periodic crystal-structure approximation, i.e., a periodic 3D boundary condition, is imposed. To chase the motion of particles, a time discrete form of the equation of motion is solved according to Verlet algorithm (Δt = 2 femto-second (fs)). Most of MD simulations were carried out by an *NPT* ensemble, where *N* (particle number in the system), *P* (pressure) and *T* (temperature) are constant. Pressure and temperature were controlled by scaling supercell lengths and particle velocities (or sometimes Nosé thermostat), respectively [11]. During the simulation, the instantaneous position and velocity of the particles were recorded at a constant time interval. Based on the accumulated data, the thermo-physical properties, e.g., diffusion coefficient and thermal conductivity, of the target system are obtained.

4. Diffusion Behavior of Fluorite-Based Oxides

4.1. Determination of Diffusion Coefficient with MD Simulations

The self-diffusion coefficient of each atom is determined based on the time evolution of the mean square displacements (MSD). This quantity can be easily determined by recording the atomic positions regularly during the simulation. The MSD function at time t corresponds to the average square distance travelled by an atom between t_0 and t. It is calculated with the following equation,

$$\mathrm{MSD} = \left\langle \left| r_i(t) - r_i(t_0) \right|^2 \right\rangle, \tag{6}$$

where $r_i(t_0)$ is the position of the atom i at the initial time t_0, and $r_i(t)$ the position at the time t. The three-dimensional diffusion coefficients, D, are deduced from the MSD functions using the Einstein's relation expressed by;

$$D = \lim_{t \to \infty} \frac{1}{6t} \left\langle \left| r_i(t) - r_i(t_0) \right|^2 \right\rangle. \tag{7}$$

In order to obtain the reliable diffusion coefficient, the larger time steps are required for the MD calculation. Thus, the typical total time steps for one-run 5×10^5 steps (1 step = 2 fs), which correspond to infinite t in Eq. (6), are used to determine the diffusion coefficient from the asymptotic slope of the MSD function.

In general, diffusion in solids occurs via point defects. The point defect concentration is thermally increased with increase in temperature. Migration is also a thermally activated process, accelerated with increasing temperature. The diffusion coefficient can then be written in the Arrhenius equation form as follows,

$$D = D_0 \exp\left(-\frac{Q}{k_B T}\right), \tag{8}$$

Where the activation energy for diffusion, Q, is the sum of the formation energy and the migration energy of the defect, D_0 a pre-exponential factor independent of the temperature, k_B the Boltzmann constant, and T the temperature in Kelvin (K).

4.2. Actinide Oxides

4.2.1. Bulk Diffusion Behavior of Oxygen and Actinides

Diffusion behavior of oxygen and actinides in oxide fuels is of great importance for the sintering behavior in the manufacturing process and for the fuel performance in the operating condition, *etc*. For instance, the sintering condition is determined by the change in diffusion coefficient with the oxidation states of oxide fuels, *i.e.*, oxygen non-stoichiometry of the system. In the operating condition, oxygen and actinides re-distributions also occur due to the large temperature gradient in the fuel pellet. In addition, the oxygen non-stoichiometry lowers the thermal conductivity as well as the melting point and influences the physico-chemical behaviors of fission products. Furthermore, the non-uniform distribution of actinides alters the axial burn-up profile. Since all these are generally very complex non-isothermal coupled oxygen-actinide cation diffusion processes in nonstoichiometric UO_{2+x} which are beyond the scope of our MD-simulations study, here, the isothermal bulk diffusion behaviors of oxygen and actinides are discussed mainly focusing on the stoichiometric UO_2.

Figure 2 shows the Arrhenius plot of oxygen bulk diffusion coefficients calculated with Yakub potential in Eq. (3) [18], together with the MD result obtained by Potashnikov et al. [19]. For comparison, the experimental data are plotted as well [20-23]. The oxygen diffusion coefficients calculated in the both MS studies for the perfect crystal (5 × 5 × 5 unit cells) exhibit a common feature that there eixst at least two kinds of different diffusion-mechanism state (or regime) ; namely, (1) a super-ionic state just below the melting point; (2) the usual point-defect diffusion state in the lower temperature range. In the former super-ionic state, as oxygen ions are considered to be in the so-called 'anion (oxygen) sublattice-melt' state, these diffuse very quickly, and the activation energy, which most plausibly corresponds to the oxygen migration energy, is quite low. The activation energy obtained by Yakub potential is less than 1.5 eV, and that obtained by Potashnikov is ~0.97 eV [19], respectively. To investigate the oxygen diffusion path, the average oxygen density was calculated and shown in Figure 3. The volumetric data show the oxygen density where the RGB (Red-Green-Blue) colors correspond to the density where R: high and B: low. As seen in this figure, oxygen moves to <001> and <111> directions. From the neutron scattering experiment [24,25], the diffusion mechanism is suggested to be via the enhanced anion Frenkel disorder or Frenkel defect clusters. On the other hand, Yakub et al. argued that the oxygen self-diffusion coefficient deviated from the simple Arrhenius law, due to the increasing mutual interactions between defects, and that the activation energy (or the slope of the Arrhenius plot) gradually decreased from around the onset of the super-ionic state to melting point [18]. From a visual inspection in Figure 3, the oxygen ions are supposed to be moving around there via various diffusion mechanisms including vacancy, interstitial and interstitialcy *etc* mechanisms.

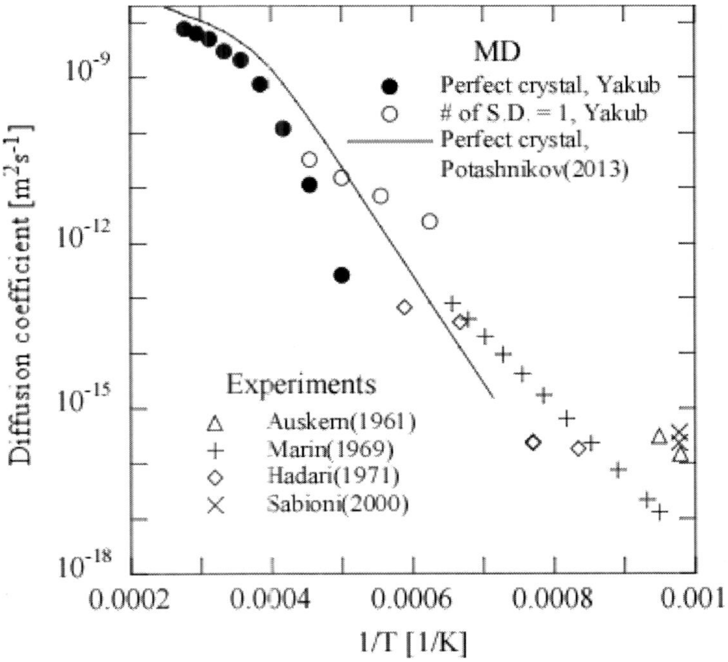

Figure 2. Oxygen bulk diffusion coefficient of UO_2 obtained by MD calculations [18,19] in comparison with experimental data [20-23]. S.D. stands for Schottky defect.

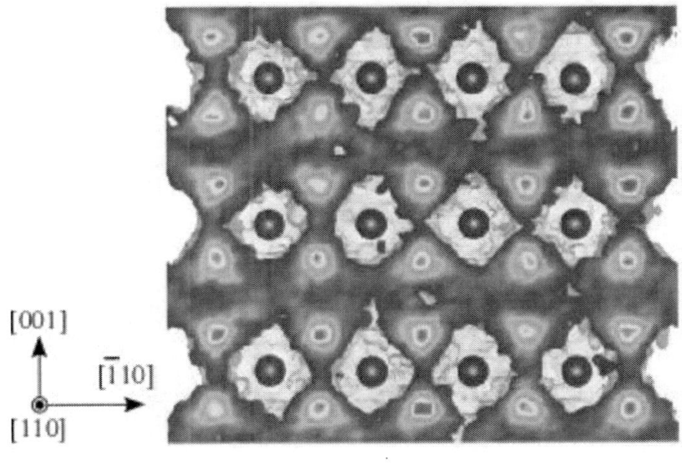

Figure 3. Average oxygen density in UO_2 supercell at 3400 K (the super-ionic state). The cross-section corresponds to the (110) plane. RGB colors indicate the O density (R: high; B: low), which was calculated by averaging the position of O ions over 40 ps, and blue balls stand for U atoms.

For the oxygen self-diffusion behavior in the usual (or normal) diffusion state at lower temperature, the activation energy estimated in our simulation [17] was 5.9 eV. Potashnikov et al. estimated it to be 3.88 eV [19]. Figure 4 shows the simulated oxygen trajectories in the stoichiometric UO_2 at 2000 K. Four and eight oxygen ions almost simultaneously move to the respective nearest neighbor (NN) sites in Figure 4(a) and (b), respectively. Potashnikov et al. also pointed out the similar ring mechanism. On the other hand, the activation energy experimentally obtained by Marin and Contamin was 2.6 eV [22], and it has been considered to be most reliable and representative for the oxygen-self diffusion in the stoichiometric UO_2. Matzke also recommended their data, and suggested the anion Frenkel defect formation and the oxygen diffusion via vacancy and interstitial mechanisms [26]. Thus, there are some non-consistent part between calculated and experimental results. Potashnikov et al. pointed out that such inconsistency is an essential problem in the simulation under the periodic boundary condition (PBC) and tried two different approaches in their paper [19]: (1) the simulation with the Schottky defects under the PBC; (2) the nano-crystal simulation under the isolated boundary condition (IBC). For the first approach, the formation reaction of the Schottky defect is expressed as:

$$\text{null} \leftrightarrow V_M + 2V_O \tag{9}$$

where V_M is the metal vacancy, V_O the oxygen vacancy. During MD calculation under the PBC with no surfaces, Schottky defects cannot be generated in the super-cell. So to evaluate the effect of Shottky defect, related ions are removed for the initial condition, which means that the Schottky defect in this condition behaves like an extrinsic defect. The result obtained by Potashnikov et al. showed that the diffusion coefficient was over two orders greater than that obtained for the perfect crystal and that the diffusion activation energy was about 1 eV [19], which means that the first approach cannot solve the above inconsistency. Our simulation with the Schottky defect (S.D. = 1) also shows the same tendency (see Figure 2). The second approach is the simulation with the free surface (i.e., isolated in vacuum). Using the octahedral nano-crystal under the IBC, they obtained the MD diffusion coefficients and the activation energy (2.9 eV) comparable to the experimental data below 1800 K, and observed the formation of Frenkel anion pairs and the diffusion mechanism via vacancy and interstitial migrations [19]. Therefore, the simulation technique using the nano-crystal under the IBC might be an answer to obtain the reliable self-diffusion coefficient for oxygen in the normal difusion state of the stoichiometric UO_2.

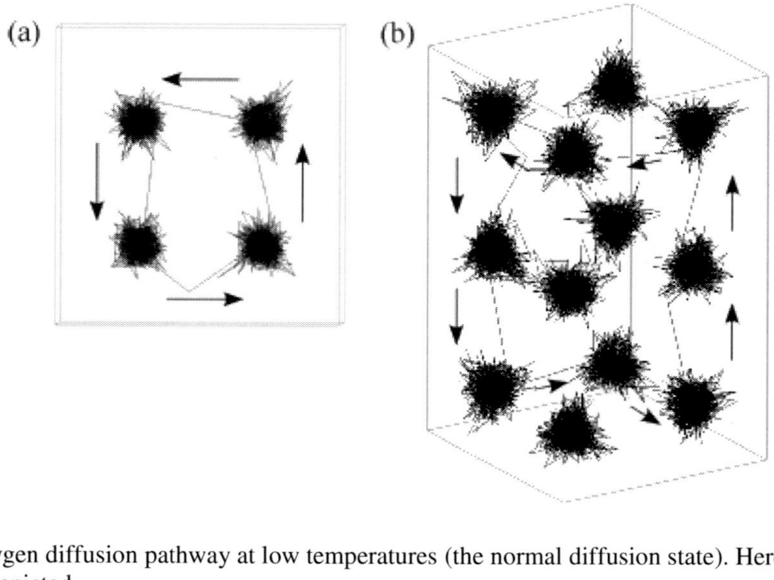

Figure 4. Oxygen diffusion pathway at low temperatures (the normal diffusion state). Here, uranium ions are not depicted.

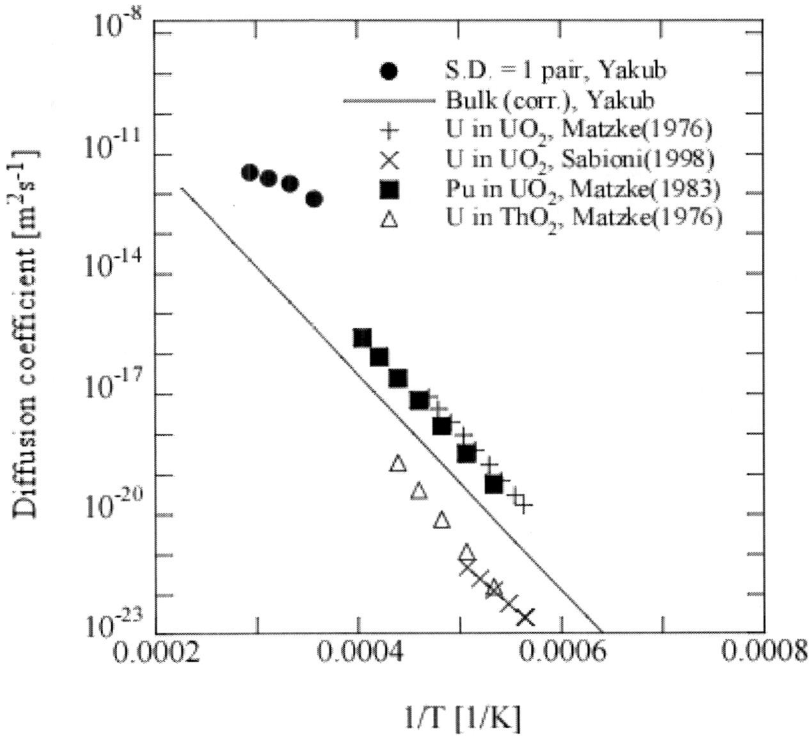

Figure 5. Uranium bulk diffusion coefficients in UO_2 obtained by MD calculations with Yakub potential function in comparison with experimental data [26-28].

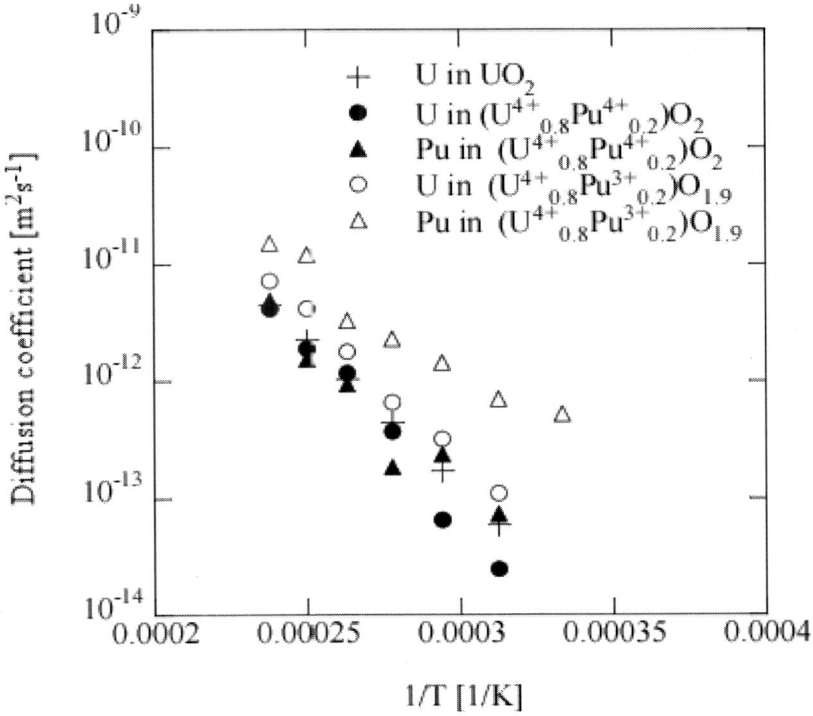

Figure 6. O/M dependence of U and Pu diffusion coefficients in MOX.

Diffusion coefficients of actinides are much smaller than that of oxygen, making it difficult to evaluate their diffusion behaviour in the MD simulation. In the MD simulation of uranium diffusion, only extrinsic vacancy diffusion was found in high temperature range. Figure 5 shows the MD-simulated uranium bulk diffusion results obtained by Yakub potential in Eq. (3), together with experimental data [26-28]. Unlike oxygen bulk diffusion, however, the activation energy of lattice diffusion of uranium in Eq. (7) could not be obtained for the perfect crystal. As a result, MD data with the Schottky defect gives only migration energy of uranium vacancy. In addition, the result thus obtained depends on the defect concentration. So, to evaluate the inherent self-diffusion coefficient of uranium, both the pre-exponential factor and the activation energy relevant to the defect formation and the uranium migration should be required. Here, the former roughly corresponds to the pre-exponential factor obtained for the diffusion coefficient with S.D. = 1, and the latter can be obtained according to $\Delta G_S - \Delta G_{FO} + \Delta H^m_{U,V}$ [26], where ΔG_S is the formation energy of the Schottky defect, ΔG_{FO} the formation energy of the oxygen Frenkel defect, the migration energy (enthalpy) of the uranium vacancy. The calculated result extrapolated to the lower temperature is comparable with experimental data (Figure 5).

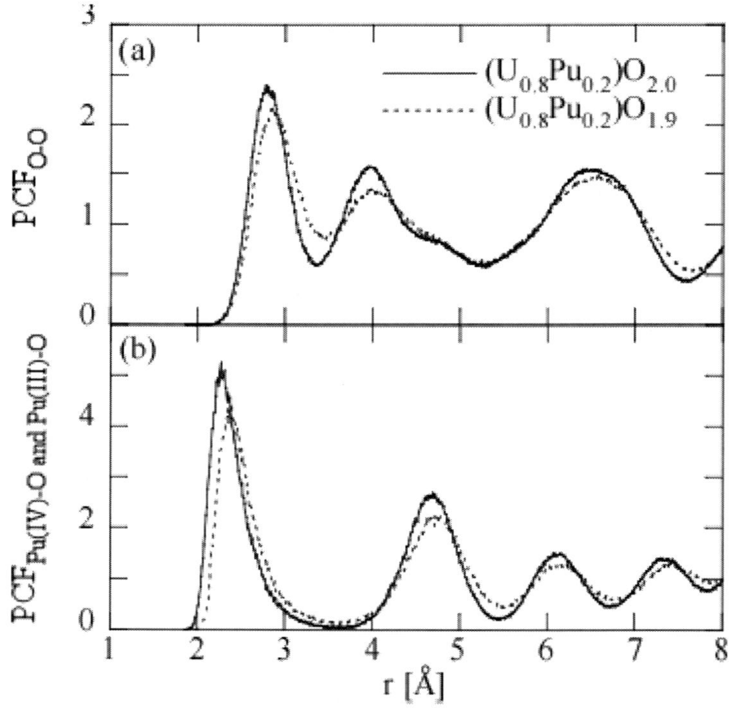

Figure 7. MD-calculated pair-correlation functions of O-O and Pu-O in MOX.

Figure 6 shows the Arrhenius plot of U and Pu diffusion coefficients in MOX at two O/M ratios obtained for the supercell (8 ×8 ×8 unit cells). This supercell includes 5 pairs of Schottky defects and these diffusion coefficients were calculated with Arima potential in Eq. (2) instead of Yakub potential, because Arima potential provides the potential parameters not only of U^{4+} but also of Pu^{4+} and Pu^{3+}. Both U and Pu ions in stoichiometric $(U_{0.8},Pu_{0.2})O_2$ have almost comparable diffusion coefficients with that of U ion in UO_2. Thus, the effect of

Pu^{4+} addition on actinide diffusion is small for the O/M ratio = 2.0. To evaluate the O/M dependence, hypo-stoichiometric $(U_{0.3}Pu_{0.2})O_{1.9}$ was provided, where ionic valence of all additive Pu was +3. The diffusion coefficient of Pu^{3+} for the O/M ratio = 1.9 is found to be larger than either those of U^{4+} and Pu^{4+} for the O/M = 2.0. The effective ionic radius of Pu^{3+} is larger than those of U^{4+} and Pu^{4+}, and some oxygen vacancies are introduced with addition of Pu^{3+} to keep the electrical neutrality. In this hypo-stoichiometric MOX, the supercell length therefore increased and the distances between Pu^{3+}-O^{2-} and O^{2-}-O^{2-} obtained from the pair correlation functions (PCFs), reflecting the local crystal structure, also increased (Figure 7). In the fluorite structure, an actinide ion exists in the oxygen cube (CN = 8), as shown in Figure 1. Hence, Pu^{3+} ions is supposed to migrate more easily than tetravalent actinide ions because of its enlarged oxygen cube. Though, it would be also possible to postulate that the lower-valent Pu^{3+} has the lower electrostatic migration barrier energy than Pu^{4+}. To perform more accurate and reliable simulations, further development of the potential function, e.g., charge variable potential model, is required especially for non-stoichiometric oxide systems.

4.2.2. Grain Boundary Diffusion Behavior of Oxygen and Actinides

As these nuclear fuel oxides are usually used in polycrystalline (either powder or pellet) form, their grain boundary structure and its associated diffusion behaviour are also important, e.g., in sintering and FP migration behaviour. Their some representative grain-boundary structures were modelled by the coincidence site lattice (CSL) theory [29]. For example, Figure 8 shows the snapshot of the supercell including Σ41 (091) / [100] structure at 300 K. At the centre and both sides of the supercell, the disordered phases which correspond to GB interfaces can be observed.

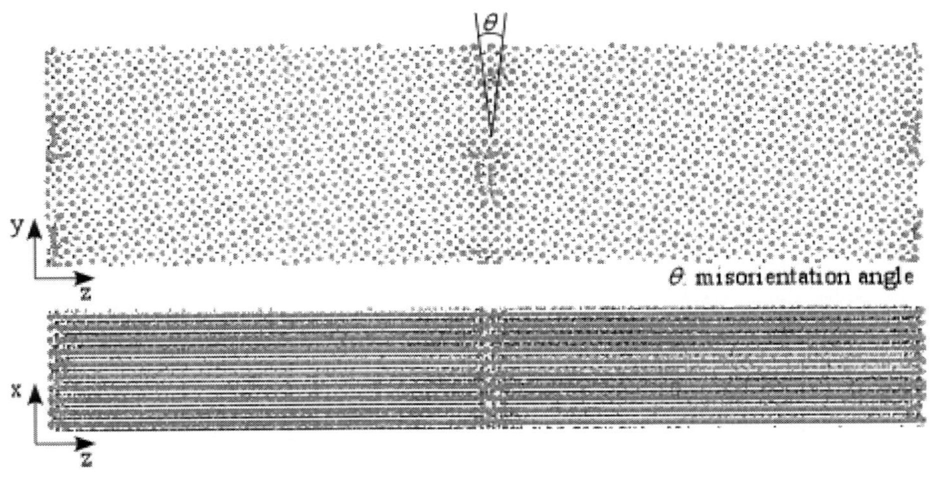

Figure 8. Σ41GB structure of UO_2. Blue: U; green: O.

Figure 9 shows oxygen GB diffusion coefficients obtained by Arima potential in Eq. (2). The GB diffusion coefficient increases with the misorientation angle. The GB diffusion coefficients are larger than bulk ones and the GB activation energies are relatively low. Originally the GB structure has the crystalline imperfection. Thus, oxygen ions quickly diffuse in the GB region. On the other hand, there are no reliable experimental data for oxygen GB diffusion. So, in Figure 9, reported surface diffusion data line is drawn [26].

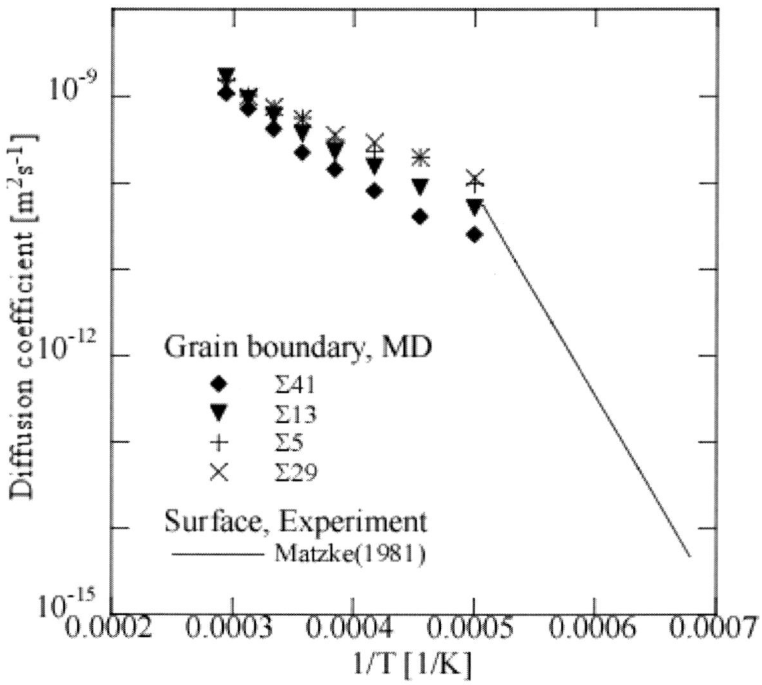

Figure 9. GB diffusion coefficients of oxygen in UO_2 obtained by MD calculations in comparison with experimental surface diffusion data [26].

In Figure 10, MD-simulated diffusion coefficients of uranium in GBs were compared with experimental data [27,30,31]. Our simulation data show that diffusion coefficients of GBs are much larger than that of bulk (Figure 5), and in addition, the extrapolations from MD results are almost comparable with experimental data of Reynolds [30] and Sato [31]. The calculated activation energy for uranium diffusion shows almost the same trend as that for oxygen diffusion. To understand the GB-structure dependence of the diffusion coefficient, the interfacial potential energy of each GB structure was calculated. The GB potential energy is given by

$$E_{\text{GB potential}} = \frac{E_{\text{GB supercell}} - E_{\text{Perfect supercell}}}{2A}, \tag{10}$$

where $E_{\text{GB supercell}}$ is the total potential energy of the supercell including GBs, $E_{\text{Perfect supercell}}$ is that of perfect crystal, A is the area of the GB interface. Since each GB supercell includes two GB interfaces in its inside, the denominator is multiplied by 2.

As an example, Figure 11 shows the GB energy of AmO_2, which has the same fluorite structure as UO_2 and is an important minor actinide (MA) in the nuclear fuel recycle. Indeed AmO_2 was found to show a closely similar MD-simulation behavior to UO_2. The GB potential energy initially increases with an increase of the misorientation angle, but turns to decrease slightly beyond the misorientation angle ~ 37° for Σ 5 GB, showing the relatively stable nature of Σ 5 GB structure. Experimental GB structure data for polycrystalline sintered UO_2 pellets are quite limited. So, the above MD result is an important first-step to clarify the effects of GB structure on material properties of fluorite oxides.

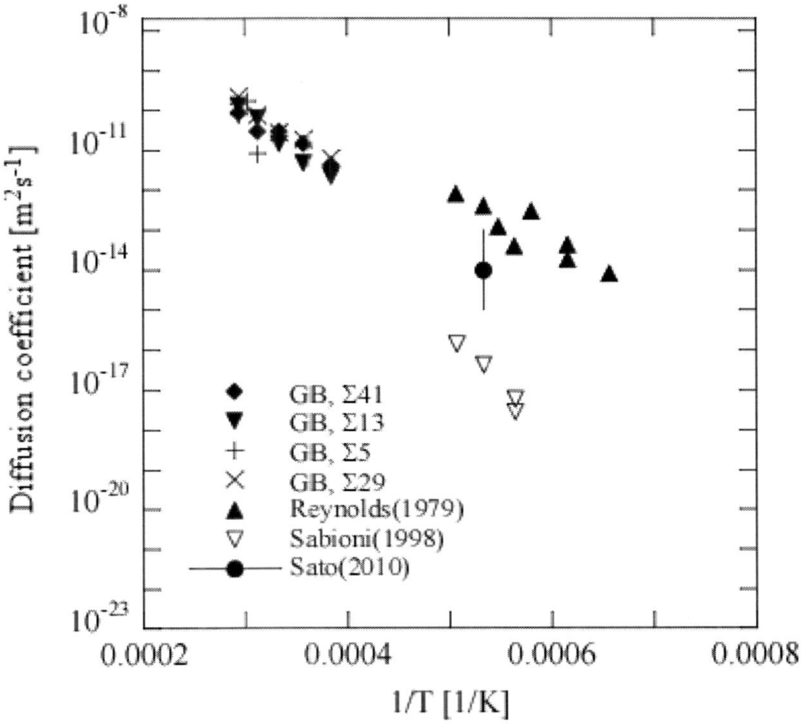

Figure 10. Uranium GB diffusion coefficients obtained for UO_2 supercell with CSL grain boundaries and their experimental literature cata [27,30,31].

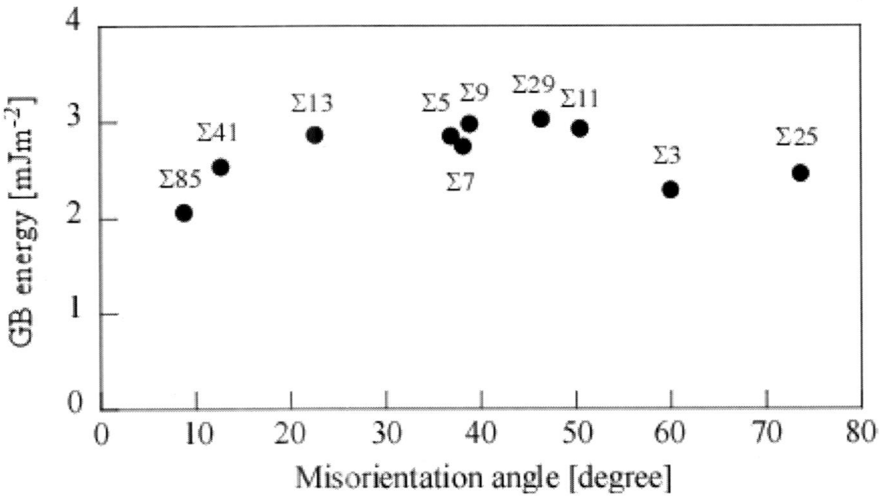

Figure 11. GB potential energy of AmO_2 as a function of misorientation angle.

4.3. Yttria-Stabilized Zirconia

Oxygen diffusion behavior in stabilized zirconia strongly depends on its lattice defect structure determined by aliovalent additive elements. MD simulations of yttria-stabilized zirconia (YSZ) were performed using Born-Mayer-Huggins type potential function in Eq. (1) by calculating the pair correlation function (PCF) which represents the local lattice structure. The PCF of each ion pair in YSZ, i.e., Zr^{4+}-Zr^{4+}, Y^{3+}-Y^{3+}, Zr^{4+}-Y^{3+}, Zr^{4+}-O^{2-}, Y^{3+}-O^{2-} and O^{2-}-O^{2-}, can provide requisite information for discussing the oxygen diffusion behavior. In Figure 12 all these PCFs are plotted as a function of distance between the respective ion pairs for YSZs with 8 and 24 mol% yttria [15]. The cation-cation nearest neighbor distances, Zr^{4+}-Zr^{4+}, Y^{3+}-Y^{3+} and Zr^{4+}-Y^{3+}, were almost independent of the yttria content. The cation-anion nearest neighbor distances, Zr^{4+}-O^{2-} and Y^{3+}-O^{2-}, decreased with increase in the yttria content. On the contrary, the anion-anion nearest neighbor distance, O^{2-}-O^{2-}, increased with the yttria content. The extended X-ray-absorption fine structure (EXAFS) measurements showed that distances between the cation-cation and cation-anion do not change much or decrease a little with increasing the yttria content [32]. Ionic radii of Zr^{4+}, Y^{3+} (at CN=6) and O^{2-} (at CN=4) are 0.72, 0.90 and 1.38 Å, respectively [33], the added Y^{3+} ion being larger than Zr^{4+}. Only in terms of ionic radii, these PCF results cannot be therefore explained well, for with Y^{3+} substitution for Zr^{4+} oxygen vacancies are introduced into the anion sublattice. Our previous study of YSZ showed that, when oxygen vacancy exists in the nearest neighbor site, O ions are inclined to move toward it due to the local-structural relaxation effects resulting from Columbic (plus elastic) interactions such as the repulsive one between O ions and the attractive one between Zr^{4+} and O^{2-} ions [15]. As pointed out by Li and Hafskjold [34], the nearest neighbor O-O distance therefore expands a little with Y^{3+} substitution, and such expanded O-O distance caused by oxygen-vacancy formation can be observed as a hump in PCF around $r \sim 3$ Å for YSZ with 24 mol% yttria in Figure 12(f).

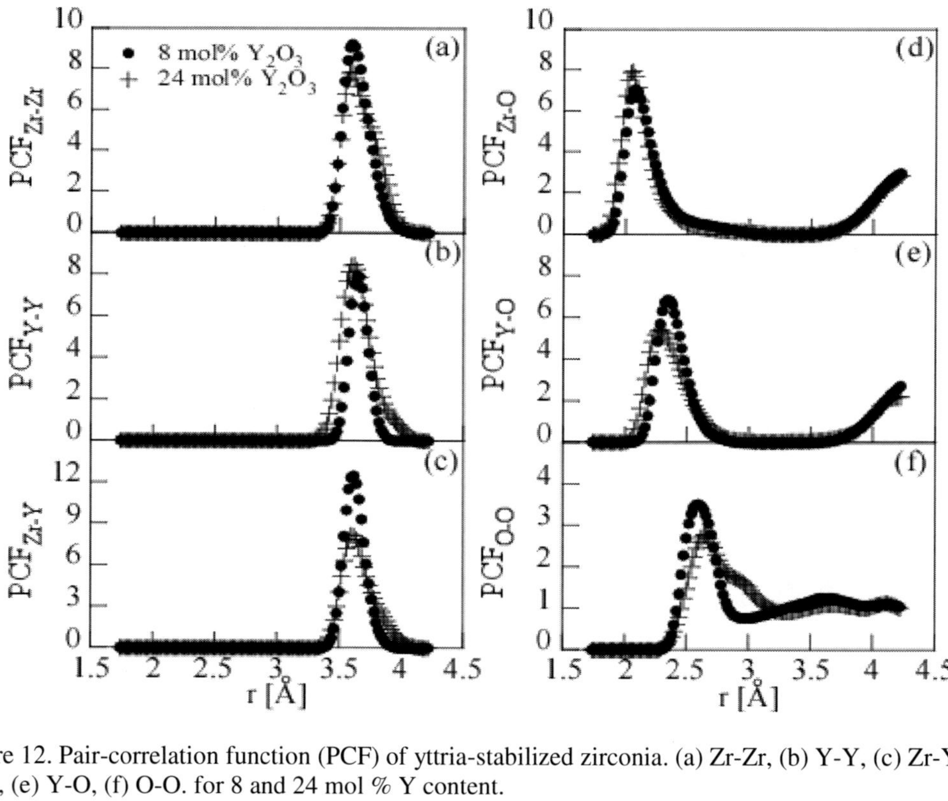

Figure 12. Pair-correlation function (PCF) of yttria-stabilized zirconia. (a) Zr-Zr, (b) Y-Y, (c) Zr-Y, (d) Zr-O, (e) Y-O, (f) O-O. for 8 and 24 mol % Y content.

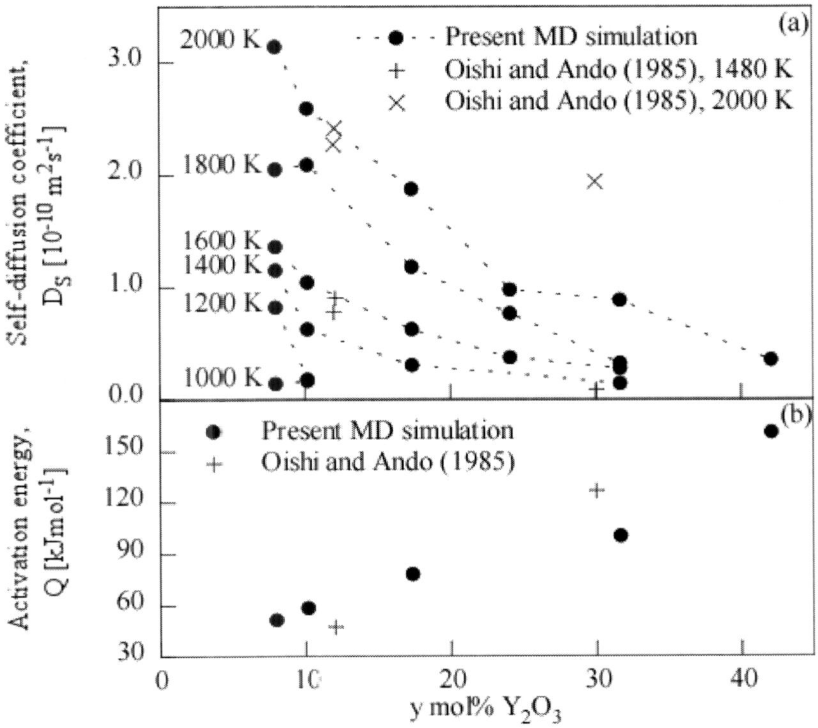

Figure 13. Self-diffusion coefficient of oxygen (a) and its activation energy (b) in YSZ.

Obviously, the self-diffusion coefficient of oxygen in this representative oxide-ion conductor YSZ is much greater than those of Zr and Y ions. In Figure 13(a), its simulated values with Eq. (7) are plotted as a function of yttria content for temperatures from 1000 to 2000 K. The oxygen self-diffusion coefficient of increases with temperature, but, decreases with increasing yttria content at constant temperature. Experimental data also show the similar tendency [35]. Figure 13(b) shows the activation energy plotted against yttria content. The activation energy increases nearly linearly with the yttria content. In the simplest case oxygen vacancy on the tetrahedral site would exchange the position with the nearest neighbor O ion through the mid-saddle point between the cations that compose the tetrahedron. As the yttria content increases, the case that the O ion would migrate through the midpoint between Zr^{4+}-Y^{3+} or Y^{3+}-Y^{3+} rather than that between Zr^{4+}-Zr^{4+} is expected to increase. The cation-cation distance scarcely changes or decreases only a little, as shown in Figure 12. However, Y^{3+} is ~20% larger than Zr^{4+}. Consequently, the O ion is forced to pass through the narrower diffusion path. The activation energy of oxygen self-diffusion therefore increases, thereby decreasing the self-diffusion coefficient itself with increasing the yttria content.

5. Thermal Conductivities of Fluorite-based Oxides

5.1. Methodology for Determining Thermal Conductivity

In the equilibrium system, the thermal conductivity is calculated from the transport coefficients of the phenomenological transport equations composed of the charge current, energy current and the cross-coupling term. The macroscopic charge and energy fluxes ($\vec{J}_Z$ and $\vec{J}_E$) are given by [36-38],

$$\vec{J}_Z = -L_{ZZ} T \nabla\left(\frac{\mu_Z}{T}\right) - L_{ZE} T \nabla(T), \tag{11}$$

$$\vec{J}_E = -L_{EZ} T^2 \nabla\left(\frac{\mu_Z}{T}\right) - L_{EE} T \nabla(T). \tag{12}$$

In the above equations, $L_{\alpha\beta}$ (α, $\beta = Z, E$) is the phenomenological coefficient and μ_Z is the electrochemical potential. The Onsager reciprocal relation gives $L_{ZE} = L_{EZ}$, so that the thermal conductivity can be expressed by

$$\kappa = L_{EE} - \frac{L_{EZ}^2}{L_{ZZ}} T. \tag{13}$$

According to the Green-Kubo relationship based on the fluctuation-dissipation theorem, such phenomenological coefficients, which relate to non-equilibrium phenomena, can be calculated under a thermal equilibrium condition. Therefore, in the EMD scheme, the phenomenological coefficients are given by the time integrals of the auto-correlation functions of microscopic fluxes, and $\vec{j}_E$,

$$L_{ZZ} = \frac{1}{3Vk_B T} \int_0^\infty dt \left\langle \vec{j}_Z(t) \cdot \vec{j}_Z(0) \right\rangle, \tag{14}$$

$$L_{EZ} = \frac{1}{3Vk_B T^2} \int_0^\infty dt \left\langle \vec{j}_E(t) \cdot \vec{j}_Z(0) \right\rangle, \tag{15}$$

$$L_{EE} = \frac{1}{3Vk_B T^2} \int_0^\infty dt \left\langle \vec{j}_E(t) \cdot \vec{j}_E(0) \right\rangle. \tag{16}$$

Here k_B is the Boltzmann constant and V is the cell volume for the simulation, and where;

$$\vec{j}_Z(t) = \sum_{i=1}^{N} z_i e \vec{v}_i, \tag{17}$$

$$\vec{j}_E(t) = \sum_{i=1}^{N} \left[\frac{m_i v_i^2}{2} + \frac{1}{2} \sum_{j \neq i} U(r_{ij}) \right] \vec{v}_i + \frac{1}{2} \sum_{i=1}^{N} \vec{v}_i \cdot \vec{r}_{ij} \vec{F}_{ij}, \tag{18}$$

where m_i is the mass of the ith particle and v_i is its velocity. The last term, including the Coulomb potential ($U(r_{ij})$: see Eqs. (4) and (5)) in Eq. (18), is relatively complex compared to the Lennard-Jones potential [39-41]. This term has been already formalized for binary ionic systems, e.g., NaCl, KCl, CaF_2, UO_2, *etc.* by several researchers [37,38,42]. In our study, the algorithm proposed by Motoyama was incorporated into the MXDORTO program [42], and MD simulations with the Busing-Ida type potential function in Eq. (2) have been performed.

Each transport coefficient is given by an integral of the auto-correlation function (ACF) of the corresponding fluxes. For $Er_{0.05}Y_{0.1}Pu_{0.2}Zr_{0.65}O_{1.925}$ at 2000 K, ACFs and their integrals of energy, charge, and coupling fluxes are plotted as a function of time in Figure 14. Here, the ACFs plotted are normalized ones (NACFs) relative to the value of ACF at 0.0 ps. All NACFs approach to almost zero within ~ 0.3 ps. Simultaneously, each transport coefficient approaches its final value. Especially, L_{EZ} in (b) clearly takes a negative value at this temperature, which indicates strong coupling between energy and charge currents.

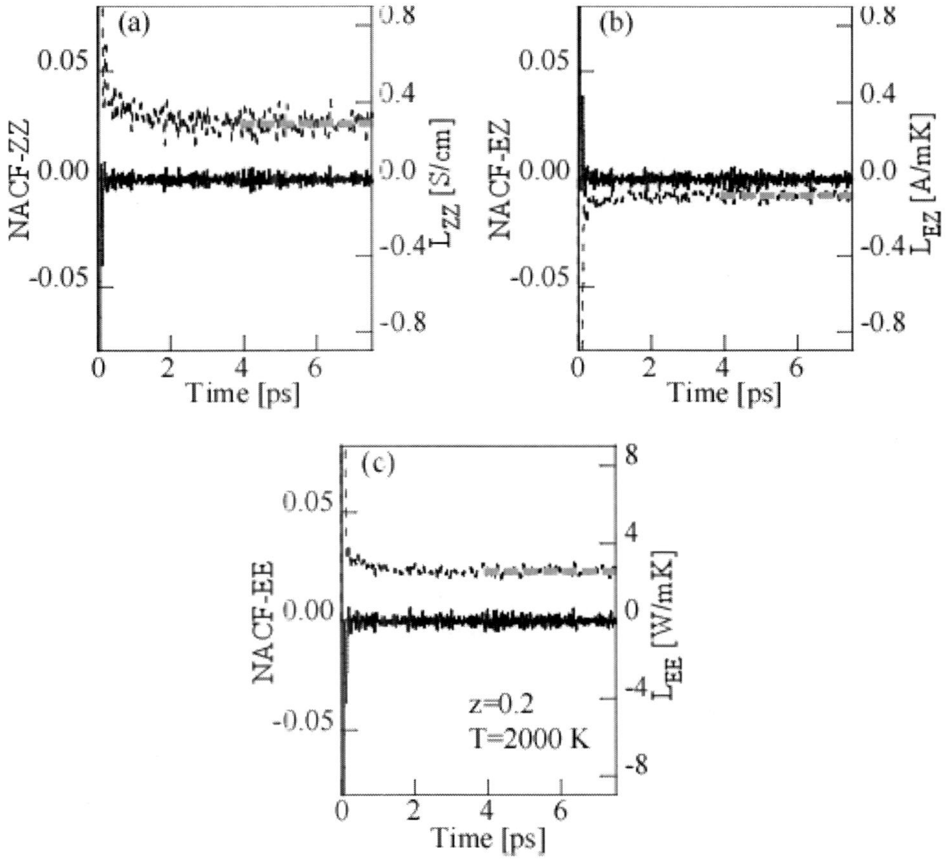

Figure 14. Phenomenological transport coefficients (grey) and auto-correlation functions (black) of IMF $Er_{0.05}Y_{0.1}Pu_{0.2}Zr_{0.65}O_{1.925}$ with Pu content z = 0.2 at T = 2000 K.

According to the homogeneous non-equilibrium MD (NEMD) method proposed by Evans [43], the dynamics of the particles in the simulated cell are governed by the following equations of motion:

$$\frac{d\vec{q}_i}{dt} = v_i , \tag{19}$$

$$\frac{d\vec{p}_i}{dt} = \vec{F}_i + \vec{\vec{D}}_i \vec{F}_{ext}(t), \tag{20}$$

Where q_i and p_i are the generalized coordinate and momentum of the particle i, v_i the velocity, the force, the perturbed external force coupled with the tensor parameter. When $\vec{F}_{ext} = 0$, the system is under the thermal equilibrium condition. The time-derivative of Hamiltonian that is the internal energy under adiabatic condition is, therefore, given by

$$\left(\frac{dH}{dt}\right)_{ad} = \vec{F}_{ext}(t)\sum_{i=1}^{N}\frac{\vec{\vec{D}}_i \cdot \vec{p}_i}{m_i} = \vec{F}_{ext}(t)\cdot\vec{j}_E, \tag{21}$$

Where J_E is the heat current. In the linear response theory, if the physical observable is the heat current, then the time-averaged heat current can be expressed with Eq. (22) via Eqs. (19-21).

$$\left\langle \vec{j}_E(t) \right\rangle_t = \frac{\vec{F}_{ext}}{k_B T}\int_0^{\infty} dt \left\langle \vec{j}_E(t)\cdot\vec{j}_E(0)\right\rangle_{eq}, \tag{22}$$

where each particle is perturbed in one direction by time-independent $\vec{F}_{ext}$. On the other hand, in the equilibrium system where the perturbed field is zero, the thermal conductivity is given by the Green-Kubo relationship (see Eq. (16)). As a result, Eq. (22) is re-written as

$$\kappa = \lim_{F_{ext}\to 0}\lim_{t\to\infty}\frac{\left\langle \vec{j}_E(t)\right\rangle}{VTF_{ext}}. \tag{23}$$

Therefore, the thermal conductivity is given as the proportional constant between the time-averaged heat current and the perturbed external force field in the present NEMD.

Figure 15 shows the time-averaged heat flux as a function of F_{ext} for UO_2. For small F_{ext} values, the heat flux increases linearly with F_{ext} at each temperature. But, it abruptly increases at a certain critical F_{ext} value, which means the breakdown of the linear response theory. In such a case, temperature cannot be controlled and increases with time. In addition, this figure shows that the maximum F_{ext} value attainable in the linear-response regime decreases with decrease in temperature. It means that the perturbed heat flux is very sensitive to F_{ext} value at low temperature. In other word, the temperature of the system is more and more sensitively influenced by the applied F_{ext} value with decreasing temperature, so that the temperature control becomes difficult at low temperature. Once such allowed maximum F_{ext} can be known with reasonable accuracy to avoid the nonlinear increase of heat flux with F_{ext}, thermal conductivity of the system can be determined precisely and efficiently. In principle, it is preferable for F_{ext} to be as small as possible for the linear response theory to be applicable. However, if F_{ext} is too small, its effect on perturbed heat flux will become also small. As a consequence and compromise, a modestly large value of F_{ext} within the linear response region is required to precisely determine the thermal conductivity of the system.

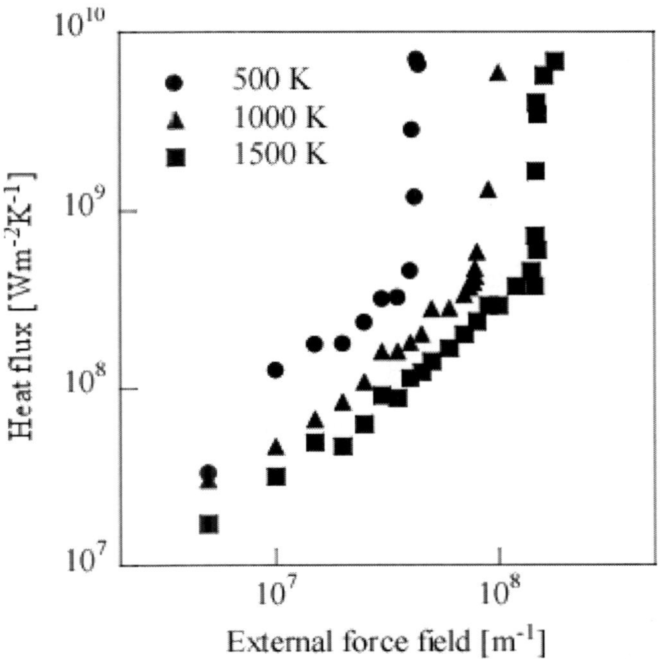

Figure 15. Heat flux as a function of external force field for UO_2.

5.2. Thermal Conductivity of Hyper-Stoichiometric UO_{2+x}

The MD-simulated thermal conductivity of UO_{2+x} is plotted as a function of temperature in Figure 16, together with experimental results [44]. The experimental thermal conductivity data decrease with increasing both temperature and oxygen nonstoichiometry, *x*, but seems to increase slightly at high temperatures due to the contribution from the conduction electrons not considered in the classical MD simulations dealing with only the phonon-conduction contribution. For $x > 0$, excess oxygens scatter the phonons and decrease the thermal conductivity. For the stoichiometric UO_2 (x=0), the thermal conductivity calculated with NEMD is relatively large at low to middle temperatures (< 1400 K). Moore et al. showed that the phonons were also scattered by the magnetic moment of U and its contribution was independent on temperature above 80 K [45]. In addition, its contribution was much larger than that from scattering by grain boundaries. Therefore, the thermal conductivities of single crystal and polycrystalline UO_2 were supposed to be comparable in the wide range of temperature [45]. Considering the contribution of phonon scattering by magnetic moment of U and unexpected lattice defects (e.g., impurities), the thermal conductivity calculated with NEMD seems to be overestimated for the stoichiometric UO_2 at low to middle temperature range in comparison with experimental data of Lucuta [44]. As to the difference between EMD and NEMD results, these should not be different in principle, if both the size of supercell and the total number of calculation step are large enough. In the present case, the supercell consisting of 3 × 3 × 3 unit cells is used. The effects of these two factors on the calculated EMD and NEMD results are not clear at present. Especially in case of $x = 0$ and low temperature, such effects seem serious, so that the NEMD thermal conductivity may be overstimated [46]. For UO_2, the thermal conductivities calculated with Eq. (16), which is an integral of auto-correlation function of heat currents, converged extremely slowly with integration time and its fluctuation was estimated to be approximately 30 % at $T \leq 700$ K. At higher temperature, the fluctuation was conservatively estimated to be 20-25 %.

As shown in Figure 16, MD calculated-thermal conductivity can be obtained in the temperature range comparable with experimental one unlike oxygen and uranium diffusion cases, because the thermal diffusion originates from lattice vibration. However, both low (< room temperature) and very high temperature (> ca. 2000 K) ranges are exceptions, because the quantum effect and the radiation conduction should be considered, respectively.

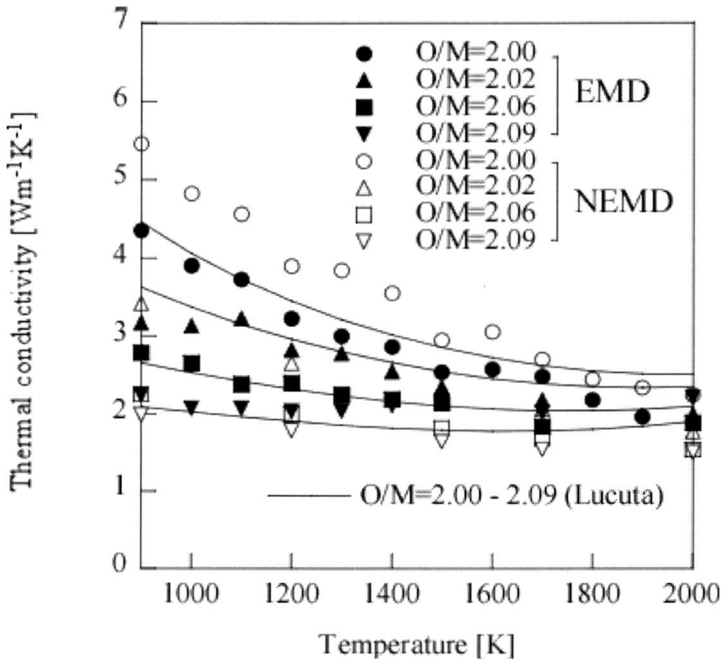

Figure 16. Thermal conductivity of hyper-stoichiometric uranium oxides UO_{2+x}.

To identify the detailed structure of oxygen defect cluster in hyper-stoichiometric UO_{2+x} (Figure 1(b)), which has direct close connection with the present thermal-conductivity problem, still remains as a largely-unresolved challenging question even today. Willis proposed the V_O : O' : O" = 2 : 2 : 2 -type and related cluster models in his early studies [7]. Furthermore, a more complex oxygen cuboctahedron (COT) cluster model is proposed based on neutron-diffraction [47-50] and EXAFS [51] experiments. Geng et al. showed the stability of COT clusters by a first-principle calculation [51,52]. Conradson et al. proposed an oxygen cluster associated with oxo group by EXAFS measurements [53], and also the split di-interstitial and related clusters are porposed in recent first-principle calculations [54]. In our MD calculation with the Busing-Ida-type potential function, both COT-o and -v clusters (in Figure 17) have been observed in $UO_{2.09}$ when such complex Willis clusters were placed as the initial state of supercell. Figure 17(a) shows the snapshot of $UO_{2.09}$ supercell annealed at 1000 K, wherein (b) COT-o and (c) COT-v clusters are found in it. Whereas the primitive Willis 2 : 2 : 2-type cluster can never be found in this system. Thus, COT-o and -v clusters could be more stable than Willis cluster at high temperature.

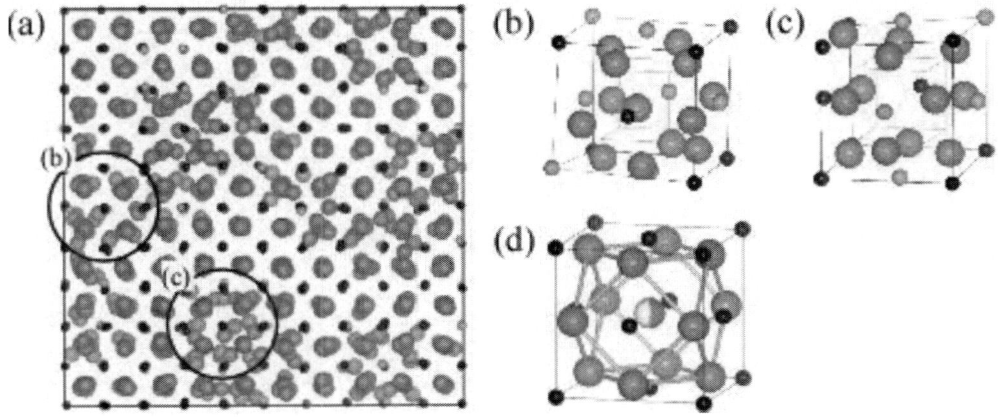

Figure 17. Oxygen defect structures found in $UO_{2.09}$ supercell. (a) Snapshot of supercell annealed at 1000 K; (b) COT-o cluster; (c) COT-v cluster: (d) the ideal COT-o (or v) cluster. Here, COT: cub-octahedron consisted of 12 O^{2-} ions, COT-o (or v) means oxygen interstitial Oi (or vacancy Vo) exists at the center.

5.3. Thermal Conductivity of Hypo-Stoichiometric (U, Pu)O_{2-x}

Figure 18 shows the thermal conductivity of stoichiometric $U_{1-y}Pu_yO_{2.0}$ crystals with y = 0.0-0.3. The MD-simulated thermal conductivity was almost constant independent of Pu content, y. At low temperature, the MD results are somewhat higher than the estimated (Gibby [55]) and experimental data (Goldsimith [56]) reporting the tendency that the thermal conductivity slightly decreases with increasing y from 0.0 to 0.3. On the other hand, Duriez concluded that the thermal conductivity of stoichiometric $U_{1-y}Pu_yO_{2.0}$ solid solution was independent of Pu content y from 0.05 to 0.15 [57]. Philipponneau and Inoue also reported that Pu content had no significant effect on the thermal conductivity for $U_{1-y}Pu_yO_{2.0}$ [58, 59]. Its influence seems to be even smaller at high temperature. It may therefore be concluded that Pu^{4+} ion as a lattice defect for the host UO_2 crystal little influences the thermal conductivity of the stoichiometric solid solution within Pu content y range from 0.0 to 0.3, although the ionic radius and the mass both differ a little between Pu^{4+} and U^{4+} ions.

The thermal conductivities of hypo-stoichiometric $U_{0.8}Pu_{0.2}O_{2-x}$ solid solutions with y = 0.2 and various x (i.e. O/M ratio) were plotted as a function of temperature in Figure 19. For the stoichiometric solid solution (x = 0), the thermal conductivity tends to decrease with increasing temperature, whereas for the hypo-stoichiometric solid solutions, its value as well as its temperature dependence becomes steadily smaller with increasing x. This means that the lower the temperature larger the effect of lattice defect (i.e., oxygen vacancy) to the thermal conductivity. At high temperature, Umklapp process is the dominant mechanism of thermal conductivity. Both the EMD- and NEMD-simulated thermal conductivities are in good agreement with experiment data except for the stoichiometric $U_{0.8}Pu_{0.2}O_2$ [57]. For stoichiometric $U_{0.8}Pu_{0.2}O_2$ the simulated ones gave the significantly larger thermal conductivity than the experimental curve in the low to middle temperature range. This might be due to the contribution from additional phonon scattering by the U^{4+} magnetic moment [45] that has not incorporated in the present MD simulations.

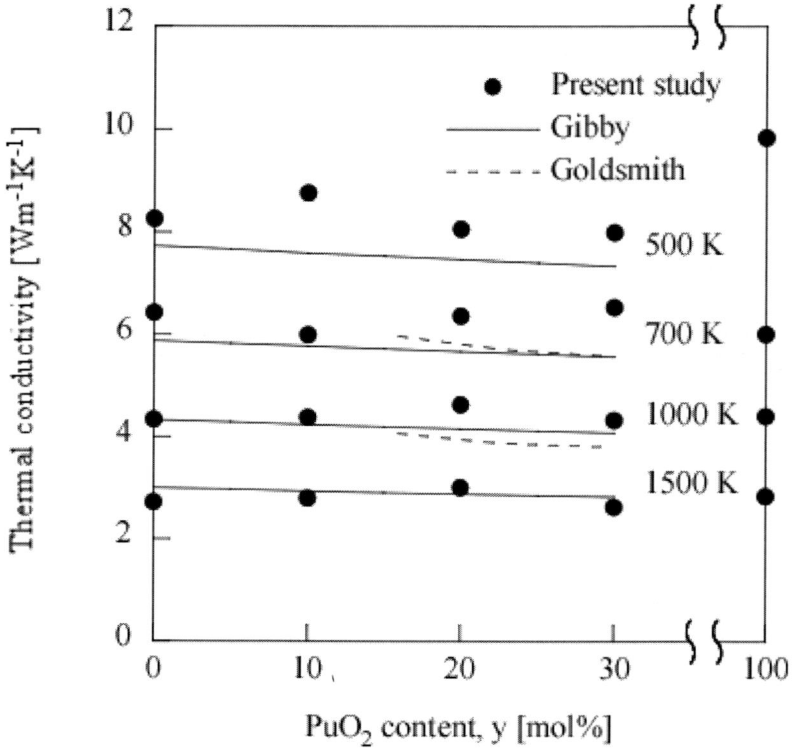

Figure 18. Thermal conductivity of stoichiometric $U_{1-y}Pu_yO_{2.0}$ as a function of *y*.

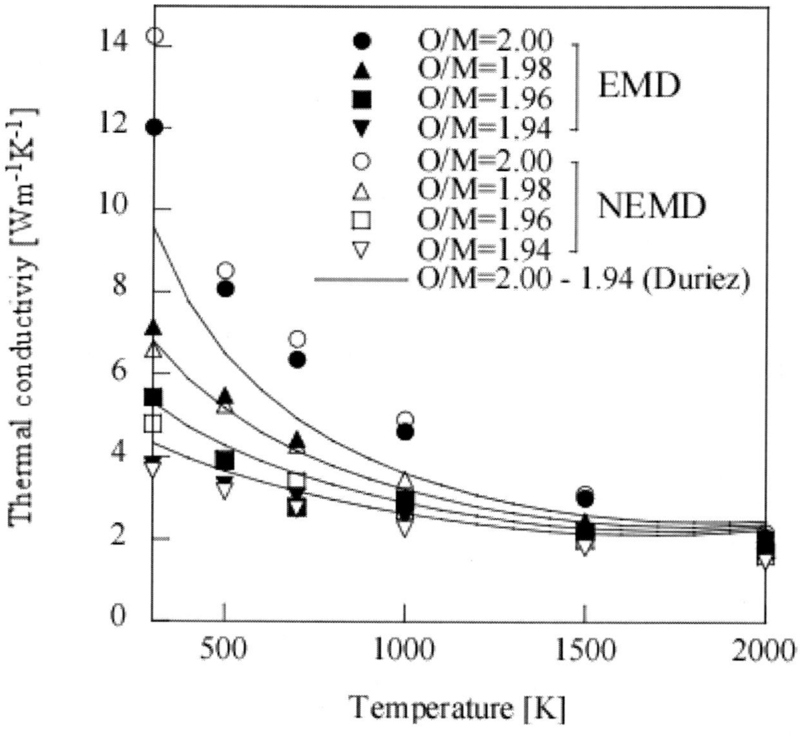

Figure 19. Thermal conductivity of hypo-stoichiometric $U_{0.8}Pu_{0.2}O_{2-x}$ solid solutions.

5.4. Thermal Conductivity of Yttria-Stabilized Zirconia and IMFs

The EMD-simulated thermal conductivity of yttria-stabilized zirconia, $Y_yZr_{1-y}O_{2-y/2}$ (y = 0.148, 0.185) is shown in Figure 20, where, the results from NEMD simulation for a heterogeneous system (Shelling [60]) and experimental data for $Y_{0.179}Zr_{0.821}O_{1.9105}$ single crystal [61] are also plotted. In our EMD simulations the thermal conductivity is calculated with Eq. (13), including the cross-coupling term between energy and charge currents. Our EMD results show that the thermal conductivity slightly decreases with increase in temperature. A similar trend is seen in NEMD [60] and experimental data [61]. Though not shown in Figure 20, the thermal conductivity of polycrystalline yttria-stabilized zirconia (cubic, 15 wt% Y_2O_3: $y = 0.16$) was measured by Raghavan [62] and was determined to be ~2.4 $W\cdot m^{-1}\cdot K^{-1}$ at $T \leq 1073$ K, in good agreement with our results at $y = 0.185$. In addition, Figure 20 shows that our calculated thermal conductivity slightly decreases with increasing y, although the calculation error is relatively large. This indicates that the amount of oxygen vacancy which increases with Y doping suppresses the thermal conductivity of the system.

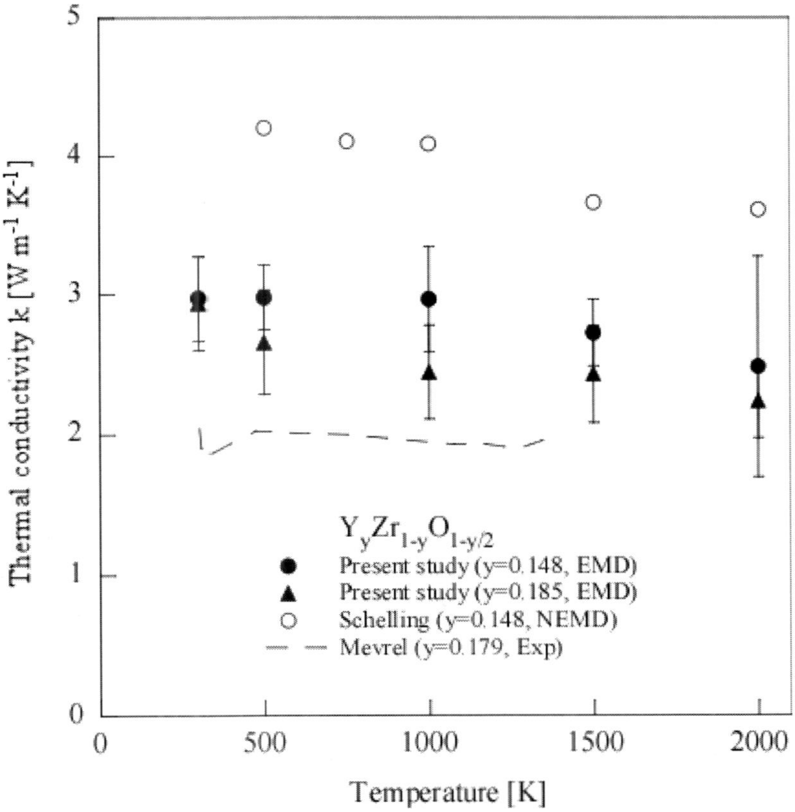

Figure 20. Thermal conductivity of yttria-stabilized zirconia.

Figure 21 shows the thermal conductivity of $Er_{0.05}Y_{0.1}Pu_zZr_{0.85-z}O_{1.925}$. The present EMD results of Pu-doped zirconia show the similar temperature dependence to yttria-stabilized-zirconia (Figure 20), and the calculated thermal conductivity decreases with an increase of Pu content, z. In addition, they are in good agreement with experimental data [63]. In Figure 21, the thermal conductivity of stoichiometric PuO_2 with no extrinsic oxygen vacancies obtained in our previous EMD simulation is also plotted [64]. The thermal conductivity of PuO_2 decreases steeply with increasing temperature, approaching to those of Pu-doped IMFs at the highest temperature range around 2000 K. These results indicate that in close parallel with MOX case (Figure 19) Umklapp phonon scattering process (i.e., intrinsic thermal resistance) dominates at high temperature range, while the impurity (i.e., oxygen vacancy) scattering process makes a major contribution for lowering the thermal conductivity in the low temperature range.

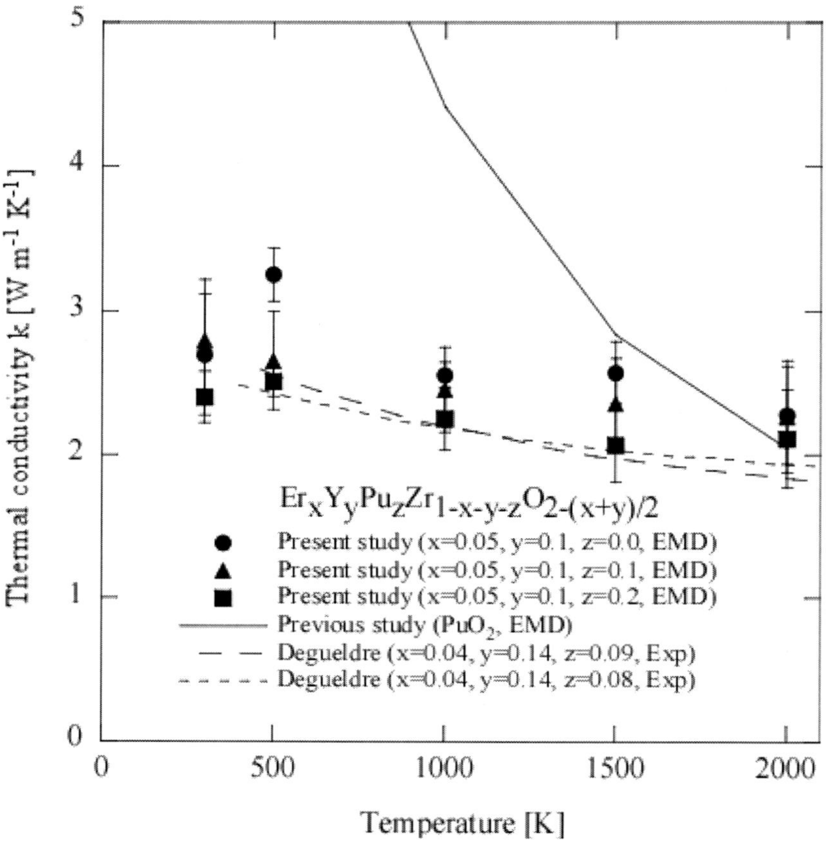

Figure 21. Thermal conductivity of zirconia-based IMF.

6. MELTING POINT OF UO_2

6.1. Methodology for Determining Melting Point

Two types of MD simulation were tested for evaluation of melting point. One was the one-phase simulation (OPS), and the other was the two-phase simulation (TPS) [5,65,66]. For OPS, the supercell at the initial state was prepared as the crystalline solid, and the MD calculation was subsequently performed at a desired temperature and pressure (*NPT* ensemble).

After enough time elapsed to reach the equilibrium state of the system, a determination was made whether the system was liquid or still solid based on the density and enthalpy of

UO_2 and visual information of the crystal structure. On the other hand, in TPS, crystalline solid and liquid phases coexisted in the supercell at the initial state [5,65,66]. Both solid and liquid phases were equilibrated for 20,000 steps (= 40 ps) either at 300 or 4,000 K. Both cells of the same size were subsequently placed side-by-side as the initial supercell. For instance, Figure 22(a) shows the supercell at the initial state 8 × 8 × 8 × 2 unit cells for TPS. Here, the last digit of 2 in the cell size means number of phases. The other processes of TPS were the same as those of OPS.

The configuration of the initial supercell with Schottky defects was made as follows: One Schottky defect pair consists of one uranium vacancy and two oxygen vacancies (Eq. (9)). One uranium ion was randomly removed from its cation sublattice site, and then the two nearest neighbor oxygens were removed from its anion sublattice site.

This procedure was repeated until the number of pairs of the Schottky defect reached the desired level. The solid and liquid cells of the initial supercell in TPS were thermally equilibrated independently, and MD calculations with Schottky defects were performed at desired temperature and pressure for the joining cell. For all the calculations, we varied the temperature at an interval of 50 K. Therefore, the error of the melting point due to this procedure is estimated to be ±25 K, which is drawn as 'error bar' in the melting point plotted in the following Figures 24 and 25.

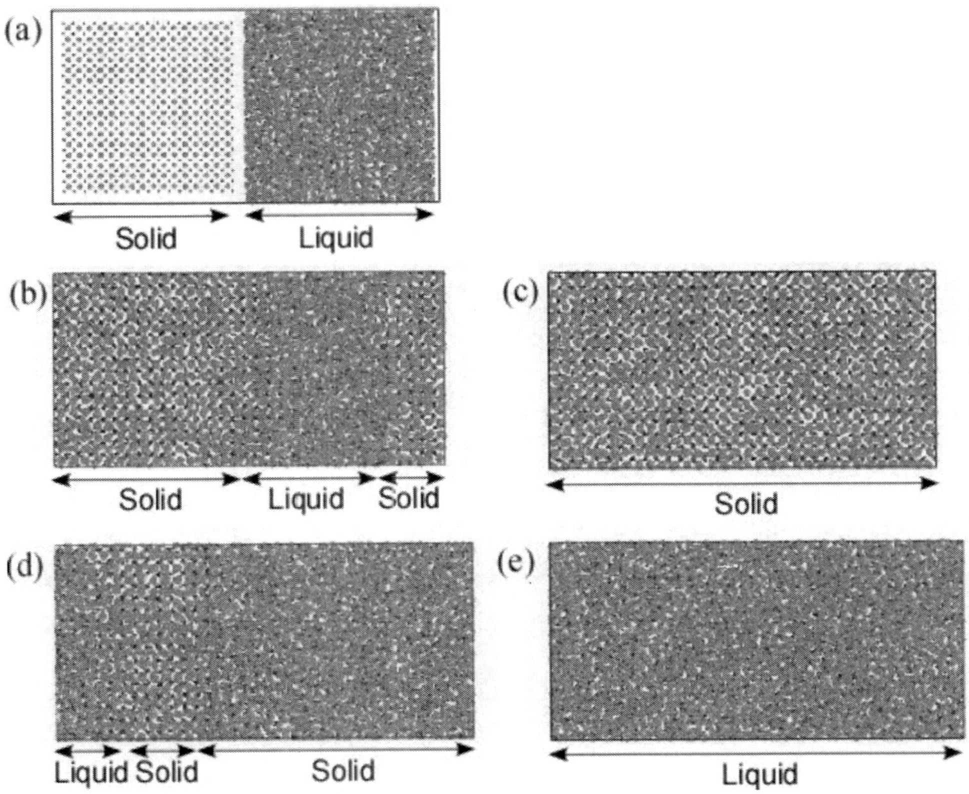

Figure 22. Crystallization and melting behavior of UO_2 observed in two-phase simulation (TPS). (a) initial state of supercell; (b) and (c) show the supercells at 3150 K for crystallization, where elapsed times are 800 and 1200 ps, respectively. (d) and (e) show the supercells at 3200 K for melting, where elapsed times are 120 and 200 ps, respectively.

6.2. One-Phase Simulation (OPS)

Melting point calculations in OPS in *NTP* ensemble were performed for the supercell of 3 × 3 × 3 fluorite unit cells using Basak [67], Arima [17] and Yakub [18] potentials. The calculated values of densities of UO_2 are plotted as a function of temperature in Figure 23, together with experimental data [68-71]. It is found that each calculated density decreases gradually with increase in temperature and drops abruptly at the melting point. Melting points (T_{MP}) obtained by Yakub, Basak and Arima potentials are 3625 K, 4225 K and 4550 K, respectively. These results indicate that the melting point becomes higher with the increase in the degree of ionicity of the employed potentials. In comparison with INSC (International Nuclear Safety Center) recommended data of 3120±30 K [68], the melting points obtained in OPS are found to be much higher for Basak and Arima potentials, even the least-ionic Yakub potential giving the closest but still significantly higher T_{MP}.

In addition, OPS has a drawback that is revealed in the solidification behavior; when the supercell was thermally equilibrated at 4000 K (in liquid state) and then cooled gradually down to 300 K, the resultant UO_2 did have not the crystalline but the amorphous phase.

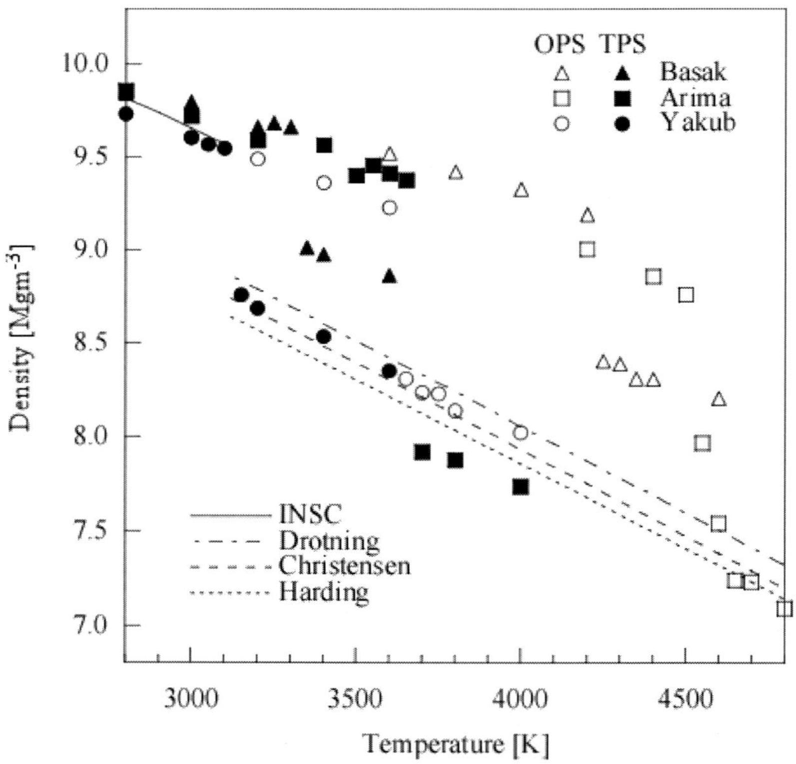

Figure 23. UO_2 density vs. temperature plot obtained by OPS and TPS, together with experimental data.

6.3. Two-Phase Simulation (TPS)

First, we performed TPS in the *NTP* ensemble for the supercell consisting of solid and liquid phases where each cell size was 3 × 3 × 3 unit cells. UO_2 densities obtained by TPS are also shown in Figure 23. TPS gives melting point of 3125, 3325 and 3675 K for Yakub, Basak and Arima potentials, respectively, which are all much lower than those respective obtained in the corresponding OPS. Yakub potential gives a comparable value with the INSC recommended data [68]. This difference in the derived T_{MP}s between OPS and TPS primarily stems from the fact that the initial supercell of TPS has two different kinds of phases of solid and liquid UO_2 at the initial state, and consequently the disorder in the liquid phase can propagate easily into the solid phase at lower temperatures. On the contrary, the initial supercell of OPS does not contain disordered phase which acts as a trigger for melting, and results in too high a melting point calculated.

Next, we investigated the dependence of melting point on the supercell size which was varied from 3 × 3 × 3 × 2 to 10 × 10 × 10 × 2 unit cells using Yakub potential. The small supercell size is desirable for efficient calculation, but the large size is better for precise calculation. The results show that the calculated melting point increases slightly with the supercell size and the value obtained for the largest cell was estimated to be 3175±25 K. Therefore, a supercell size greater than 7 × 7 × 7 × 2 unit cell is enough to precisely evaluate the melting point. For TPS with a supercell of 8 × 8 × 8 × 2 unit cells, the atomic configurations at 3150 and 3200 K are visually shown in Figures 22(b)-(e): In snapshot (b) at 800 ps at 3150 K, the solid phase expands into the liquid region. After 1200 ps in (c) the supercell completely became solid. In contrast, at 3200 K, the liquid phase still grows at 120 ps in (d), and the supercell is completely melted at 200 ps in **(e)**.

Melting behavior of UO_2 relates to the U and O diffusion. Yakub potential is thus judged to give the greater diffusion coefficients of both ions than Arima potential, as seen in Figure 23, consequently giving the lower melting temperature. At present, Yakub potential seems to be most suitable for evaluation of thermal and transport properties of UO_2 by MD simulation.

6.4. Pressure Dependence of Melting Point

The pressure dependence of melting point was investigated to clarify the relative phase stability of solid and liquid UO_2. All calculations were performed in the *NPT* ensemble for 8 × 8 × 8 × 2 unit cells using Yakub potential. Pressure was changed from 0.1 MPa (~1 atm) to 3000 MPa (~3 × 10^4 atm). Figure 24 shows that the melting point of UO_2 linearly increases with applied pressure. This result leads to the following T_{MP} – pressure (p) relationship;

$$T_{MP}\ [\mathrm{K}] = (3178 \pm 20) + (11.5 \pm 1.1) \times 10^{-2} p, \tag{24}$$

where p is pressure in MPa. Manara et al. [72] have experimentally determined the melting point of UO_2 as a function of pressure in 10 - 250 MPa range, and derived the relationship; T_{MP} [K] = 3147 + 9.29 × 10^{-2} p. Both the melting point at 0 MPa and the slope of dT_{MP} / dp obtained from our TPS are a little larger than those of the latter.

The Clausius-Clapeyron equation gives the classical thermodynamic relationship between melting point and pressure of the system as follows,

$$\frac{dT_{MP}}{dp} = \frac{T_{MP}\Delta V_m}{\Delta H_{fusion}}, \tag{25}$$

where ΔV_m is the molar-volume change on melting (fusion) and ΔH_{fusion} the corresponding enthalpy change of the system. The left-hand side of Eq. (25) is already known from the above-derived T_{MP} – pressure (p) relationship in Eq. (24), and the right-hand side can be independently evaluated. To do this, we used the MD results obtained for the largest two supercells (8 × 8 × 8 × 2 and 10 × 10 × 10 × 2 unit cells): These give the average $\Delta V_m = 2.50 \times 10^{-6}$ $m^3 \cdot mol^{-1}$and $\Delta H_{fusion} = 60.38$ $kJ \cdot mol^{-1}$. Consequently, the right-hand side of Eq. (25) was estimated to be 13.13×10^{-2} $K \cdot MPa^{-1}$, in reasonable agreement with the dT_{MP}/dp value of 11.5×10^{-2} $K \cdot MPa^{-1}$ in Eq. (24). So, the present MD simulations are judged to be thermodynamically consistent, confirming that the Clausius-Clapeyron equation indeed holds here.

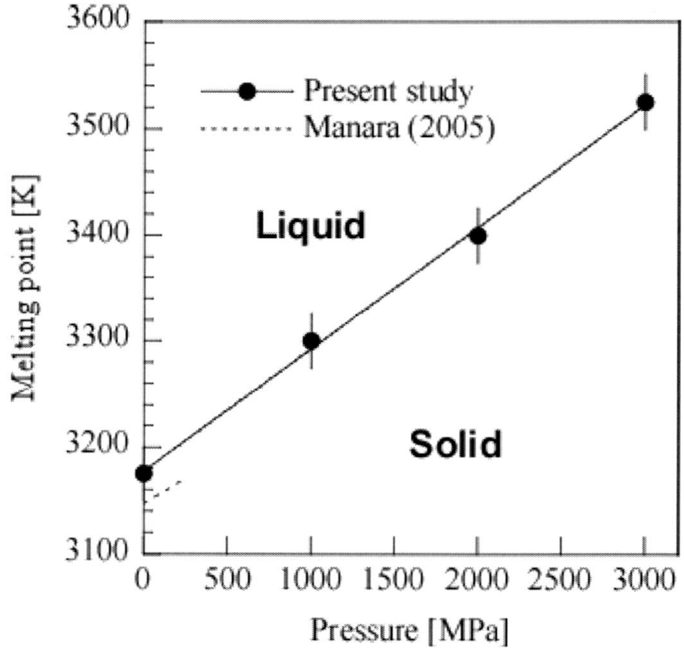

Figure 24. Melting point of UO_2 as a function of pressure in TPS.

6.5. Effect of Schottky Defects

The effect of Schottky defects on melting point was investigated by TPS with Yakub potential. As described in 4.2, Schottky defect in UO_2 consists of one uranium vacancy and two oxygen vacancies (Eq. (9)). A maximum of 150 pairs of Shottky defects corresponding to a maximum 7.3 % (= {150 / (8 × 8 × 8 × 4)} × 100, where 4 in the denominator is a number of U ions in the unit cell) defect concentration in Figure 25 was introduced into the 8 × 8 × 8 unit cells. These defect concentrations may seem to be somewhat too high; however, two rationalizations for such high defect concentration in UO_2 in the vicinity of T_{MP} appears possible: Firstly, as afore-argued in 4.2 for its high-temperature oxygen diffusion behavior shown in Figure 2, its oxygens are most likely in a highly-disoredred 'sublattice melt state', probably greatly enhancing the cation-sublattice disorder too, and second, under high irradiation, especially in the rim region of UO_2 fuels, the occurrence of such highly-disoredred both cation and anion state seems increasingly more realistic [73,74]. This study therefore also stands as a preliminary trial attempt for investigating such highly-defective state. It is found in Figure 25 that the melting point of UO_2 decreases with an increase in the concentration of Schottky defects. Kurosawa pointed out that the melting of ionic crystals is closely related to the amount of lattice defects as well as to the strength of lattice vibrations [75]. It can be interpreted that the Schottky defect which locally disorders the crystallinity of the system causes the lowering of melting point. In real UO_2 crystals, Schottky defects exists a few % at high temperatures such as the pre-melting state. From MD simulations, the melting point is, therefore, anticipated to be a little lower than 3175 K.

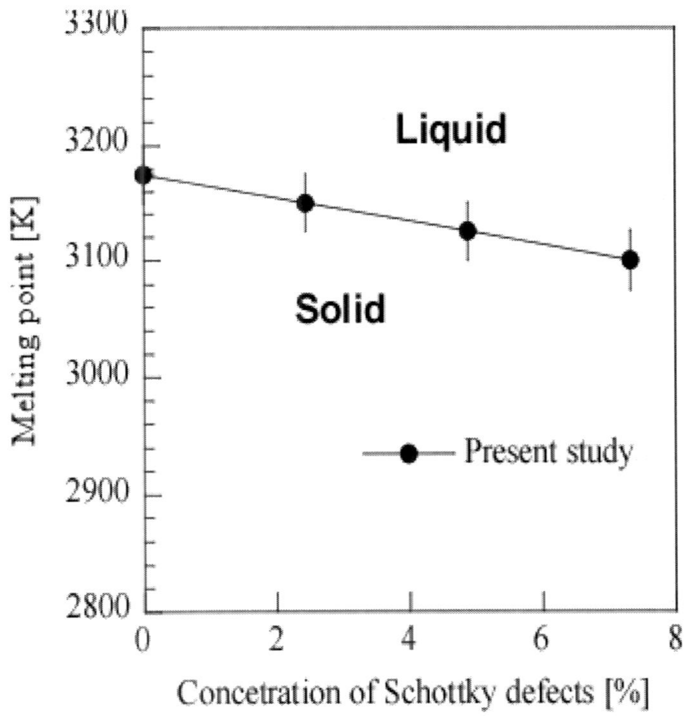

Figure 25. Melting point of UO_2 as a function of concentration of Schottky defects.

Conclusion

Fluorite-based oxides covered here, i.e., actinide oxides and zirconia-based oxides are widely used in nuclear energy field and are exposed to severe irradiation environments such as very high temperature, steep temperature gradient, and high radiation field and so on. Notably, these materials have a wide variety of metal as well as oxygen compositions and contain various lattice defects. Therefore, these systems are expected to exhibit diverse material properties dependent on each specific and their combined synergetic conditions. In order to understand such complicated behavior at the atomic level in a complimentary manner with their experimental investigations (often not easily done), molecular dynamics (MD) simulations are of particular value and are finding increasing applications in these days. In this chapter, the author has attempted to demonstrate some of such practical applicability of MD simulations using classical two-body interatomic potential functions by successfully describing some of their basic atomic diffusion, heat conduction and melting *etc.* properties. Of course, as is above apparent, the success attained here is only to quite a limited extent, and the more elaboration and extension both in methodology and application is to be awaited in the future. For example, in the current classical MD regime, it is difficult to identify the more exact forms and nature of (possibly existing) various complex defect-cluster structures in the fluorite lattice, and the interatomic potential model should be improved, e.g., to a variable charge model. Needless to say, in parallel the first-principle MD should be further developed and leveraged. In addition, non-equilibrium MD calculations, e.g., under thermal gradient and radiation field, should be developed to directly simulate and compare with the actual fuel behavior both in operation and in severe accident condition.

REFERENCES

[1] B.J. Alder and T.E. Wainwright, *Phys. Rev.* (1962) 127, 359.
[2] D.C. Rapaport, *Computer Phys. Comm.* (2006) 174, 521.
[3] G.V. Lewis and C.R.A. Catlow, *J. Phys. C: Solid State Phys.* (1985) 18, 1149.
[4] K. Govers, S. Lemehov, M. Hou and M. Verwerft, *J. Nucl. Mater.* (2007) 366, 161.
[5] K. Govers, S. Lemehov, M. Hou and M. Verwerft, *J. Nucl. Mater.* (2008) 376, 66.
[6] D.R. Olander, "Fundamental aspect of nuclear reactor fuel elements", TID-26711-P1, (2nd Ed.), Technical Information Center ERDA (1976).
[7] B.T.M. Willis, *Acta Cryst.* (1978) A34, 88.
[8] K. Momma and F. Izumi, *J. Appl. Crystallogr.* (2008) 41, 653.
[9] IAEA, "Viability of inert matrix fuel in reducing plutonium amounts in reactors", IAEA-TECDOC-1516 (2006).
[10] T. Arima, S. Yamasaki, K. Idemitsu and Y. Inagaki, *J. Nucl. Mater.* (2008) 376, 139.
[11] K. Hirao and K. Kawamura, "Material design using personal computer", Shokabo, Tokyo (1994).
[12] H.W. Brinkman, W.J. Briels and H. Verweij, *Chem. Phys. Lett.* (1995) 247, 386.
[13] Y. Ida, *Phys. Earth Planet. Inter.* (1976) 13, 97.
[14] P.P. Ewald, *Ann. Phys.* (1921) 64, 253.
[15] T. Arima, K. Fukuyo, K. Idemitsu and Y. Inagaki, *J. Mol. Liq.* (2004) 113, 67.
[16] T. Arima, S. Yamasaki, K. Yamahira, K. Idemitsu, Y. Inagaki and C. Degueldre, *J. Nucl. Mater.* (2006) 352, 309.
[17] T. Arima, K. Yoshida, K. Idemitsu, Y. Inagaki and I. Sato, *IOP Conf. Series: Mater. Sci. Eng.* (2010) 9, 012003.
[18] E Yakub, C. Ronchi and D. Staicu, *J. Chem. Phys.* (2007) 127, 094508.
[19] S.I. Potashnikov, A.S. Boyarchenkov, K.A. Nekrasov, A.Ya. Kupryazhkin, *J. Nucl. Mater.* (2013) 433, 215.
[20] C.S. Sabioni, W.B. Ferraz and F. Millot, *J. Nucl. Mater.* (2000) 278, 364.
[21] B. Auskern and J. Belle, *J. Nucl. Mater.* (1961) 3, 267.
[22] J.F. Marin and P. Contamin, *J. Nucl. Mater.* (1969) 30, 16.

[23] Z. Hadari, M. Krouppand Y. Wolfson, *J. Appl. Phys.* (1971) 42, 537.
[24] K. Clausen, W. Hayes, J.E. Macdonald, R. Osborn, *Phys. Rev. Lett.* (1984) 52, 1238.
[25] K.N. Clausen, M.A. Hackett, W. Hayes, S. Hull, M.T. Hutchings, J.E. Macdonald, K.A. Mcewen, R. Osborn, U. Steigenberger, *Phys. B* (1989) 156 & 157, 103.
[26] Hj. Matzke, in *Nonstoichiometric Oxide*, edited by O.T. Sørensen (Academic, New York, 1981) pp. 156-232. Hj. Matzke, *J. Phys.* (1976) C7, 452.
[27] C.S. Sabioni, W.B. Ferraz and F. Millot, *J. Nucl. Mater.* (1998) 257, 180.
[28] Hj. Matzke, *J. Nucl. Mater.* (1983) 114, 121.
[29] H. Ogawa, *Mat. Trans.* (2006) 47, 2706.
[30] G.L. Reynolds and B. Burton, *J. Nucl. Mater.* (1979) 82, 22.
[31] I. Sato, K. Tanaka and T. Arima, *IOP Conf. Series: Mater. Sci. Eng.* (2010) 9, 012005.
[32] P. Li, I.W. Chen and J.E. Penner-Hahn, *Phys. Rev.* (1993) B48, 10074.
[33] R.D. Shannon and C.T. Prewitt, *Acta Cryst.* (1969) B25, 925.
[34] X. Li and B. Hafskjold, *J. Phys.: Condens. Mat.* (1995) 7, 1255.
[35] Y. Oishi and K. Ando, "Transport in non-stoichiometric compounds", Plenum Press, New York (1985) p189.
[36] P. Sindzingre and M.J. Gillan, *J. Phys.: Condens. Matter* (1990) 2, 7033.
[37] P.L.D Lindan and M.J. Gillan, *J. Phys.: Condens. Matter* (1991) 3, 3929.
[38] N. Galamba, C.A. Nieto de Castro and J.F. Ely, *J. Chem. Phys.* (2004) 120, 8676.
[39] B. Bernu and J.P. Hansen, *Phys. Rev. Lett.* (1982) 48, 1375.
[40] C. Pierleoni, G. Ciccotti and B. Bernu, *Europhys. Lett.* (1987) 4, 1115.
[41] C. Pierleoni and G. Ciccotti, *J. Phys: Condens. Matter* (1990) 2, 1315.
[42] S. Motoyama, Y. Ichikawa, Y. Hiwatari and A. Oe, *Phys. Rev.* (1999) B 60, 292.
[43] J. Evans and G.P. Morriss, "Statistical Mechanics of Nonequilibrium Liquids", Academic press, London, 1990.
[44] P.G. Lucuta, Hj. Matzke and I.J. Hastings, *J. Nucl. Mater.* (1996) 232, 166.
[45] J.P. Moore and D.L. McElroy, *J. Am. Cerm. Soc.* (1971) 54, 40.
[46] P.K. Schelling, S.R. Phillpot and P. Keblinski, *Phys. Rev.* (2002) B65, 144306-1.
[47] D.J.M. Bevan, L.E. Grey, B.T.M. Willis, *J. Solid State Chem.* (1986) 61, 1.
[48] F. Garrido, R.M. Ibberson, L. Nowicki, B.T.M. Willis, *J. Nucl. Mater.* (2003) 322, 87.
[49] R.I. Cooper and B.T.M. Willis, *Acta Cryst.* (2004) A60, 322.
[50] F. Garrido, A.C. Hannon, R.M. Ibberson, L. Nowicki and B.T.M. Willis, *Inorg. Chem.* (2006) 45, 8408.
[51] Y. Geng, Y. Chen, Y. Kaneta, M. Iwasawa, T. Ohnuma and M. Kinoshita, *Phys. Rev.* (2008) B77, 104120.
[52] Y. Geng, Y. Chen, Y. Kaneta and M. Kinoshita, *Phys. Rev.* (2008) B77, 180101.
[53] S.D. Conradson, D. Manara, F. Wastin, D.L. Clark, G.H. Lander, L.M. Morales, J. Rebizant and V.V. Rondinella, *Inorg. Chem.* (2004) 43, 6922.
[54] D.A. Andersson, J. Lezama, B.P. Uberuaga, C. Deo and S.D. Conradson, *Phys. Rev.* (2009) B79, 024110.
[55] R.L. Gibby, *J. Nucl. Mater.* (1971) 38, 163.
[56] L.A. Goldsmith and J.A.M. Douglas, *J. Nucl. Mater.* (1972) 43, 225.
[57] C. Duriez, J.P. Alessandri, T. Gervais and Y. Philipponneau, *J. Nucl. Mater.* (2000) 277, 143.
[58] Y. Philipponneau, *J. Nucl. Mater.* (1992) 188, 194.
[59] M. Inoue, *J. Nucl. Mater.* (2000) 282, 186.

[60] P.K. Schelling and S.R. Phillpot, *J. Am. Ceram. Soc.* (2001) 84, 2997.

[61] R. Mévrel, J.C. Laizet, A. Azzopardi, B. Leclercq, M. Poulain, O. Lavigne and D. Demange, *J. Euro. Ceram. Soc.* (2004) 24, 3081.

[62] S. Raghavan, H. Wang, R.B. Dinwiddie, W.D. Porter and M.J. Mayo, *Scripta Mat.* (1998) 39, 1119.

[63] C. Degueldre, T. Arima and Y.W. Lee, *J. Nucl. Mater.* (2003) 319, 6.

[64] T. Arima, S. Yamasaki, Y. Inagaki and K. Idemitsu, *J. Alloys Compd.* (2005) 400, 43.

[65] B. Belonoshko and L.S. Dubrovinsky, *Geochim. Cosmochim. Acta.* (1995) 59, 1883.

[66] K. Harafuji, T. Tsuchiya and K. Kawamura, *J. Appl. Phys.* (2004) 96, 2501.

[67] B. Basak, A.K. Sengupta and H.S. Kamath, *J. Nucl. Mater.* (2003) 360, 210.

[68] International Nuclear Safety Centre (INSC) Material Properties Database, Argonne National Laboratory, Illinois, USA, http://www.insc.anl.gov/matprop/.

[69] W.D. Drotning, "Proceedings of the 8th Symp. On Thermophysical Properties" Gaithersburg, Maryland, June 15-18, 1981, National Bureau of Standard (1981).

[70] A. Christensen, *J. Am. Ceram. Soc.* (1963) 46, 607.

[71] J.H. Harding, D.G. Martin and P.E. Potter, Commission of European Communities Report EUR 12402 (1989).

[72] D. Manara, C. Ronchi, M. Sheindlin, M. Lewis and M. Brykin, *J. Nucl. Mater.* (2005) 342, 148.

[73] R.A. Jackson and C.R.A. Catlow, *J. Nucl. Mater.* (1985) 127, 161.

[74] H.Y. Geng, Y. Chen, K. Kaneta and M. Kinoshita, *J. Alloys and Compd.* (2008) 457, 465.

[75] T. Kurosawa, *J. Phys. Soc. Jpn.* (1957) 12, 338.

In: New Research Trends of Fluorite-Based Oxide Materials ISBN:978-1-63117-350-9
Editors: Akio Nakamura and Junichiro Mizusaki

Chapter 10

FORMATION AND MORPHOLOGY OF UO_2 IMAGE CRYSTAL

***Hiroyuki Serizawa*[*1], *Ohisi Yuji*[2], *Yoshiyuki Kaji*[1] *and Shinsuke Yamanaka*[2]**

[1]Nuclear Science and Engineering Directorate, Japan Atomic Energy Agency, Ibaraki, Japan

[2]Department of Nuclear Engineering, Graduate School of Engineering, Osaka University, Osaka, Japan

ABSTRACT

This chapter gives a review of our recent study on image crystal observed in face-centered-cubic (FCC) fluorite-type dioxide of uranium, UO_2. Image crystal is a shape controllable negative crystal which has been observed in various minerals. The negative crystals formed in a large single-crystal body have attracted interest as expensive jewelry because of their mysterious appearance and rarity. Nonetheless, for many scientists, such cavity has been nothing else than a volume defect which occurs in single crystals by chance. We have found recently that image crystal is formed in UO_2 when helium (He) is enclosed in the gas bubble. its shape changing dramatically with the condition of He precipitation such as annealing temperature and He inner pressure. Truncated octahedron-, octa-triacontahedron- and pentacontahedron-type negative crystals were observed. Our study clarifies that the growth process of the image crystal can be described by the step free energy model rather than the attachment energy model. We also touch briefly the bearing of these new findings for UO_2 application as nuclear fuel. Our recent research showed that the helium precipitated in the surface region of poly-crystalline UO_2 forms blister of which lid had been broken away to form shallow basin on the surface with hexagonal fringe. Though the phenomena is interesting from scientific standpoint of view by itself, it should be noted that the variation of surface morphology affect the densification behavior in the sintering process or swelling.

* Corresponding author: E-mail address: serizawa.hiroyuki@jaea.go.jp.

1. INTRODUCTION

Although many researchers had reported the presence of a polyhedral cavity, called negative crystal [1-3], in various natural minerals, as well as in nuclear oxide fuel UO_2 [4-10], their formation mechanism had been unclear, for such kind of cavity, usually found by chance, has been scarcely regarded as a serious research target. It has been known for long time that a polyhedral microscopic void is formed in UO_2 matrix by the heat treatment at high temperatures above 2273 K [6]. (See Figure 3.) The shape of the void coincides with the equilibrium shape of the matrix, which is called Wulff shape after G. Wulff who originally gave a simple geometric construction for the shape of a crystal at equilibrium [11]. The morphological studies [6,7] have demonstrated that the equilibrium shape of voids formed in UO_2 matrix is a truncated octahedron bounded by the {111} and {001} facets. Whereas, we have recently got an impressive field emission scanning electron microscopy (FE-SEM) photograph incidentally during another investigation aiming at reducing the high radiotoxicity of minor actinides (MAs = Np and Am, *etc.*) accumulated through nuclear power plant operation.

Figure 1 shows the FE-SEM image that caught our interest. The figure shows the fracture surface of a single-crystal UO_2 that was heat treated in helium at 90 MPa (900 atm) at 1473 K, followed by annealing at 1573 K for 1 h. The cavity in the centre of the figure is bounded by well-grown facets. Moreover, the cavity has the shape of a complex but well-defined polyhedron. Precipitated fission product (FP) gas such as xenon (Xe) or krypton (Kr) is well known to form spherical gas bubbles [13-17], and this figure implies that helium has an unusual ability to form symmetrical shaped cavities. The difference in the behavior of these gases is attributable to the disparity of diffusion coefficient and solubility. The gas precipitated in the matrix moves by reiterating re-dissolution and re-precipitation along with concentration gradient. As is well known, the diffusion coefficient and solubility of helium is much higher than those of xenon and krypton. So, it is reasonable to consider that helium accumulated in the gas bubble can easily re-dissolves into the matrix and move toward the surface of the particle where the activity of helium is zero. Thus, the inner pressure of the helium gas bubble easily decreases by the high temperature annealing. On the other hand, precipitated xenon or krypton remains behind at the precipitated position to form a cavity which is called FP gas bubble. Because of their low solubility, xenon and krypton tend to precipitate and clump together rather than dissolves into the matrix. Since these gases precipitated in the cavity cannot escape from the gas bubble, the inner pressure of the FP gas bubble might be high. As is described later, the shape of the cavity is closely related with the inner pressure. By image analysis using computer code Shape [18], we had successfully modeled a truncated octahedron-, octa-triacontahedron-, pentacontahedron-types of image crystals observed in matrix. However, a cuboctahedron-type image crystal which is considered to be formed in the shape change process from the truncated octahedron-type to the pentacontahedron-type was not found in the sample matrix. We therefore concluded that the pentacontahedron-type image crystal, of which the relative surface area (the value of the surface area normalized by that of the Wulff-shaped image crystal with the same volume) is much smaller than that of the cuboctahedron, is actually formed in the transformation process.

Figure 1. FE-SEM photograph of image crystal formed in the single-crystal UO_2 particle [12].

The Castell who investigated the Wulff shape negative crystal of UO_2 made an interesting suggestion on this point [6]: Although researchers generally had not considered the cavity as being filled with gas at any significant pressure, gas at appropriate pressure would be enclosed within the cavity and affect the cavity's shape. In the reference, he mentioned that the issue remains a point of the further investigation. Considering every factor, we conjectured that the helium enclosed in the image crystal affects the formation of the complex symmetrical shape, and Figure 1 precisely reflects the influence of the inner pressure on the shape of the cavity. We named such shape-controllable cavity bounded by facets "image crystal". Not to mention, surface and interfacial phenomena are ubiquitous and constitute a topic of current utmost interest in materials research, such as nucleation, wetting and crystal-faceting *etc.*. Of particular merit is that the measurement of the concaved facet allows us to study impurity-free pure surfaces, avoiding the problem of contamination [19-24]. In this chapter, for more in-depth understanding of this intriguing gas/solid interface problem, our recent study on the growth process of the image crystal formed in UO_2 [11,26] is reviewed, also briefly touching its nuclear-technological implication.

2. Formation and Shape-Change Mechanism of Image Crystal

2.1. Concept of Image Crystal

Crystal structure of UO_2 is fluorite (F)-type face centered cubic (FCC). The lowest energy faces are {111}, {011} and {001} [25,26]. The crystal habit of UO_2 takes the form of a octahedron or truncated octahedron. Figure 2 shows single crystal UO_2 particles prepared by a chemical vapour transport method using UO_2 powder as the source material and tellurium tetrachloride ($TeCl_4$) as the transport reagent [27] as follows:

$$UO_2(s) + 2Cl_2(g) \leftrightarrow UCl_4(g) + O_2(g).$$

Figure 2. Optical photograph of single-crystal UO_2 sample prepared and used in this study [12].

Two types of particle can be seen: One is an octahedron-type single crystal which consist of {111}. The other is a truncated octahedron-type single crystal which is surrounded by {111} and {001}. The figure indicates that the {111} and {001} surface energies are sufficiently close. The model of truncated octahedron-type single-crystal UO_2 is given in Figure 3 [6,28]. Casetll described that this type of structure is Wulff shape of UO_2 and the boundary of equilibrium voids within UO_2 will also adopt this shape. The {111} to {001} area ratio is highly sensitive to the surface energy ratio. The Wulff-shaped void is sometimes formed in UO_2 matrix too when the single-crystal is prepared from the poly-crystalline powder through high temperature heat treatment. On the other hand, we found that a spherical gas bubble formed by a helium precipitation transforms to polyhedral-shaped cavity by the heat treatment. We infused helium into the single-crystal and the poly-crystalline UO_2 by a hot isostatic pressing method. The infusion conditions were as follows: 1473 K for 100 h under a helium pressure of 91 MPa. Both the heating and the cooling rates were 20 K / min. The cross sections of the single-crystal sample before and after helium treatment differ significantly, as shown in Figures 4(a) and (b), respectively. The cross section of the sample before helium infusion is clean; no visible defect is observed, whereas many round cavities are observed after helium infusion in Figure 4(b). The figure indicates that the cavities were formed during high pressure heat treatment in helium. Recent study using Knudsen effusion mass spectrometry showed that the amount of helium dissolved in the single-crystal UO_2 was 4×10^{-2} cm^3 (STP) / gUO_2 [29]. The solubility of helium decreases as the temperature decreases [30], and the diffusion coefficient of helium in UO_2 is sufficiently high for it to move in the grain at high temperature greater than 1273 K. Therefore, super-saturated helium would easily diffuse along a concentration gradient toward the grain surface, where the equilibrium helium concentration should be low [29,31], during the cooling process. However, when the sample is cooled very rapidly, a substantial amount of helium that cannot escape from the grain inside would clump together to form gas bubbles. Thus, the cavities seen in Figure 4(b) are judged to have been formed by precipitated helium. The cavity-creation mechanism had been discussed by Kashibe et al. [14]. They examined the formation of FP gas bubble in UO_2 and found that enhanced coarsening of bubbles was observed near the grain boundary. They interpreted that a sufficient amount of vacancy is supplied from external vacancy sources such as free surfaces, grain boundaries or irradiation-induced sub-grain boundaries for huge bubbles etc. However, as is apparent in Figure 4(b), such kind of coarsening is not observed here, leaving the mechanism hardly clear and unresolved.

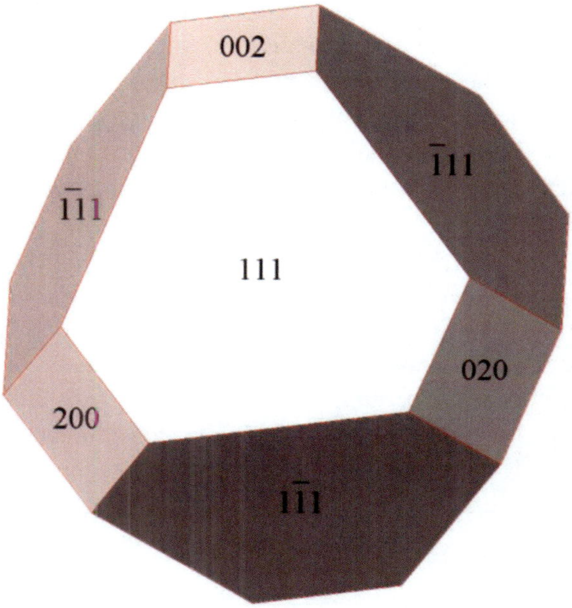

Figure 3. The truncated octahedron crystal habit or Wulff shape of UO_2.

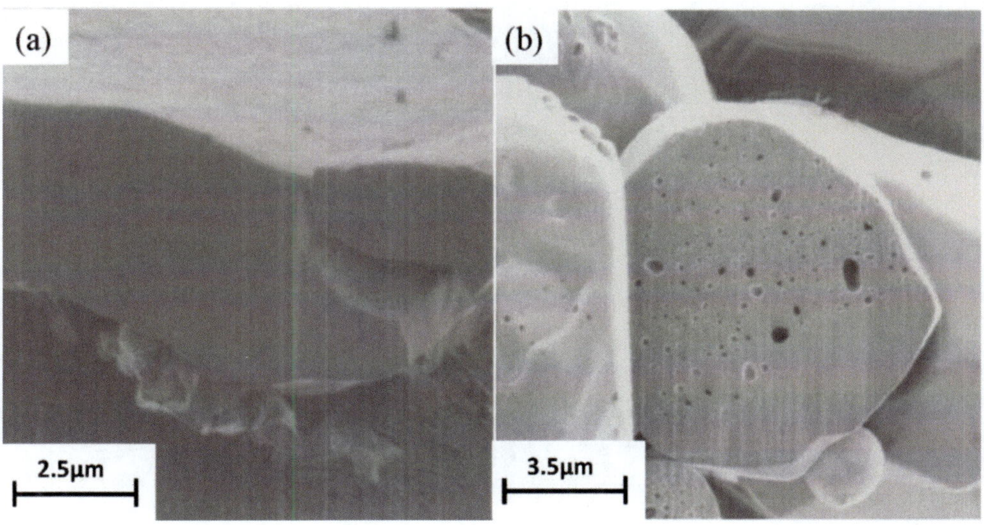

Figure 4. Cross section of the single-crystal UO_2 sample examined by secondary ion mass spectrometry (SIMS). (a), Before helium infusion. (b), After helium infusion. [12].

The typical fracture surfaces of the samples annealed at temperatures greater than 1473 K in helium at atmospheric pressure (1atm = 0.101395MPa) are shown in Figure 5. The results of area analysis of the cavities formed on the fracture surface are also summarized in Table 1. It is apparent in Figure 5 that the well-grown larger-size cavities are much more abundant in single crystal UO_2 than in polycrystalline one, and in addition, the cavity density in the former itself is about ten times as high as that in the latter. The fourth column of Table 1 shows in turn that the total avergae diameter of the cavity in the latter poly-crystalline sample is overall slightly larger than that in single-crystal sample, though the distinction between them is not clear since the scattering of the data is so large. As seen in the eighth column, the ratio of the area occupied by the cavity in the single-crystal is evidently larger than that in the poly-crystalline sample. The cross sections of the single-crystal samples in the figure are {111} cleavage surfaces. The lines running across the images are a result of the cleavage process. The symmetrical shape of the cavity, called image crystal, is at a glance attractive morphologically.

Although the bright symmetrical features varied depending on the heat treatment condition, the image crystals that formed in the single crystal appeared to align with their edges parallel to a certain direction that would be related to the crystal orientation of the fracture surface. Image crystal is also seen on the fracture surface of the poly-crystalline sample, but obviously much less in number. We ascribe this difference to the absence / presence of grain boundary [11]: During the annealing, healium dissolved in the polycrystalline UO_2 was able to be released more easily from the matrix to the surface of each small grain, i.e., grain boundary. Namely, the growth of the cavity containing helium is suppressed in the polycrystalline UO_2 by the quick escape of helium to the grain boundary.

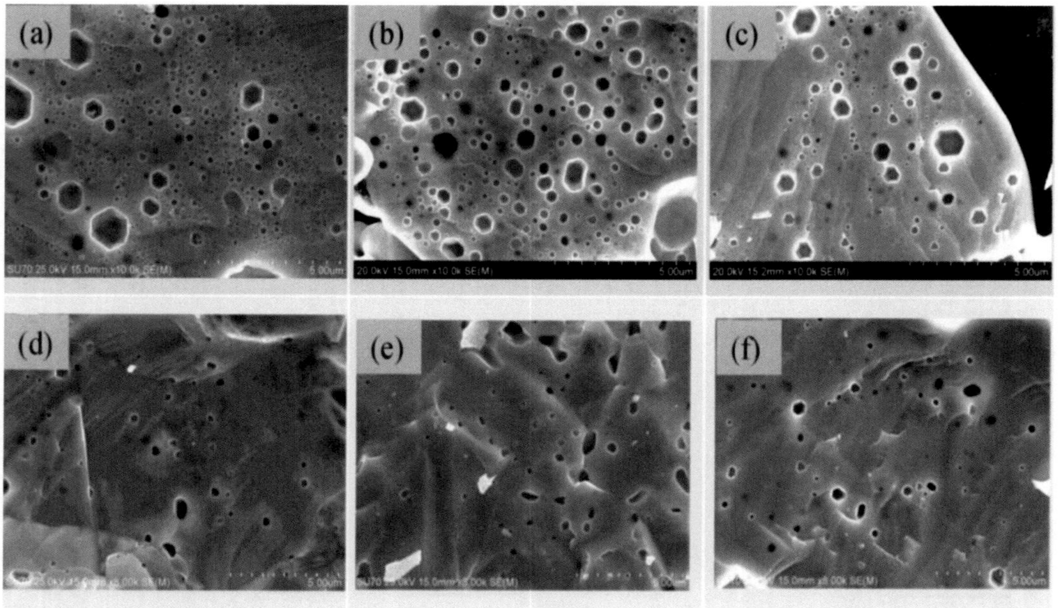

Figure 5. FE-SEM images of fracture surfaces of annealed UO2 after helium treatment at 1473 K Upper halves for single-crystal UO2 annealed for 1 h; (a) at 1573 K, (b) at 1773 K, (c) at 1973 K. Lower halves for poly-crystalline UO2 annealed for 1 h; (d) at 1573 K, (e) at 1773 K, (f) at 1973 K .

Table 1. Information on the cavity on the fracture surface examined by Image analysis [12]

Sample	Heat treatment (K)	Detected number of the cavity	Mean size of the cavity (μm)	Standard deviation (μm)	Area of field (μm^2)	Density $x10^{12}$ (n/m^2)	Acavity** /Area of field
Poly-crystal UO_2	B. H.*	139	0.369	0.229	441.8	0.31	0.0336±0.130
	1573	76	0.474	0.435	441.8	0.17	0.0303±0.0256
	1773	90	0.394	0.27	441.8	0.2	0.0248±0.0117
	1973	90	0.492	0.484	441.8	0.2	0.0387±0.0375
Single-crystal UO_2	B. H.*	341	0.225	0.138	141.6	2.41	0.0957±0.0360
	1573	592	0.19	0.136	141.6	4.18	0.119±0.0607
	1773	518	0.218	0.134	141.6	3.66	0.137±0.0516
	1973	352	0.193	0.17	141.6	2.49	0.0727±0.0564

*Before heat treatment; ** Area occupied by cavity.

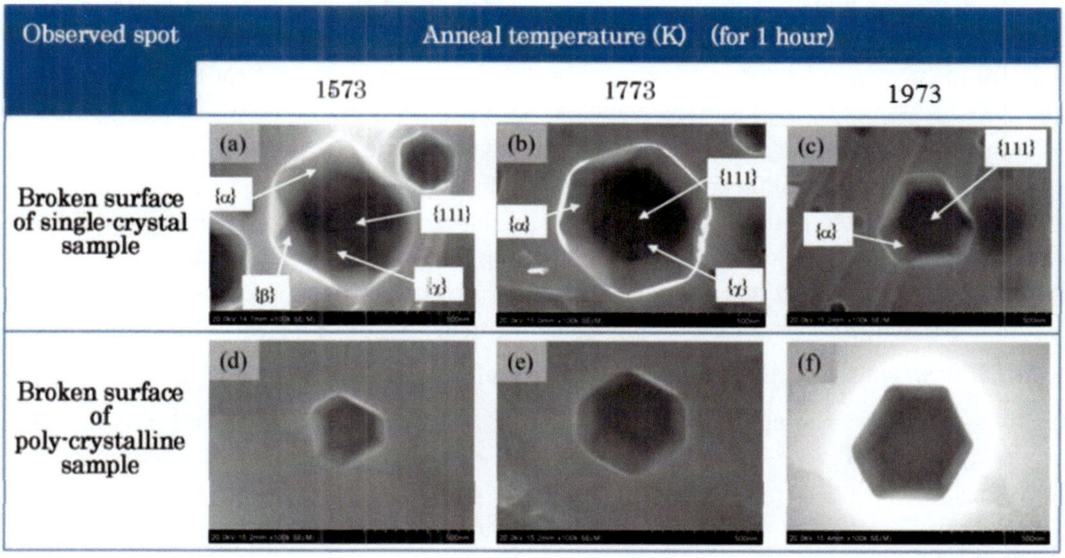

Figure 6. FE-SEM images of typical shape of image crystals, showing almost identical features between single- and poly-crystalline samples. [12].

2.2. Structure analysis of image crystal

In this section we discuss the transformation mechanism of the image crystal from the stand point of facet growth which depends on the surface energy or step free energy. Figure 6 shows typical shape of the image crystals at high magnification. Their faceted nature is clearly visible. The FE-SEM images of the single-crystal samples show the {111} cleavage surface, and the bottom face of the image crystals, which is almost parallel to the cleaved surface, is indexed as {111}. The shape of the image crystal formed in the single-crystal sample changes depending on the annealing temperature. On the other hand, all of the image crystals formed in the poly-crystalline sample shown in the lower-side figure is composed of the facets analogous to the image crystal formed in the single crystal at 1973 K.

In the figure, the formation of other facets with higher Miller indices, {001}, {110} and {311}, is observed on the surface of the image crystal. According to the characteristics of crystals with the fluorite $Fm\overline{3}m$ symmetry, the morphological importance (MI) of the facets follows the sequence $MI_{\{111\}} > MI_{\{001\}} > MI_{\{110\}} > MI_{\{311\}}$ [7].

In general, The MI sequence is considered to be determined by surface formation energy. In case of UO_2, two different types of energies for this, surface free energy and attachment energy have been reported by Abramowski et al. [25]. The surface free energy is defined as the energy required for cleaving that surface from a bulk crystal. The surface with the lowest surface free energy is the most stable one, which corresponds to a cleavage surface. Whereas, the attachment energy is the energy released when a growth slice of thickness d_{hkl} is attached to the surface from infinity [32]. Since energy is gained when matter is attached to the surface, the value of the attachment energy is negative. The surface with the highest attachment energy is the most favorite surface; the material tends to be added on the adjacent surfaces with the lower attachment energy, consequently enhancing the formation and growth of the surface with the highest attachment energy. Abramowski et al. evaluated these energies distinctly through ab initio calculation. However, according to them the surface free energy of UO_2 for the preferred interatomic potential is calculated to be 2.72 Jm^{-2} for the {001} surface and 1.27 Jm^{-2} for the {111} surface, which gives a surface energy ratio of 2.14, as compared with the experimentally determined one of 1.42 ± 0.05 reported by Castell [6]. Castell pointed out some problems on the theoretical treatment of the surface free energy and showed that the surface energy of the {001} face should be less than that calculated by Abramowski et al.. Consequently, the reliable surface formation energy to predict the sequence of MI for UO_2 lattice planes available at this moment is only the attachment energy calculated by them. In addition, MI sequence predicted by the attachment energy coincides with that of the crystal with $Fm\overline{3}m$ symmetry. So, hereafter, we often refer to attachment energy in discussing the relative surface morphological stability relationship. The model of polyhedral shaped image crystals shown in Figure 7 was constructed on the basis of the attachment energy and FE-SEM images. The pentacontahedron (a) consists of the {111}, {100}, {110} and {311} facets. The truncated octahedron (c) consists of the {111} and {001} facets; the latter is similar to the Wulff shape reported by Castell [6]. The polyhedron (b) is an octa-triacontahedron bounded by facets indexed as {111}, {001} and {311}. These models indicate that facets {α}, {β} and {χ} in Figure 6 are indexed as {001}, {110} and {311}, respectively.

Figure 8(a) and 8(b) again show FE-SEM photographs of the image crystal formed at 1773 K in a further-magnified scale; the emphasis is on the step bunching and growth of the {110} facets. In Figure 8(a), a full-grown {311} facet appears, whereas the {110} facet is almost invisible. Figure 8(b) is a high-magnification FE-SEM image of area A of Figure 8(a), where four {311} facets meet in the centre of the image. Extensive steps and terraces appear around the small face that corresponds to the {110} facet. The complex nature of the steps is particularly evident in the region where the {110} facet meets two {311} facets. The rising sides of the steps are geometrically constrained to be close to the {111} and {001} facet orientations. These results imply that $MI_{\{311\}} > MI_{\{110\}}$, which disagrees with the morphology predicted by the attachment energy, since the attachment energy of {110} is higher than that of {311} [25]. However, considering the fact that the image crystal in Figure 8 is seen during the growth process and would be in far from equilibrium state, the contradiction is not particularly surprising.

Here, we consider the critical size of nucleus on the surface and introduce another function called step free energy (SFE) model [33]. When a nucleus grow on the surface, such as the step limiting a terrace, the total free energy of step, *F*, is the line integral of a step line tension, γ, for the inside of the plane normal n = (cos θ, sin θ) to the step:

$$F = \int_{step} \gamma(\theta) dl, \tag{1}$$

The SFE is regarded as a 2D or 3D analogue of the surface free energy [34]. A relationship between the growth rate and the 2D nucleation is discussed by Sweegers et al. on the basis of a connected net analysis [35]. According to them, the size of the critical nucleus depends on the driving force for crystallization and on the surface free energy of the nucleus. They showed that the smaller the step free energy, the smaller the size of the critical nucleus. Thus, the smallest step free energies achievable for any nucleus on a specific surface determine the actual growth rate for the orientation. In brief, the SFE model predicts that the growth rate of the faces is not proportional to the attachment energy, but is determined by the step free energy and the relevant 2D growth mechanism. Therefore, the value of the SFE is a direct indicator of mass transport rates close to equilibrium [36,37]. Although the theoretical value of the SFE in UO_2 has not yet been reported, the growth rate is known to increase as the SFE decreases because the critical nucleus becomes smaller. Experimental results indicate that the MI of the faces in the growth process follows the sequence $MI_{\{111\}} > MI_{\{001\}} > MI_{\{311\}} > MI_{\{110\}}$; therefore, the SFE magnitude should be in the order of $SFE_{\{111\}} < SFE_{\{001\}} < SFE_{\{311\}} < SFE_{\{110\}}$.

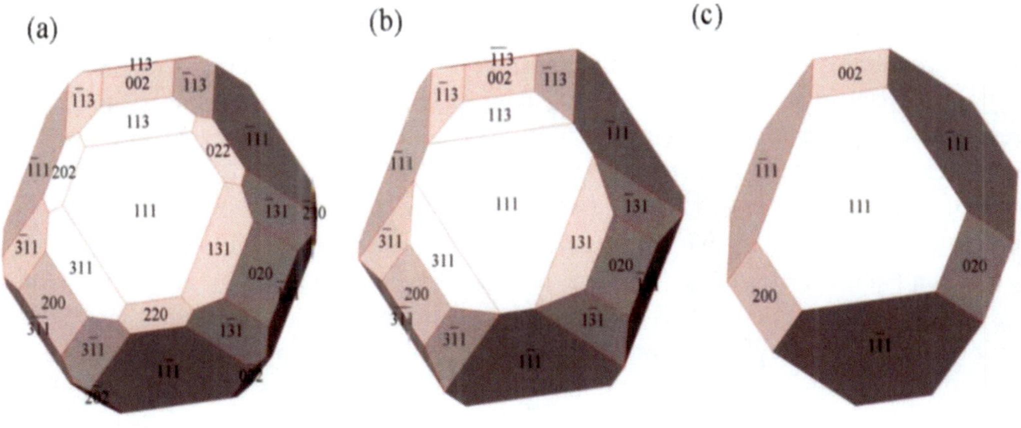

Figure 7. Evaluated shapes of image crystals. (a), (b) and (c) are models of the image crystals formed by annealing UO_2 single crystal at 1573 K, 1773 K and 1973 K shown in Figures 6(a), (b) and (c), respectively [12].

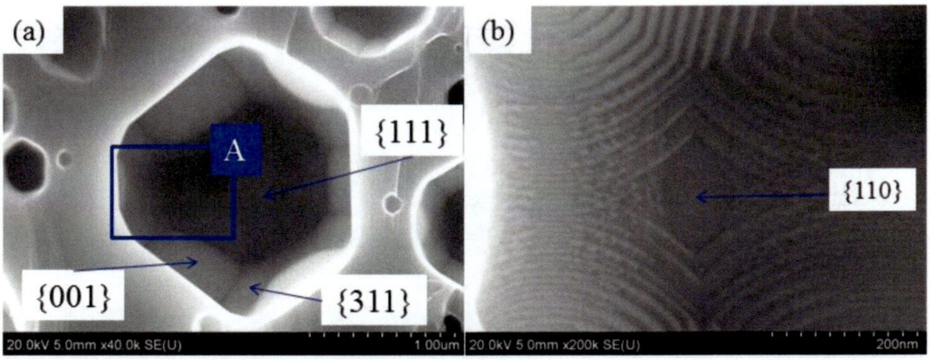

Figure 8. Bunching and growth of facets with higher indices. (a) FE-SEM image of image crystal observed in single-crystal UO_2 annealed at 1773 K. (b) High-magnification image of field of view A shown in Figure 8(a).

2.3. Transformation Mechanism of Image Crystal

It should be noted that the shape of the image crystal formed in poly-crystalline UO_2 differs from that in single-crystal UO_2 even if the annealing temperature was the same. As mentioned previously, this is attributed to the difference in grain size. A recent study on helium diffusivity in UO_2 showed that the release behaviour of helium in both single-crystal and poly-crystalline samples can be successfully analysed using an equivalent sphere model [38,39]; both samples give almost identical intra-grain bulk diffusion coefficient of helium within experimental scatter [31]. This result implies that the diffusion coefficient of helium at the grain boundary is so high that helium in the poly-crystalline sample, which consists of small grains, is easily released through the grain boundary. Hence the concentration of helium in the poly-crystalline sample tends to be lower than that in the single-crystal sample when the heating condition is the same. The results of the image-crystal area analysis described in 2.1 accord with this argument. As shown in Table 1, the area ratio occupied by the cavity of the single-crystal is evidently much higher than that of the poly-crystalline sample, which means that a larger amount of helium was precipitated in the single crystal. As a result, the cavity-area ratio of the single-crystal sample annealed at 1973 K is smaller than those annealed at lower 1573 and 1773 K, for the diffusion coefficient of helium in UO_2 grain should increase with increasing temperature. Namely, the decrease in the cavity-area ratio at higher temperature reflects the decrease in the amount of helium in the grain because dissolved helium can move faster and escape to the grain surface more easily. Since the activity of helium in the image crystal is considered to be approximately equal to that in the matrix, the inner pressure of the cavity formed in the single-crystal sample should be higher than that in the poly-crystalline sample. The release of the dissolved helium by heat treatment changes the activity of helium in the matrix, thereby decreasing the inner pressure of the cavity. When the rate of mass transport of UO_2 is sufficiently high, preferable faces grow on the wall of the cavity to form image crystal with equilibrium shape.

Unfortunately, we cannot assert whether equilibrium was attained in our series of experiments. However, as Figure 6 shows, the shape of the image crystal in the poly-crystalline sample is almost the same for annealing temperatures ranging from 1573 to 1973 K, whereas the image crystal in the single crystal was exceedingly deformed with increasing annealing temperature, which strongly suggests that the structure of the image crystal depends crucially on the inner pressure rise at higher annealing temperature.

Despite zealous efforts [6,25,26], little had been known about the morphology of UO_2. Abramowski et al. examined the morphology of UO_2 by first-principle calculation [25,26]. As mentioned formerly, they calculated the surface energies to be 2.72 Jm^{-2} for the {001} plane and 1.27 Jm^{-2} for the {111} plane. The surface energy ratio calculated with these values is 2.14, which is much larger than the experimentally determined value of 1.42 ± 0.05. To this point we will return shortly later. In addition, as pointed out by Castell [6], a STM study on UO_2 (001) surface revealed that the surface has a maze-like (1 x 1) domain structure on different terraces. However, such complex structure is not taken into account in the theoretical model of Abramowski et al. [40]. Accordingly, the only available knowledge regarding the surface energy of UO_2, is the surface free energy ratio $g_{\{001\}}/g_{\{111\}}$, so that it seems difficult to quantitatively discuss the morphology of image crystal in terms of surface free energy data. To get rid of this difficulty, in the following analysis, we assume that the image crystal is deformed only by mass transport, which means that once it forms, its volume does not change throughout the growth process. The validity of this assumption is also inspected and discussed in this section.

First, we examined the truncated octahedron-type image crystal observed in the single-crystal UO_2 annealed at 1973 K (see Figure 7c). We denote the total surface areas of the {111} and {001} facets as $s_{\{111\}}$ and $s_{\{001\}}$, respectively. Considering the geometric symmetry, the distances from the centre of the truncated octahedron to each surface, $d_{\{111\}}$ and $d_{\{001\}}$, are given as a function of the total surface areas of the respective facets as

$$d_{\{111\}} = \frac{\sin\theta_\beta\sqrt{s_{\{001\}}}}{2\sqrt{2}} + \left\{(\sqrt[4]{3})\sqrt{\left(\frac{3\sqrt{3}s_{\{001\}}}{32} + \frac{s_{\{111\}}}{8}\right)} - \frac{1}{4}\sqrt{\frac{3s_{\{001\}}}{2}}\right\}\sin\theta_\alpha, \quad (2)$$

$$d_{\{001\}} = 2\left\{(\sqrt[4]{3})\sqrt{\left(\frac{3\sqrt{3}s_{\{001\}}}{32} + \frac{s_{\{111\}}}{8}\right)} - \frac{1}{4}\sqrt{\frac{3s_{\{001\}}}{2}}\right\}\sin\theta_\beta, \quad (3)$$

where θ_α and θ_β indicate interfacial angles of the adjacent surfaces (111) and (−111), and (111) and (001), respectively. Using these parameters, the volume, V of the cavity, is expressed as

$$V = 4\sqrt{3}d_{\{111\}}{}^3 - 4\left(\sqrt{3}d_{\{111\}} - d_{\{001\}}\right)^3. \quad (4)$$

Because the numerical values of $d_{\{111\}}$ and $d_{\{001\}}$ vary with the size of the image crystal, we selected the ratio $d_{\{001\}}/d_{\{111\}}$, denoted as r_d, as the primary parameter in the following analysis. When the volume and r_d are fixed, the shape of the truncated octahedron is uniquely identified. The average ratio

$$r_{d\,exp.} = \left(\frac{d_{\{001\}}}{d_{\{111\}}}\right)_{exp.} = 1.27 \pm 0.067. \quad (5)$$

has been obtained as the measured ratio for truncated octahedron-type image crystal [11]. As mentioned above, Castell examined the surface free energy ratio between the {111} and

{001} facets of UO_2 and estimated the ratio $g_{\{001\}} / g_{\{111\}}$ as 1.42 ± 0.05 by measuring the surface area ratio of the Wulff-shaped voids [6]. Moreover, it is known that for two faces with the respective surface free energies γ_1 and γ_2, the distances d_1 and d_2 between the parallel planes $h_1k_1l_1$ and $h_2k_2l_2$ will be [7,41]

$$d_1/d_2 = \gamma_1/\gamma_2 \,. \tag{6}$$

Therefore, $d_{\{001\}}/d_{\{111\}}$ for the truncated octahedron-type image crystal with the Wulff shape is just 1.42 ± 0.05. Here we define the Wulff-shaped image crystal with a fixed size that is used as a mould for making things: Assuming that the equilibrium values, $d_{eq.\{001\}}$ and $d_{eq.\{111\}}$, are 1.42 and 1 μm, respectively, for which $d_{\{001\}} / d_{\{111\}} = 1.42$, the volume $V_{eq.}$ is calculated to be 6.807 μm^3 in Eq. (4).

Next, we consider the variations in $d_{\{001\}}$ and $d_{\{111\}}$ with decreasing r_d while the volume calculated by Eq. (4) is maintained at the above 6.807 μm^3 throughout the transformation. The r_d dependence of $d_{\{001\}}$ and $d_{\{111\}}$ and shape change of the image crystal is shown in Figure 9. As the {001} facet grows, the surface area increases with a dramatic decrease in $d_{\{001\}}$, which results in a simultaneous increase in $d_{\{111\}}$ with the contraction of the {111} facet. When the r_d decreases from 1.42 to 0.8660, firstly a pentacontahedron-type image crystal appears, and then this further transforms to a cuboctahedron-type, in which the two sets of facets come to form two triangles and two squares, meeting at a single point.

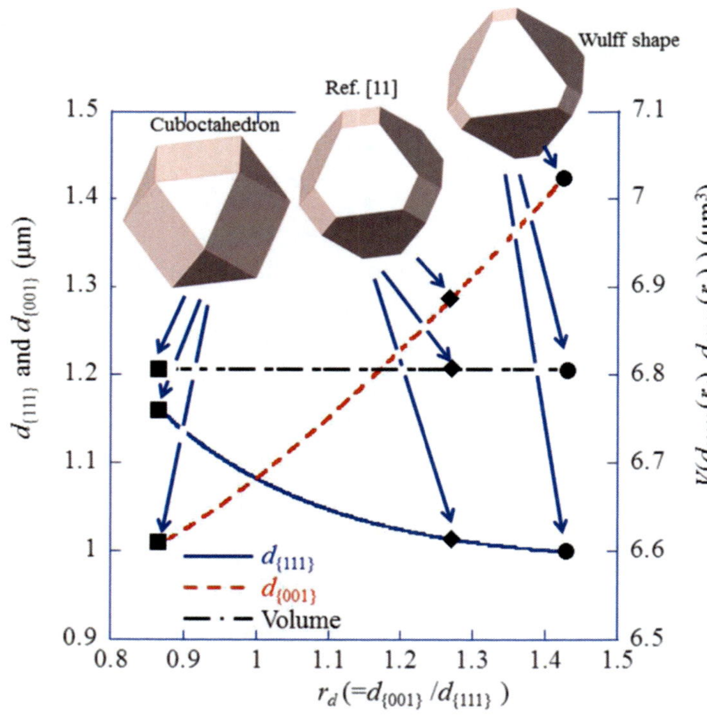

Figure 9. Dependence of $d_{\{100\}}$ and $d_{\{111\}}$ on r_d when the volume is kept constant: White plane {111}, colored plane; {001}[12].

Because of the outstanding development of the {001} facet, the image crystal becomes more rounded than the Wulff shape. Using these two parameters, $d_{\{111\}}$ and $d_{\{001\}}$, the relationship between the shape and the surface free energy can be clarified. However, since the absolute values of the surface free energies for the {111} and {001} facets are not available, we use the relative value normalized by the surface free energy of the Wulff-shaped image crystal having the same volume. The areas of the two facets, $s_{\{111\}}$ and $s_{\{001\}}$, are expressed as functions of $d_{\{001\}}$ and $d_{\{111\}}$, respectively, as follows:

$$s_{\{111\}} = 12\{\sqrt{3}d_{\{111\}}{}^2 - (\sqrt{3}d_{\{111\}} - d_{\{001\}})^2\}, \tag{7}$$

and

$$s_{\{001\}} = 12\{\sqrt{3}d_{\{111\}} - d_{\{001\}}\}^2. \tag{8}$$

The overall average surface free energy of the image crystal with the Wulff shape, denoted as $\Gamma_{eq.}$, is defined as

$$\Gamma_{eq.} = \frac{s_{\{111\}eq.}\gamma_{\{111\}} + s_{\{001\}eq.}\gamma_{\{001\}}}{s_{\{111\}eq} + s_{\{001\}eq}} = \frac{18.76\gamma_{\{111\}} + 1.169\gamma_{\{001\}}}{19.929}, \tag{9}$$

where $\gamma_{\{hkl\}}$ is the surface free energy of the {*hkl*} facet per unit area. Although $\gamma_{\{hkl\}}$ is generally a function of temperature, we treat it as a constant assuming that the surface entropy of the image crystal with well-grown facets is small enough for the temperature dependence to be negligible. Whereas, the surface free energy of the truncated octahedron-type image crystal of which $d_{\{001\}}/d_{\{111\}}$ equal to r_d, $\Gamma(r_d)$, is given as follows

$$\Gamma(r_d) = \frac{s_{\{111\}r_d}\gamma_{\{111\}} + s_{\{001\}r_d}\gamma_{\{001\}}}{s_{\{111\}r_d} + s_{\{001\}r_d}}, \tag{10}$$

where subscript r_d indicates that the value is that of the truncated octahedron with $d_{\{001\}}/d_{\{111\}} = r_d$. Thus, the normalized value by $\Gamma_{eq.}$ is expressed as

$$\frac{\Gamma(r_d)}{\Gamma_{eq}} = \frac{19.929\left\{s_{\{111\}rd} + s_{\{001\}rd}\frac{\gamma_{\{001\}}}{\gamma_{\{111\}}}\right\}}{\left(18.76 + 1.169\frac{\gamma_{\{001\}}}{\gamma_{\{111\}}}\right)\left(s_{\{111\}rd} + s_{\{001\}rd}\right)}. \tag{11}$$

The dotted line in **Figure 10** shows the dependence of $\Gamma(r_d)/\Gamma_{eq.}$ given by Eq. (11) on r_d. The surface free energy of the truncated octahedron-type image crystal increases consistently along the dotted line as the shape changes continuously from the Wulff shape to a cuboctahedron. The solid circle and the error bar represent experimental data and the standard deviation evaluated in the area analysis. Although the data in Ref. [12] involve some uncertainty, they agree well with the calculated line. This seems to substantiate the validity of the premise of the analysis that the shape change or transformation of image crystal occurs primarily by mass transport.

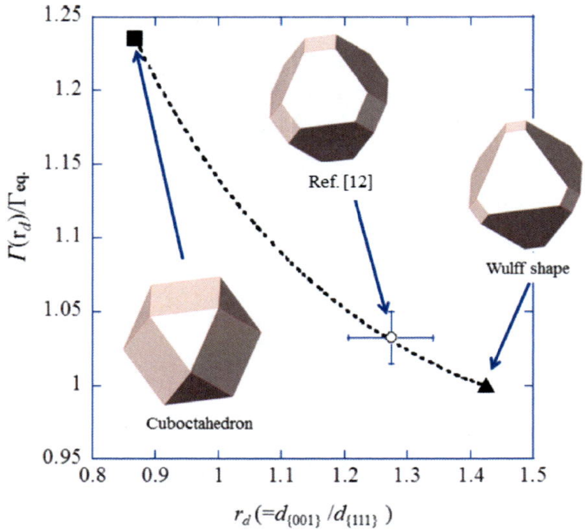

Figure 10. Dependence of surface free energy on r_d when the volume of the octahedron is kept constant [12].

The mechanical force balance on a gas-filled bubble embedded in a solid medium had been discussed by Olander [14,42]. For the equilibrium bubble in stress-free solids, the force balance takes the form of $p = 2\gamma / R$, where p, γ and R represent the inner gas pressure, surface free energy and bubble radius, respectively. If we consider two spherical gas bubbles having different surface energies but the same volume, the ratio of the inner pressures, p_1 / p_2, is equal to that of the surface free energies, γ_1/γ_2. We cannot adopt this relationship because the image crystal is not a sphere with a homogenous surface free energy. However, we can regard the overall average surface free energy of the image crystal as a homogeneous one and use this relationship to analyse the inner pressure in truncated octahedron-type image crystals. Then, we can consider the vertical axis of Figure 10 as the ratio of the inner pressure $p(r_d) / p_{eq.}$. As the figure shows, the $\Gamma(r_d) / \Gamma_{eq.}$ value of the image crystal in our work is about 1.03, its shape is known to be deformed in the direction of increasing the surface free energy so as to balance the higher inner pressure of the helium enclosed in the image crystal.

At first we considered that the image crystal would change the shape continuously from the truncated octahedron to the cuboctahedron with increasing helium inner pressure. However, the cuboctahedron-type image crystal has not yet been identified in reality [12].

Table 2. Relative surface areas of facets bounding the pentacontahedron-type image crystal (Figure 7(a)) formed in single-crystal UO_2

Facets	(s{hkl}/s{111})exp.	s{hkl}(μm²)	s{hkl }*/s{111}*	d{hkl}*/d{111}	No. faces
{111}	1	6.382	1	1	8
{001}	0.1633 ± 0.0227	1.043	0.1634	1.080	6
{110}	0.2148 ± 0.0571	1.365	0.2139	1.129	12
{311}	1.435 ± 0.4008	9.160	1.435	1.059	24

Instead of the cavity with cuboctahedron, a pentacontahedron-type image crystal formed in the matrix. To clarify the transformation mechanism, we examined the structure of this complicated polyhedron using a computer code SHAPE [18]. The $s_{\{hkl\}}/s_{\{111\}}$ value obtained by this analysis and the structural parameters simulated are summarized in Table 2.

The distance parameter $d_{\{hkl\}}^*/d_{\{111\}}$ in the Table was determined so that $s_{\{hkl\}}^*/s_{\{111\}}^*$ agreed with $(s_{\{hkl\}}/s_{\{111\}})_{exp}$, e.g., in Eqs. (2) and (3). The shape of the pentacontahedron is specified uniquely using these parameters. The total surface area of this pentacontahedron with a volume of 6.807 μm^3 is 17.95 μm^2; this value is obtained by summing the surface areas of all the facets listed in the third column of the Table.

The variation in the total surface area of the image crystal resulting from the transformation in its shape is shown in Figure 11. The dotted line is the theoretical value calculated using Eqs. (7) and (8). The solid circle and the error bar represent the experimental data and the standard deviation in shape analysis.

The pentacontahedron-type image crystal labelled (d) was drawn using the structural parameters listed in Table 2. The surface area decreases with decreasing r_d but increases in the region where $r_d \leq 1$, whereas the relative surface area of the pentacontahedron-type image crystal is much smaller than that of the truncated octahedron (a) and cuboctahedron (e). The surface free energy of the truncated octahedron-type image crystal increases with decreasing r_d, as shown in Figure 10. However, the decrease in r_d is limited; our experimental results indicate that the lowest value is 1.08, which is the value for truncated octahedron denoted by (c) in Figure 11. When the helium inner pressure becomes much higher, the shape of the polyhedron is supposed to change to the facets with higher indices with a complex polyhedron and lower surface area rather than the cuboctahedron with a larger surface area. Namely, the increase in the inner pressure promotes the formation of the facets with much higher indices, making the image crystal more round. All these results imply that the round shape of the cavity formed just after the helium treatment (Figure 4(b)) is a result of the high helium inner pressure. Then, it is conjectured that the image crystal is formed with decrease in the inner pressure caused by the helium release during the subsequent heat treatment.

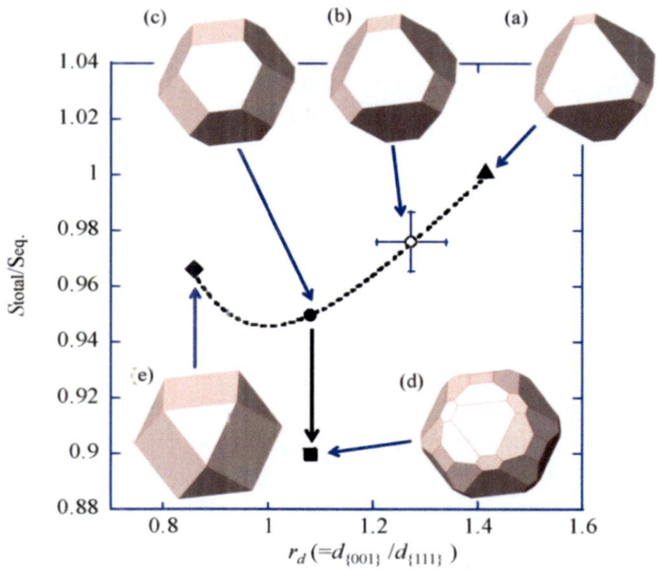

Figure 11. Dependence of surface area on r_d when the volume of the octahedron is kept constant [12].

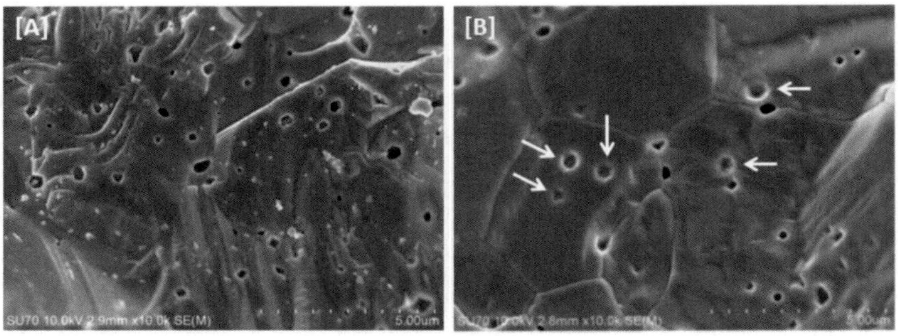

Figure 12. FE-SEM images of surface microtexture of poly-crystalline UO_2 [A] as-prepared [B] after helium infusion [28].

3. Variation of Surface Morphology with Formation of Image Crystal

We have recently found an interesting phenomenon related to the formation of image crystal [28] that shallow basins with hexagonal fringe are formed on the surface of helium-infused poly-crystalline UO_2 particle, as shown in Figure 12: The microstructure of the surface varies drastically with helium infusion.

The surface of particle shown in Figure 12[A] is a broken surface of a sintered pellet. So, the morphology of the surface is rough, and numerous open pores are observed. In contrast, Figure 12[B] shows that the surface of UO_2 was smoothed out by the surface diffusion during helium infusion plus the accompanying high temperature heat treatment. A complex wrinkle is seen on the particle surface, which is a step produced in the process of the surface diffusion accelerated by the high temperature heat treatment; the atoms on the surface migrated according to the Gibbs-Thomson effect (the reduction of local chemical potential due to nanoscale curvature), and changed the morphology. Though the presence of the grain boundary became obvious in Figure 12[B], it is not clear whether the morphology of the grain boundary changed during the helium infusion process or not, for the flesh broken surface in Figure 12[A] is so rough that the grain boundary is not visible in FE-SEM analysis. According to Sattonnay et al., at temperatures above 873 K, implanted helium nucleates to form small gas bubbles in UO_2 within 15 min [43]. The sample was not quenched after helium infusion in our or his experimental study. The cooling of the furnace from 1473 K to 873 K required approximately 30 min, which is enough for precipitated helium to form gas bubble.

It is known that the blistering occurs by gas precipitation in solids. The idea of gas-driven blister growth on metal surface was first proposed by Evans et al. [44,45]. Kuri et al. showed that similar phenomenon is also observed near the surface region of ceramic (yttria-stabilized zirconia) [46]. Much of the research on helium in UO_2 has been conducted by French researchers [43,47-49]. Their studies clarified the fundamental behavior of helium in UO_2, such as the lattice location of the dissolved helium, temperature dependence of the cluster size, formation of gas bubbles. It has also been shown by lattice location experiments that He atoms occupy octahedral interstitial positions in the UO_2 lattice [47]. In addition, they clarified that the surface of the implanted UO_2 flakes away when the local helium concentration reaches approximately 1 at. % [49]. Thus, it seems reasonable to conclude that the basins observed in Figure 12[B] are the traces of blisters whose lids were forced open, which causes the roughness of the surface, particularly on the edge of the ruptured blister [44].

The interesting point here is the shape of the fringe that surrounds the basins. If the basin is indeed a trace of the blister, i.e., a ruptured open mouth of the image crystal near the surface area [26], its shape would have similar ordered polyhedral structure to that of the latter. (The shape of the fringe relates to the formation of the image crystal near the surface area [28].) As mentioned in 2.2, the shape of the negative crystal found in polycrystalline UO_2 was that of a truncated octahedron [12].

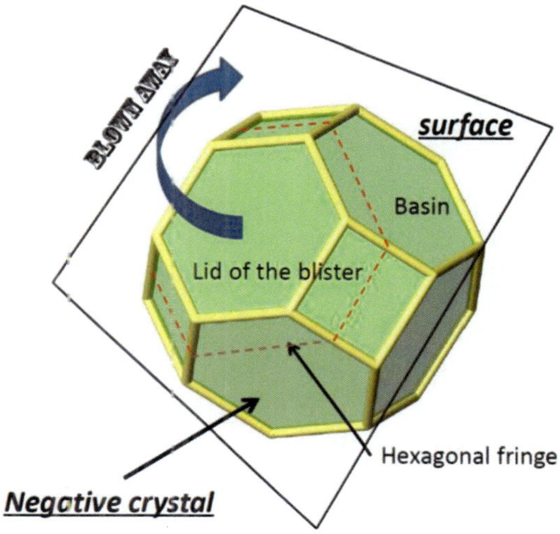

Figure 13. Schematic expression of hexagonal-fringe blister formation on the surface of poly-crystalline UO_2 particle [28].

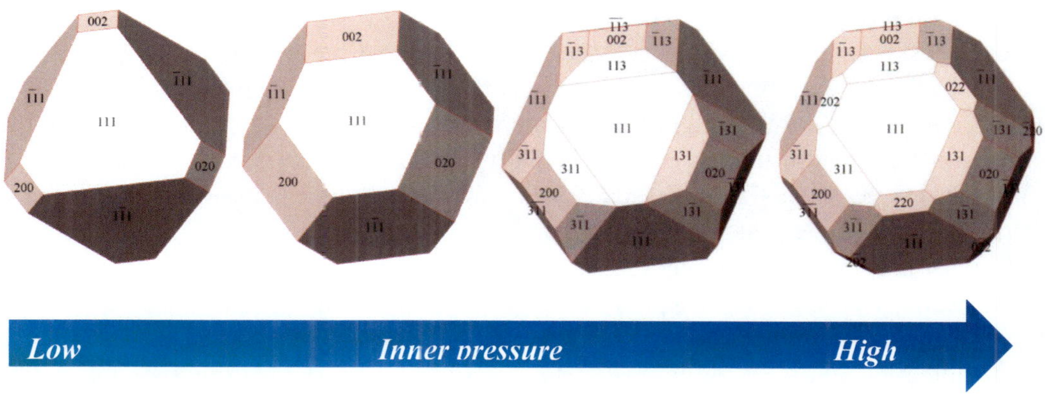

Figure 14. Transformation of image crystal formed in FCC-type UO_2.

The FE-SEM image of truncated octahedron-type image crystal appeared on the broken surface of polycrystalline UO_2 is given in Figures 6(d)-(f), whose detailed original (i.e., non-ruptured) structure being given in Figure 7(c). As can be seen in these figures, the cavity is surrounded by {100} and {111} facets. If the image crystal is formed in the vicinity of the surface and the surface ruptures, as seen in Figure 13, the shape of the fringe would be hexagonal corresponding to the section of the truncated octahedron, just as observed on the broken surface in Figures 6(d)-(f).

It is noteworthy that image crystal is formed without the helium release process. As discussed in 2.1, in case of single-crystal UO_2, the shape of the observed cavity just after helium infusion was round (see Figure 4(b)), its shape changing by mass transport through the high temperature heat treatment after helium infusion. On the other hand, the presence of the blister with hexagonal fringe implies that the image crystal has been formed in the poly-crystalline UO_2 without such high temperature heat treatment. This difference can also be explained in the same manner as in 2.1. The helium captured in the near-surface cavity can easily escape to the free surface and decrease its inner pressure to the level suitable for forming truncated octahedron type image crystal; the temperature at which helium infusion was performed is high enough for the mass transport to occur.

CONCLUSION

Though 'negative' or here-renamed 'image' crystal is occasionally seen in many crystal masses, little attention has been paid on such polyhedral cavity, for it had been regarded as a peculiar defect observed by chance. Our recent study on UO_2 clarified that the shape of image crystal undergoes orderly transformation depending on the inner gas pressure of the cavity. When gas is trapped in the gas bubble, the facets bounding the cavity grow. Basically, the shape is determined by the balance between the inner pressure and the surface energy. In case of FCC fluorite-type UO_2, it has a truncated-octahedron Wulff shape bounded by {111} and {001} facets. With increasing the inner pressure, the area of {001} facet increases.

With further increase in the inner pressure, {113} and {110} facets appeared in this order. However, the evaluation of morphological importance (MI) of facets follows the sequence $MI_{\{111\}} > MI_{\{001\}} > MI_{\{110\}} > MI_{\{311\}}$. The phenomena indicate that not only the attachment energy or the surface free energy but also the step free energy have to be taken into account to explain transformation mechanism, as schematically summarized in Figure 14. We clarified that the the step free energy is the one most important factor that governs the morphology of the FCC-type single-crystal UO_2 in its growth process, which creates a stir in the field of the crystal growth. The formation of cuboctahedron-type image crystal has not yet been reported: Assuming that the shape transformation occurs only through mass transport with no volume change, we have shown that this is because the pentacontahedron-type image crystal with the smaller surface area and hence the smaller overall surface energy than that of the cuboctahedron-type one is formed preferentially. The shallow basin is sometimes observed on the surface of helium infused FCC-type UO_2 particle. The shape of the fringe is hexagonal. The section of the image crystal formed in poly-crystalline UO_2 particle is polyhedral, and its shape resembles that of the fringe of the basin. It is thus hypothesized that the basin is the trace of the ruptured blister of the negative or image crystal formed in the vicinity of the grain surface. The formation of gas bubbles in a fuel matrix is well known to induce swelling of the pellet. When helium produced by the α-decay of minor-actinide (MA) is accumulated in raw powder of MA-MOX fuel, helium may precipitate in the grain-boundary region to form gas bubbles. Because this phenomenon will seriously affect densification during the sintering process, the formation of gas bubbles poses a non-trivial problem with respect to the quality control (QC) of nuclear fuel when MA-MOX is adapted as a standard fuel for fast-breeder reactors (FBRs). This example illustrates one practical aspect of above-reviewed basic image-crystal study in nuclear-fuel technology. (QC is expected to become much more difficult when raw powder is stored for a prolonged period before the sintering process.)

REFERENCES

[1] Y. Furukawa and S.J. Kohata,. *Cryst. Growth* (1993) 129, 571-581.
[2] A.J. Gratz, and P. Bird, *Geochim. Cosmochim. Acta* (1993) 57, 965-976.
[3] A.J. Gratz and P. Bird, *Geochim. Cosmochim. Acta* (1993) 57, 977-989.
[4] M.J. Buerger, *Am. Mineral.* (1932) 17, 228-233.
[5] M.J. Buerger, *Am. Mineral.* (1932) 17, 391-395.
[6] M.R. Castell, *Phys. Rev.* (2003) B68, 235411-235417.
[7] W.V. Lierde, *J. Mater. Sci.* (1970) 5, 527-531.
[8] T.R. Padden, Westinghouse Electric Corp. (1960).
[9] R.G. Robins, R.S. Wilks and B.T.J. Bradbury, *Nucl. Mater.* (1962) 5, 262-263.
[10] S. Takahashi, T. Kubota, H. Doi and H. Uchikoshi, *J. Atom. Ener. Soc. Japan* (1960) 2, 73-77.
[11] G. Wulff, *Zeitschr. Krystallogr. Mineral.* (1901) 34, 449-530.
[12] H. Serizawa, J. Matsunaga, Y. Haga, K. Nakajima, M. Akabori, T. Tsuru, Y. Kaji, S. Kashibe, Y. Ohisi and S. Yamanaka, *Cryst. Growth Des.* (2013) 13, 2815-2823.
[13] M.E. Cunningham, M.D. Freshley and D.D. Lanning, *J. Nucl. Mater.* (1992) 188, 19-27.
[14] S. Kashibe, K. Une and K. Nogita, *J. Nucl. Mater.* (1993) 206, 22-34.
[15] J. Spino, D. Baron, M. Coquerelle, A.D. Stalios, *J. Nucl. Mater.* (1998) 256, 189-196.
[16] L.E. Thomas, C.E. Beyer and L.A. Charlot, *J. Nucl. Mater.* (1992) 188, 80-89.
[17] K. Une, S. Kashibe and A. Takagi, *J. Nucl. Sci. Technology* (2006) 43, 1161-1171.
[18] E. Dowty, *Am. Mineral.* (1980) 65, 465-471.
[19] H. Aizawa and M. Tsukada, *Phys. Rev.* (1999) B59, 10923-10927.
[20] S. Hasegawa, N. Sato, I. Shiraki, C.L. Petersen, P. Bøggild, T.M. Hansen, T. Nagao and F. Grey, *Jpn. J. Appl. Phys.* (2000) 39, 3815-3822.
[21] T. Hirahara, I. Matsuda, M. Ueno and S. Hasegawa, *Surface Science* (2004) 563, 191-198.
[22] C. Schmidt and J. Ulrich,. in: M.Mazzotti (Ed.) *18th International Symposium on Industrial Crystallization*, AIDIC, Milan, Italy (2011) 96-97.
[23] C. Schmidt and J. Ulrich, *J.Cryst. Growth* (2012) 353, 168-173.
[24] X.X.Y. Liu and P. Bennema, *Phys. Rev.* (1996) B53, 2324-2335.
[25] M. Abramowski, R.W. Grimes and S. Owens, *J. Nucl. Mater.* (1999) 275, 12-18.
[26] M. Abramowski, S.E. Redfern, R.W. Grimes and S. Owens, *Surf. Sci.* (2001) 490, 415-420.
[27] J.-C. Spirlet, *J. Nucl. Mater.* (1989) 166, 41-47.
[28] H. Serizawa, J. Matsunaga, N. Shirasu, K. Nakajima, S. Kashibe and Y. Kajia, *J. Asian Ceram. Soc.* (2013) 1, 289-295.
[29] K. Nakajima, H. Serizawa, N. Shirasu, Y. Haga and Y. Arai, *J. Nucl. Mater.* (2011) 419, 272-280.
[30] F. Rufeh, D.R. Olander and T.H Pigford, *Nucl. Sci. Eng.* (1965) 23, 335-338.

[31] K. Nakajima, N. Shirasu, Y. Haga, H. Serizawa and Y. Arai, presented in *GLOBAL*, Makuhari, Japan (2011).

[32] P. Hartman and P. Bennema, *J.Cryst. Growth* (1980) 49, 145-156.

[33] A. Pimpinelli and J. Villain,. *Surface free energy, step free energy, and chemical potential, physics of Crystal Growth*, Cambridge Univ. Press, Cambridge (1998) 23-39.

[34] D.C. Schlößer, L.K. Verheij, G. Rosenfeld and G. Comsa, *Phys. Rev.* (1999) B82, 3843 -3846.

[35] C. Sweegers, S.X.M. Boerrigter, R.F.P. Grimbergen and H. Meekes, *J. Phys. Chem.* (2002) B106, 1004-1012.

[36] C. Herring, *The use of classical macroscopic concepts in surface-energy problems*, in: Structure and Properties of Solid Surfaces, R. Gomer and C.W. Smith eds., The Univ. Chicago Press, Chicago (1953) pp5-81.

[37] K.K. Mon, S. Wansleben, D.P. Landau and K. Binder, *Phy. Rev. Lett.* (1988) 60, 708-711.

[38] A.H. Booth, Atomic Energy of Canada Limited Chalk River Project Research and Development, Chalk River, Ontario (1957) pp1-8.

[39] A.H. Booth and G.T. Rymer, Atomic Energy of Canada Limited Charlk River Project Research And Development, Chalk River, Ontario (1958) pp1-27.

[40] C. Muggelberg, M.R. Castell, G.A.D. Briggs and D.T. Goddard, *Appl. Surf. Sci.* (1999) 142, 124-128.

[41] P.G. Shewmon and W.M. Robertson,. *Variation of Surface Tension with Orientation*, in: A.S.f. Metals (Ed.) Metal Surface, Metals Park, Ohio (1963) pp67-98.

[42] D.R. Olander, *Fundamental Aspects of Nuclear Reactor Fuel Elements*, Technical Information Center, Office of Public Affairs, Energy Research and Development Administration (1976).

[43] G. Sattonnay, L. Vincent, F. Garrido and L. Thome, *J. Nucl. Mater.* (2006) 355, 131-135.

[44] J.H. Evans and B.L. Eyre,. *J. Nucl. Mater.* (1977) 67, 307-310.

[45] J.H. Evans,. *J. Nucl. Mater.* (1978) 76-77, 228-234.

[46] G. Kuri, M.D. Beli and D. Gavillet,. *Nucl. Instr. Meth. in Phys. Res.* (2006) B245, 445-454.

[47] F. Garrido, L. Nowicki, G. Sattonnay, T. Sauvage and L. Thome,. *Nuclear Instruments and Methods in Physics Research* (2004) B219-220, 196-199.

[48] F. Garrido, L. Vincent, L. Nowicki, G. Sattonnay and L. Thome,. *Nucl. Instr. Meth. in Phys. Res.* (2008) B266, 2842-2847.

[49] S. Guilbert, T. Sauvage, H. Erramli, M.-F. Barthe, P. Desgardin, G. Blondiaux, C. Corbel and J.P. Piron,. *J. Nucl. Mater.* (2003) 321, 121-128.

In: New Research Trends of Fluorite-Based Oxide Materials ISBN:978-1-63117-350-9
Editors: Akio Nakamura and Junichiro Mizusaki

Chapter 11

STABILIZED ZIRCONIA FOR ACTINIDE TRANSMUTATION AND CONDITIONING

Marcus Walter and Joseph Somers*
European Commission, Joint Research Centre,
Institute for Transuranium Elements, Karlsruhe, Germany

ABSTRACT

Stabilized zirconia is a potential material for use in the dedicated transmutation of plutonium and minor actinides (MAs) in nuclear reactors or accelerator driven systems (ADS). Due to its high aqueous durability, it can also be used for Pu and MA conditioning and long term storage. In this review a number of features of stabilized zirconia pertinent to its operational safety assessment as a nuclear fuel and as a nuclear waste form are gathered and discussed. The main issues covered include its material properties in terms of synthesis, phase formation and stability, oxygen potential, local structure, thermal (expansion, conductivity and vaporization etc) properties, radiation resistance, and aqueous durability.

1. INTRODUCTION

1.1. Transmutation of Plutonium and Minor Actinides

One crucial general issue for the public acceptance of nuclear energy relies in finding appropriate solutions for the highly radiotoxic actinides generated in nuclear reactors. Most of them are α-emitters, requiring safe disposal of the spent fuel for about 100,000 years. After several hundred years of storage, the ingestion radiotoxicity of spent nuclear fuels is dominated by plutonium (Pu) and the so-called minor actinides (MAs): neptunium (Np), americium (Am) and curium (Cm) [1]. These actinides are generated from ^{238}U by multiple neutron capture and β–decay conversions, with ^{239}Pu ($t_{1/2}$ = 24,100 years) and ^{240}Pu ($t_{1/2}$ =

* E-mail address: 1052076@web.de (M. Walter), joseph.somers@ec.europa.eu (J. Somers).

6,563 years) dominating at the beginning of fuel irradiation. With increasing irradiation time ^{238}Pu ($t_{1/2}$ = 87.7 years), ^{241}Pu ($t_{1/2}$ = 14.4 years) and ^{242}Pu ($t_{1/2}$ = 375,000 years) are also produced. Due to its short half-life (5 hours) ^{243}Pu decays to ^{243}Am ($t_{1/2}$ = 7,370 years), which is then transmuted during reactor operation to other MA nuclides such as ^{244}Cm ($t_{1/2}$ = 18.1 years) and ^{245}Cm ($t_{1/2}$ = 8500 years). Furthermore, the Am and Cm concentration in spent fuel increases when mixed uranium-plutonium (MOX) fuel is used instead of ^{235}U enriched UO_2. In addition, ^{241}Am ($t_{1/2}$ = 432 years) is formed by the β–decay of ^{241}Pu both during irradiation and the post-irradiation storage period. The only relevant Np nuclide, ^{237}Np ($t_{1/2} = 2.14 \cdot 10^6$ years), is mainly generated by the $^{238}U(n, 2n)$ reaction and subsequent β-decay of ^{237}U. A further contribution to the Np inventory increase comes from the α-decay of ^{241}Am, which becomes substantial after several hundred years of storage. During storage in a geological repository, the decay of Pu and MAs generates further new decay-product nuclides and chemical species, which alter the physical and chemical properties of the stored material. Spent fuel is sensitive to oxidative dissolution; if ground water penetrates the natural and engineered barriers in place at the repository site, the soluble hexavalent UO_2^{+2} cation is formed by oxygen uptake from ground water. This is also true for pentavalent cations, e.g., NpO_2^+ and PuO_2^+. The dissolution and migration of actinides would be further promoted in the presence of carbonate and sulfate anions due to complex formation. Consequently, strict containment of the spent fuel in the storage casks and its, as complete as possible, isolation from ground water is needed to inhibit the actinide migration to biosphere.

Partitioning and transmutation (P&T) represents an alternative way to reduce the amount of these nuclides for final disposal. Thereby, the required time for safe disposal can be reduced drastically, even to a few thousand years. For this, the spent fuel must be dissolved and the Pu and MA separated from the other spent fuel constituents. They must be then converted into a suitable form for their irradiation in reactor systems. The spent fuel reprocessing currently applied in facilities like La Hague (France) and Sellafield (United Kingdom) involves spent fuel dissolution in nitric acid, the recovery of Pu and U by liquid extraction, and the vitrification of the MAs and fission products (FPs). New safe processing technologies are still required to separate the MAs from the FPs, a challenging problem due to the similar chemical behavior of the trivalent MAs and the trivalent lanthanide FPs.

In the P&T scenario, separated Pu and MAs are recovered and then converted to their oxides for interim storage. To prepare targets and fuels for transmutation, dilution of PuO_2 and MA oxides in a matrix is needed. If the undiluted materials were used directly, the produced power in the reactor and the resulting fuel and/or target operation temperature would be too high for safe operation. A matrix is also needed to maintain their physical and chemical properties as the material undergoes chemical transformation during irradiation. For this reason, PuO_2 is blended today with UO_2 and the resulting $(U,Pu)O_{2-x}$ mixed oxide fuel (MOX) is currently inserted as part of light water reactor cores. However, the use of UO_2 as a matrix for Pu transmutation has a limited efficiency if the goal is the incineration of the Pu, as new Pu nuclei are continuously generated from ^{238}U via neutron capture. Therefore, a closed fuel cycle with multiple reprocessing of the spent fuel can be envisaged, as a means to reduce the amount of Pu and MA in the waste. As an alternative, the separated Pu and MA can be incorporated in an inert matrix, i.e., a material not containing ^{238}U, and the irradiation can be performed either in nuclear reactors or dedicated accelerator driven systems (ADS) [2-7]. In contrast to a UO_2 matrix, such inert matrix materials must be 'inert' towards neutron capture, thus greatly reducing the formation of new actinide nuclei. During irradiation, Pu and MA undergo either fission or are transmuted to nuclei with significantly shorter half-lifes. The transmutation scheme of ^{241}Am under thermal neutron irradiation is shown in Fig. 1. Although the ^{242}Am ($t_{1/2}$ = 16 hours) generated by neutron capture is fissile, it almost immediately decays to either ^{242}Cm or ^{242}Pu. Most of the ^{242}Cm ($t_{1/2}$ = 163 days) undergoes α–decay to ^{238}Pu, also leading to a significant accumulation of helium, which can cause the material to swell and therefore induce a safety-concerned change in the dimension of the target material [8,9]. In fast neutron spectra, the capture process and build up of higher actinides is less pronounced, but the total transmutation rate is much lower than for a thermal neutron spectrum [10]. In case of pure Pu transmutation, a burnable poison (BP), like Er, must be added to flatten the power density evolution during irradiation enabling a safe target operation [11]. As can be seen in Fig. 1, power flattening can also be realized by addition of Am or Np to the target, as they also capture neutrons at the beginning of irradiation [12].

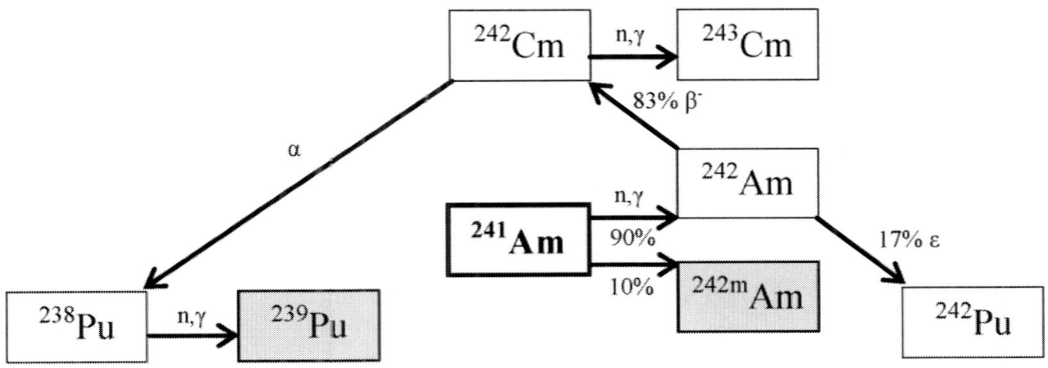

Figure 1. Transmutation scheme of ^{241}Am under thermal neutron irradiation. ^{239}Pu and ^{242m}Am are predominantly transmuted by fission.

In principal, the safety margin of inert matrix material should match or at least be close to that of the current UO_2 and MOX fuels or preferably should represent an improvement on the latter [2]. A long list of safety related criteria need to be assessed. The inert matrix should be resistant against irradiation induced damage caused by both neutrons and FPs to avoid swelling of the material or formation of microcracks, which could eventually lead to the relocation (broken chips) or to target-cladding interactions. Porosity increase could lead to undesired fission gas release. The transmutation target must be thermally stable and should have a high thermal conductivity, at least similar to UO_2, to remove heat from the target. Interactions with the cladding and the coolant should not occur. If the material is sent for direct disposal after irradiation ("once-through-then-out (OTTO)" strategy), it should be highly durable in aqueous solutions to limit release of remaining radionuclei during storage of the irradiated target. Even if actinide destruction is not complete within such an OTTO strategy, the amount of fissile Pu nuclei would be drastically reduced and the Pu isotope vector quality deteriorated [11] guaranteeing long-term proliferation resistance. Alternatively, an inert matrix, which is easily dissolved, can be chosen, if reprocessing of the irradiated target is preferred to enhance the total transmutation efficiency in a closed fuel cycle. Two types of inert matrix target are distinguished: (a) homogeneous target, in which actinides are incorporated in the inert matrix in the form of a solid solution, and (b) heterogeneous target, where Pu and MAs are included in a separate phase distributed in the inert matrix [2,13]. Candidate inert matrices have been down selected, based on the above-mentioned criteria; as a result, yttria stabilized zirconia $(Zr,Y)O_{2-x}$ (YSZ) and ZrN were regarded as the most relevant homogenous matrices, and MgO, Al_2O_3, $MgAl_2O_4$, Mo (depleted in ^{95}Mo) are considered for heterogeneous matrices to dilute actinide oxides. The total Pu and MA content of transmutation targets is in the order of 0.7 to 2.0 $g \cdot cm^{-3}$, depending on the isotope composition and the irradiation conditions. The present chapter focuses on stabilized zirconia as the most promising homogenous inert matrix material [12,14]. Though its thermal conductivity is not entirely satisfactory, its high aqueous durability and radiation resistance make stabilized zirconia a promising material for conditioning of Pu and MAs [15], e.g., for the direct disposal of Pu resulting from nuclear disarmament stock piles, derived from civilian use of nuclear energy [2,16], and for intermediate storage of separated Cm [17].

1.2. Stabilized Zirconia

Pure zirconia, ZrO_2, has a monoclinic crystal structure at room temperature and ambient pressure. It can be stabilized in tetragonal ($P4_2/nmc$) and cubic ($Fm\bar{3}m$) forms by the addition of large aliovalent dopants like Ca^{+2}, Y^{+3}, and other rare earth elements (REEs). For the transmutation of Pu and MAs, YSZ is of special interest, as both Y and Zr have low neutron cross sections. YSZ forms a cubic defect fluorite structure when the Y content exceeds 16 mol% $YO_{1.5}$ [18]. This structure persists up to the melting point at about 2700°C. The lattice parameter of stabilized zirconia $(Zr,REE)O_{2-x}$ exhibits a systematic variation depending both on the ionic radii of the dopant cations (REE) and their concentration in the solid solution [19]. As the actinide dioxides crystallize in the same structure type, they can be incorporated readily in stabilized zirconia matrices forming a solid solution. The fluorite structure is face-centered cubic (fcc), with the cations occupying the (0,0,0) and the oxygen anions (oxide ions) the (¼,¼,¼) positions (see Fig. 2 left). As is typical for solid solutions, the different cations are supposed to be randomly distributed on the same crystallographic position and each cation is coordinated by eight oxygen anions. In stabilized zirconia oxygen vacancies are created by the aliovalent dopants and therefore the average oxygen coordination number of the cation becomes less than eight. Furthermore, in YSZ these oxygen vacancies are supposed to be preferentially associated with the Zr atoms, although Y can also accept vacancies [20-22] at higher Y content.

In the case of larger trivalent dopants, as an extreme case of oxygen vacancy and host-cation association scheme, an ordered $A_2Zr_2O_7$ pyrochlore structure ($Fd\bar{3}m$) is formed in a narrow central region of the A_2O_3–ZrO_2 phase diagram. Such ordering occurs if the ionic radius of the dopant lies in a specific range (1.05–1.16 Å), which corresponds to La–Gd and Pu–Cf in the lanthanide and actinide series, respectively [23,24]. The ordering is lost at high temperature in some case, and pyrochlore transforms to a 'disordered'defect fluorite structure. This order-disorder transition temperature is lowest for $Gd_2Zr_2O_7$ and increases with the ionic radius of the A cation in the $A_2Zr_2O_7$ pyrochlore [24]. The pyrochlore structure has thus a unit cell parameter twice as large as that of the parent fluorite structure ranging from 10.5 to 10.8 Å, depending on the REE A [23]. The trivalent A atoms occupy the (½,½,½) position and the tetravalent zirconium atoms are located at the (0,0,0) position (see Fig. 2 right for A=Nd).

There are two occupied oxygen positions: 8b at (3/8,3/8,3/8) and 48f at (x,1/8,1/8), where the oxygen positional parameter, x, lies between 0.3125 to 0.375 depending on the A cation. The 48f position is relaxed in the direction of the vacant 8a position at (1/8,1/8,1/8). The Zr atoms are coordinated by six oxygen atoms (48f) at a distance of about 2.1 Å, while the trivalent A atoms are surrounded by two close apical oxygen atoms at approx. 2.3 Å (8b) and six more-distant oxygen atoms at approx. 2.6 Å (48f) [25]. As in the case of defect fluorite structure, $A_2Zr_2O_7$ pyrochlore can incorporate actinides by forming a solid solution with the actinides in trivalent, tetravalent, or mixed valence state. The most favored pyrochlore matrices for nuclear applications are $Nd_2Zr_2O_7$ or $Gd_2Zr_2O_7$. The Nd cross section for neutron capture is small enough to consider $Nd_2Zr_2O_7$ as inert matrix for Pu and MA transmutation. In contrast, the extremely high neutron cross section of ^{155}Gd and ^{157}Gd inhibits its use for transmutation targets, but, if $Gd_2Zr_2O_7$ is used as a conditioning matrix for excess weapons Pu the neutron absorption of Gd is beneficial with respect to criticality inhibition. Other ordered derivatives of the fluorite structure include δ–type $A_4Zr_3O_{12}$ ($R\bar{3}$), which is formed with smaller REEs (A=Sc,Y,Ho–Lu) and $CaZrTi_2O_7$. The latter and titania based pyrochlores are also considered as conditioning matrices for Pu and MAs [15]. However, as titania based compounds are thermally less stable than stabilized zirconia, they are not considered sufficiently safe for transmutation, and therefore not discussed further in this chapter.

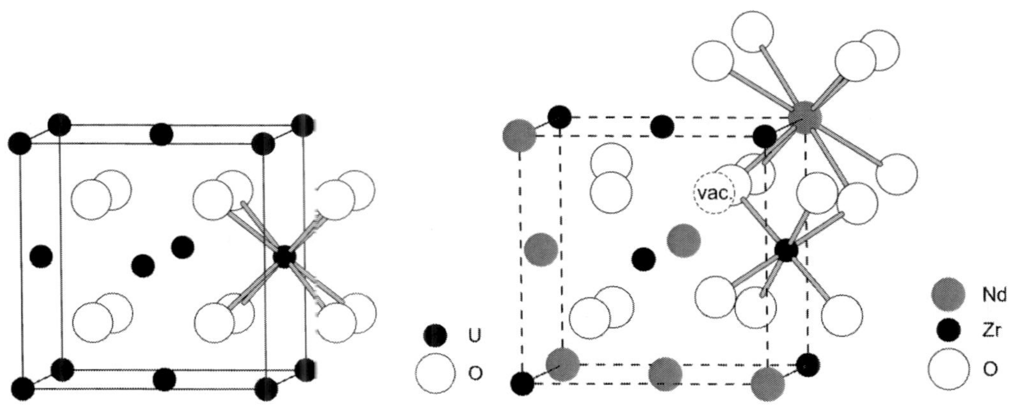

Figure 2. Fluorite (UO_2, a = 5.471 Å) and pyrochlore structure ($Nd_2Zr_2O_7$, a/2 = 5.338 Å).

1.3. Actinide Oxides

One of the most interesting features of the actinide elements is their capability to exist in a wide range of oxidation state. For lighter actinides, relevant for transmutation and conditioning, the trivalent oxidation state is the lowest known for binary oxides [26]. Sesquioxides are reported for Pu–Es. As for the lanthanide sesquioxides, hexagonal (A form, $P\bar{3}m1$), monoclinic (B form, $C2/m$), and cubic (C form, $Ia3$) modifications exist. The cubic (C) form is a derivative of the fluorite structure with doubled unit cell and with ¼ of the oxygen atoms removed, resulting in an ordered six-fold coordination of the cation. Actinide dioxides with fluorite structure (see Fig. 2 left) are found for Th–Cf. The lattice parameter is highest for ThO_2 (a = 5.597 Å [26]) and systematically decreases with atomic number of the actinide to 5.310 Å for CfO_2 (actinide contraction). Due to the more stable nature of dioxides some actinide sesquioxides (Pu_2O_3, Am_2O_3, and Bk_2O_3, etc) are susceptible to oxidation if stored under air. On the contrary, AmO_2, CmO_2, and CfO_2 are susceptible to reduction when heated, posing safety issues related to fuel pin cladding interactions. Oxygen is released from CmO_2 and CfO_2 heated in air at temperatures higher than 300°C, whereas AmO_2 is stable in air up to 900°C. Furthermore, δ–type intermediate phases, Am_7O_{12}, Cm_7O_{12}, and Cf_7O_{12}, have also been identified. Binary oxides of pentavalent Pa, U, and Np exist. In addition, U_4O_9, U_3O_7, U_3O_8, and UO_3 exist. Both Np_2O_5 and UO_3 decompose once heated above 400°C in air. At 800°C only NpO_2 and U_3O_8 are stable under air. It should also be noted that higher actinide oxidation states can be found in ternary oxides. For example, Cm(IV) can be stabilized in $BaCmO_3$ [27] and Np(VI) and Pu(VI) can be stabilized as $SrNpO_4$ and $SrPuO_4$, respectively [28].

Due to their mutually similar fluorite–based crystal structures, Pu and MAs can be readily incorporated into a stabilized zirconia matrix forming solid solutions. The present chapter first of all summarizes some representative preparation methods of such solid solutions for transmutation targets and then provides an overview of chemical and physical properties of the prepared materials. For completeness, results of a limited number of irradiation experiments aiming to evaluate their operational safety are summarized briefly.

2. Synthesis of Solid Solutions and Composites

One major safety aspect of actinide handling is the high radiotoxicity of inhaled actinide particles. Consequently, the handling of actinide powders or dust-generating processes must always be performed in hermetically sealed glove boxes. Glove boxes equipped with lead glass are needed when a large amount (i.e., gram scale upwards) of Np or Pu is processed. Also Am and Cm can be handled in such shielded glove boxes if their amount in solid form does not exceed several hundred mg. In solution self shielding diminishes drastically. Although the dominant γ–emission of ^{241}Am can be easily shielded by lead due to its low energy (59 keV), the dose rate at the operator's hands is unacceptable. The problem of the shielding against external irradiation becomes more serious in the case of Cm. The most abundant nuclide ^{244}Cm in spent fuel is an α, γ–emitter, which also undergoes spontaneous fission and thereby emits a significant yield of neutrons too. Handling such highly radioactive material using manual procedures is unacceptable. Consequently, the remote handling by master-slave manipulators in shielded cells is a requisite for the synthesis of Am and Cm bearing materials. The synthesis routes in such facilities must be optimized in terms of simplification, minimization of process steps and waste, and avoiding the generation of radioactive dust which would settle on the installation surfaces and complicate the maintenance due to a high dose rate [17]. In addition, extensive pretest processing with less or non radioactive surrogate materials have to be performed to establish safe routes.

Starting with traditional powder processing and precipitation techniques, several advanced preparation techniques of stabilized zirconia doped with actinides to synthesize suitable transmutation targets will be described in the following section. Such advanced methods are also essential for small-quantity production for safety research.

2.1. Powder Metallurgy

The blending and milling of powders is often a starting point to obtain a desired solid solution. The constituent oxide powders are milled either manually by mortar and pestle, or in mechanical ball or attrition mills. The milling can be performed dry [29] or with liquid additives like acetone and ethanol to avoid agglomeration and the generation of dust [16,30]. Zinc stearate, $Zn(C_{18}H_{35}O_2)_2$, is added optionally at 0.5–2 wt% as a lubricant for improving the powder–powder and also the powder–press die wall interaction. The material is then compacted at pressures in the 80–500MPa range to yield pellets for sintering at temperatures at 1400 to 1750°C for several hours. To avoid the generation of oxygen during irradiation (FPs are in general and as a whole more noble than the actinides), sintering is performed under reducing atmosphere (e.g., Ar/H_2, or N_2/H_2 gas mixtures). The addition of several thousand ppm of H_2O vapor mitigates the reducing conditions of Ar/H_2 or N_2/H_2 gas mixtures and can adjust the oxygen/metal (O/M) ratio of actinide oxides as safety related parameter.

A drawback of powder processing is the low solid state reactivity of the oxide powders, which necessitates higher sintering temperatures and repeated heating (after crushing, re-milling and re-pelletizing) to obtain homogeneous solid solutions [31-33]. The degree of homogeneity is usually less than when other advanced methods (see below) are used. Nevertheless, it was adapted successfully for production of zirconia based Pu transmutation targets with the composition $(Zr_{0.73}Y_{0.14}Er_{0.04}Pu_{0.08})O_{2-x}$ [31]. The process can be optimized by using reactive fine stabilized zirconia powder with high specific surface area as obtained by coprecipitation techniques. Due to dust generation, powder blending and milling methods are not suitable for the synthesis of Am bearing targets and is entirely unacceptable for Cm bearing targets [17].

2.2. Coprecipitation

Obtaining a homogenous material can be more easily achieved by using coprecipitation processes. Here, solutions of different metals are prepared, mixed to the desired stoichiometry, and finally coprecipitated by ammonia [16], oxalate, ammonium-oxalate or other chemical agents [34-36], or by simple evaporation. This process has the advantage that the actinides are available as an aqueous solution after spent fuel reprocessing. However, the mass streams which can be handled are limited in the case of Pu, as water moderates neutrons and consequently reduces the critical mass of the installation. The precipitate must be aged and washed to remove unwanted anions (typically nitrates). The dried precipitate is then converted into the oxide form by calcination at temperatures in the 600–800°C range. The calcined precipitate can then be pressed into pellets and sintered under similar conditions as for the powder blending process. A homogeneous solid solution can be obtained even after such a single thermal treatment [31].

2.3. Sol–Gel Process

An incremental improvement of the coprecipitation process is gel-supported precipitation, also often simply called the sol-gel process. In this process, the mixed metal solution, modified by thickeners, is dispersed in the form of droplets, which keep an ideal spherical form once they gel. After calcination, spheres with 40–150μm diameter can be obtained by using a rotating cup atomizer [37] (see Fig. 3), whereas the use of a needle type dropper results in larger spheres of 500–1000μm after calcination. Two principal gelation methods can be distinguished. In the external gelation process droplets are dispersed in an ammonia bath, where ammonia diffuses inside the droplets causing hydrolysis and precipitation of the metal ions. Due to the presence of polymers in the feed solution, the precipitation is geometrically controlled and the near spherical shaped droplet is transformed to the gel. After settling of the spheres at the bottom of the bath, they are aged, washed to remove anions and ammonia, dried in air or by an azeotropic distillation with C_2Cl_4, and finally calcined at temperatures of 600 to 800°C. This process was initially devised for the safe production of high quality UO_2 fuel kernels for the High Temperature Reactor (HTR), but was successfully extended to the production of stabilized zirconia [37] and actinide doped stabilized zirconia [29].

The second gelation method, called internal gelation, is based on the addition of HMTA (hexamethylenetetramine $C_6H_{12}N_4$) and urea ($CO(NH_2)_2$) to the feed solution, which is dropped into a hot silicon oil bath [31,38,39] or hot C_2Cl_4 [40]. At temperatures of ~100°C, HMTA and urea decompose and the liberated ammonia causes the precipitation. The spherical shape of the precipitate is guaranteed by the surface tension between the aqueous droplet and the surrounding silicon oil. Further processing steps include aging, separation from oil, washing with kerosene and ammonia solution, and finally calcination. The calcined material originating from either external or internal gelation can then be directly pressed into pellets and sintered. Alternatively, a milling step can be included before pressing to improve the sintering process and pellet microstructure [31,39]. Pu doped stabilized zirconia has been prepared by this method and a $(Zr,Y,Er,Pu)O_{2-x}$ solid solution was obtained after sintering at 1700°C [31].

Figure 3. YSZ spheres (ca. 80 μm) obtained by sol-gel process (rotating cup atomizer).

The addition of carbon black to the feed solution is reported to improve the pellet microstructure by softening of the calcined spheres and avoiding a "blackberry" structure of pellets directly pressed from sol-gel spheres [41,42]. In contrast to conventional precipitation techniques, the spheres obtained by the sol-gel process generate no dust, certainly a desirable attribute when dealing with MA targets. Also, the material is free flowing and can be handled remotely with ease. Significant amounts of different contaminated liquids accumulate, however, and must be treated as radioactive waste or recycled [43].

2.4. Infiltration Process

One remarkable property of the spheres fabricated by the sol-gel process is their high open porosity, which can be in the order of 50 to 80 vol%. This important attribute has lead to the evolution of a combined sol-gel and infiltration route (see Fig. 4) [37] where spheres of the inert matrix compound, e.g., YSZ, are produced in the absence of actinide compounds (i.e., in a conventional chemical installation). The calcined spheres are then transferred into glove-boxes or shielded nuclear facilities, where they are infiltrated with concentrated actinide nitrate solutions until the incipient wetness point is reached. Thus, the open porosity of the spheres is completely filled, but no free solution is present between them. The actinide concentration in the final product can be adjusted by the actinide concentration in the infiltrant solution. Approximately 6–10 wt% actinides can by infiltrated by a single infiltration step. The spheres are dried after infiltration to remove moisture and calcined between 600 to 800°C to convert the nitrate to the oxide. Furthermore, the spheres can be subjected to multiple infiltration steps to reach yet higher actinide loadings. Final steps include pellet pressing and sintering. During sintering, a solid solution is formed with the actinides, e.g., $(Zr,Y,Ce)O_{2-x}$, $(Zr,Y,Pu)O_{2-x}$, [37] and $(Zr,Y,Am)O_{2-x}$ [44]. Alternatively, the infiltrated spheres can be directly sintered after calcination and could even be coated with carbon and SiC for use as fuels in the high temperature reactor (HTR) [43].

The outstanding benefit of the combined sol-gel and infiltration process is the near elimination of radioactive waste. The liquid waste generated during the sol-gel process is free of radioactivity and can be treated as conventional chemical waste. No liquid radioactive waste is produced during the actinide infiltration, although the off-gas from the calcination must be treated for removal of NO_x.

Concentrated actinide solutions can also be infiltrated into powders or granules generated by pre-compaction [45,46]. Another process variation relies on the dispersion of reactive YSZ (5–9 m^2/g specific surface area) in 2-propanol to which an aqueous Ce nitrate solution is added [47]. After drying, the material was grinded and thermally treated at 450°C in air. A $(Zr,Y,Ce)O_{2-x}$ solid solution with densities of 96.4–97.5 % of the theoretical was found after sintering pressed pellets in air. Hot isostatic pressing (HIP) at 160MPa and 1400°C was applied to sintered $(Zr,Y,Ce)O_{2-x}$ pellets to further eliminate porosity. Unfortunately, the HIP processing led to a non-homogeneous redistribution of pores and transgranular cracks [47].

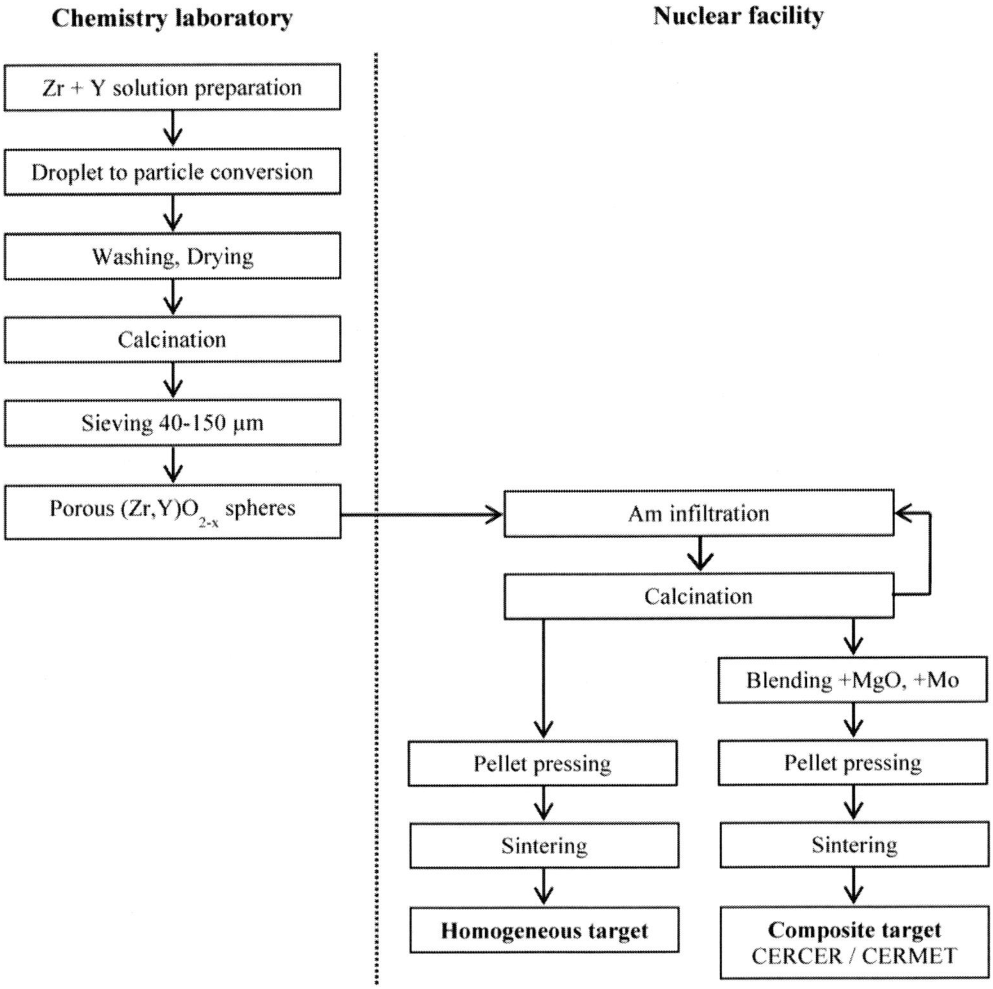

Figure 4. Combined sol-gel and infiltration route for the production of homogeneous and composite targets.

2.5. Composite Targets

A major safety related drawback of zirconia based transmutation targets is their low thermal conductivity, which results in a high pellet temperature during irradiation. The situation can be improved by heterogeneous composite targets, where the actinide doped zirconia matrix is diluted in a second inert matrix material with an inherently high thermal conductivity. To reach the desired Pu and MA content in the composite, the Pu and MA concentration in the stabilized zirconia must be higher than for a homogeneous target. If ceramic (e.g., MgO, Al_2O_3, $MgAl_2O_4$) or metal (e.g., Mo, stainless steel) are used, the composites are referred to as either CERCER (ceramic-ceramic composite) or CERMET (ceramic-metallic composite), respectively. The selected diluting material must fulfill the same criteria as an inert matrix fuel, with respect to cladding interactions and neutron absorption, etc. Although MgO (periclase) undergoes hydration in contact with water, its application in light water reactors seems feasible as Pu and MA are fixed in the stabilized zirconia matrix, which reduces the hydration-induced cracking of MgO grains [48]. The materials used for the composite should not react chemically to form undesirable phases (especially low melting point eutectics) during sintering or reactor operation. This phenomenon was observed, when Am nitrate was infiltrated into an $MgAl_2O_4$ matrix, forming a rhombohedral $AmAlO_3$ phase after sintering under Ar/H_2 instead of microdisperse AmO_2 as desired [8,49,50].

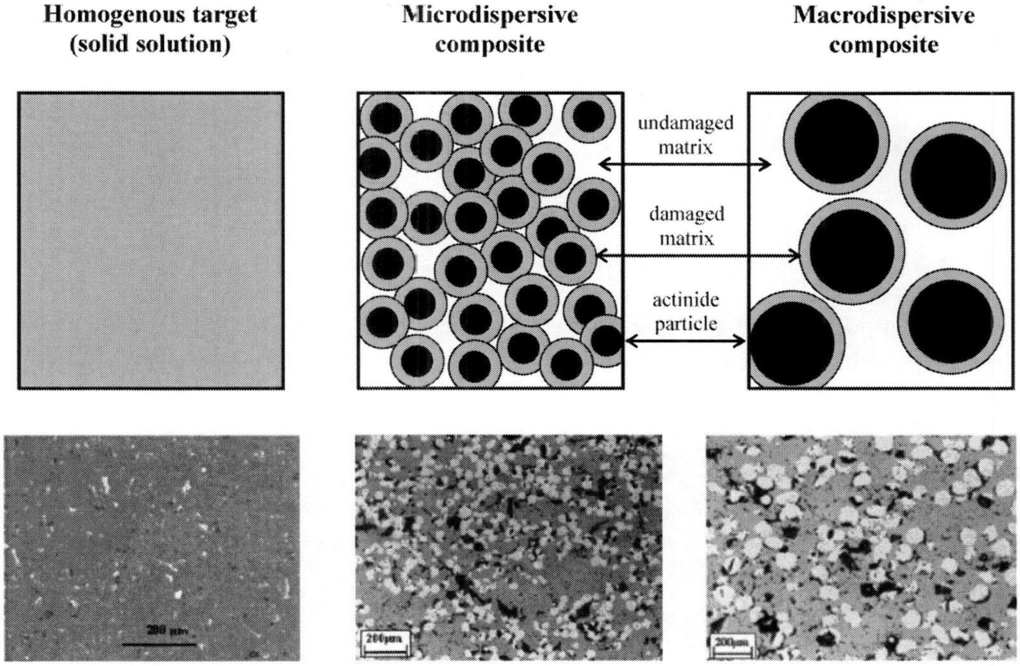

Figure 5. Schematic illustration of damage distribution caused by fission and α-decay products in a homogenous target, microdisperse, and macrodisperse composite targets. The ceramographs below [44] © (2003) Elsevier correspond to a homogeneous $(Zr,Y,Pu)O_{2-x}$ target and to $MgO–(Zr,Y,Pu)O_{2-x}$ composites with sphere size of 30–50 μm and 90–130 μm for micro- and marco-dispersed targets, respectively.

A special and important aspect of composite materials is the effective range of radiation damage caused by neutrons, FPs, and α-decay products (^4He and recoil nuclei). In a homogenous target (solid solution) the whole material is equally affected (i.e., damaged) by these particles. In contrast, radiation damage due to fission and α-decay products in composite targets occurs mainly in the particle of stabilized zirconia matrix containing Pu and MAs [8]. The properties of the blending material are affected only in a small shell extending to several μm in direct contact with the actinide bearing phase. Nevertheless, damage due to neutrons occurs in the bulk material too. Localization of the radiation damage caused by fission and α-decay products depends on the particle size and volume fraction of the actinide doped phase (see Fig. 5); it is small in microdisperse composites, where the particle size is similar to the range of fission fragments (8–10 μm [8]). In contrast, in macrodisperse targets (50–300 μm particles), the bulk of the material is not subjected to damage induced property deterioration.

This particle size effect on the safety performance during Pu and MA transmutation is currently under investigation in irradiation experiments comprising homogeneous targets and microdisperse and macrodisperse composites [29,44,51].

The production of composite fuels can be achieved by a variety of methods. Powder metallurgy is ideal for the production of microdisperse targets and has been deployed for the production of $(U,Zr,Y)O_2$ based composites with Al_2O_3 or $MgAl_2O_4$ [29], and for $(Zr,Pu,Mg)O_2$–MgO [52]. Simple evaporation of solutions of all the precursors is yet another method to achieve the goal of microdisperse targets [36,53,54]. Formation of slurries or emulsions and their controlled evaporation can also produce excellent microdisperse targets. Sieving of granulates or sol–gel produced particles is perhaps the ideal way to approach this issue, if well defined particle sizes are needed [44,51,55].

3. Material Properties

3.1.Phase Stability of Defect Fluorite and Pyrochlore Structure

When considering the incorporation of actinides in stabilized zirconia with a defect fluorite or pyrochlore structure, the phase diagrams of the parent ternary oxides, ZrO_2–CaO, ZrO_2–$REEO_{1.5}$, and ZrO_2–actinide oxides are an important starting point. In stabilized zirconia, the defect fluorite structure is usually formed at dopant (e.g., Ca, REE) concentrations exceeding 15 mol%, and in addition it often involves some ordered phases like perovskite–type $CaZrO_3$, δ–type $A_4Zr_3O_{12}$ (A=Sc, Y, Ho–Lu) and pyrochlore–type $A_2Zr_2O_7$ (A=La–Gd, Pu–Cf) around each stoichiometric composition range. At higher REE concentrations, the C- and B-type sesquioxides are also formed. The phase diagrams of ZrO_2–actinide dioxides and their surrogate ZrO_2–CeO_2 exhibit a coexistence of tetragonal and cubic solid solutions at typical heating conditions [56]. The stability of the tetragonal phase decreases to the benefit of the cubic phase with temperature and decreasing Zr concentration. Early ZrO_2–PuO_2 [57] and ZrO_2-AmO_2 [58] phase diagrams indicated a wide stability field for single fluorite type solid solutions. However, sample preparation in those studies was performed under slightly reducing conditions which lead to the formation of significant amounts of Pu(III) or Am(III) and thence stabilize the defect fluorite structure in a similar manner to REE dopants. A similar phenomenon is reported for $(Zr,Cm)O_{2-x}$ [59] and $(Zr,Am,Cm)O_{2-x}$ with Cm/Am=0.2 sintered at 1500°C in air [60], where trivalent Cm and Am stabilize the cubic structure.

In principle, one can assume that the solid solution of isostructural cubic stabilized zirconia and actinide dioxides also results in a (defect) fluorite structure. This has been shown for most trivalent and tetravalent actinides incorporated in stabilized zirconia up to approximately 30-40 mol% actinide, i.e., the range of interest for nuclear applications [30,37,45,61–65]. The incorporation of Th(IV) in stabilized zirconia is limited due to the large difference of the ionic radii [16]. Despite the low solubility of ZrO_2 in UO_2, U–rich solid solutions of single fluorite phase are reported, if sufficiently stabilized by REEs [66]. However, the miscibility of stabilized zirconia and actinide dioxides appears limited at approximately 1400°C when the Zr / actinide ratio is close to unity. The coexistence of two cubic $(Zr,Ca,U)O_2$ phases is reported between 35 and 70 mol% U at constant Zr / Ca = 0.3

(1400°C, vacuum [62]), but a single cubic phase is observed for $(Zr,Ca,U)O_{2-x}$ sintered at 1750°C under H_2 [61]. Also $(Zr_{0.425}Ca_{0.075}Np_{0.5})O_{1.925}$ sintered at 1400°C in air has two cubic phases [30], but $(Zr_{0.42}Y_{0.08}Np_{0.5})O_{1.96}$ sintered at 1600°C under Ar / H_2 exists as a single fluorite phase [65]. This temperature effect is also well known for the more intensively studied CeO_2–ZrO_2–$YO_{1.5}$ system. Although the phase diagram at 1600°C (air) shows a miscibility gap between the cubic stabilized zirconia and CeO_2 with two cubic phases occurring around Ce /Zr ~ 1 [67], more recent X-ray and neutron diffraction experiments indicate the formation of a complete solid solution above 1200°C [68]. In the case of U, the $(Zr,Ca,U)O_{2-x}$ and $(Zr,Y,U)O_{2-x}$ defect fluorite structure formed under reducing conditions can be destroyed upon oxidation in air. $(Zr,Ca,U)O_{2-x}$ is stable in air up 35 mol% U at 800°C, 20 mol% U at 1000°C [61]. and 10 mol% at least 1200°C [62]. The decomposition of $(Zr,Ca,U)O_{2-x}$ in air is associated with weight gain resulting in monoclinic ZrO_2, $CaUO_4$, and U_3O_8 [61,62]. Oxidation of cubic $(Zr,Y,U)O_{2-x}$ with up to 20 mol% U (Y/Zr=0.2) in air results in a tetragonal derivative ($P4_2/nmc$) of the fluorite structure. At U concentrations of 50 mol% and higher, $(Zr,Y,U)O_{2-x}$ decomposes during oxidation to ZrU_2O_7 and U_3O_8, but this can be suppressed by increasing the Y/Zr ratio. $(Zr_{0.17}Y_{0.33}U_{0.5})O_{2-x}$ maintains the fluorite structure when oxidized at 800°C in air [65].

Although incorporation into a pyrochlore structure is mainly considered for trivalent actinides, higher oxidation states like Np(V) and U(VI) can be incorporated in $Nd_2Zr_2O_7$ up to at least 5 mol% [69]. The solubility of Th(IV) in $Nd_2Zr_2O_7$ and $Gd_2Zr_2O_7$ is limited to less than 5 mol% Th [69,70]. Compensation with Al permits the Th solubility to reach x = 0.4 in $(Gd_{2-x}Th_x)(Zr_{2-x}Al_x)O_7$ [71]. The Pu(IV) solubility limit in $La_2Zr_2O_7$ sintered at 1500° air is about 10 mol%, but the total Pu solubility can be enhanced by partial reduction to Pu(III) in vacuum (35 mol% Pu [33]) or under Ar / H_2 (up to 50 mol% [33,72]), in agreement with an atomistic simulation, suggesting that the solubility of actinides in pyrochlores will increase in the order of U(IV) < Pu(IV) < Pu(III) [73]. The miscibility of $La_2Zr_2O_7$ and $Pu_2Zr_2O_7$ is not continuous in the Pu-rich range [33], but this could also be an artifact of incomplete reduction to Pu(III). Another important aspect is the transition from the disordered defect fluorite structure to the ordered pyrochlore structure when the ratio REE / Zr nears unity. Such a transition is observed by progressive incorporation of trivalent cations like Pu(III) [64] or Nd(III) [62] into stabilized zirconia. Sintering of $(Zr,Y,Pu)O_{2-x}$ with Y / Zr = 0.25 at 1500°C under Ar / H_2 results in a defect fluorite phase up to 30 mol% Pu, but a single pyrochlore phase is found at higher Pu contents. Ordered pyrochlore structures were also observed in $(Zr,Am,Cm)O_{2-x}$ solid solutions for 36–51 mol% actinide (at Cm/Am = 0.2 [60]). The ordering of the pyrochlore structure can be either partially (oxygen sublattice) or fully (cation and oxygen sublattice) reversed by oxidation of the trivalent actinides (Pu,Am) [23]. Oxidation at moderate temperature (≤ 800°C) introduces additional oxygen atoms, which will disorder the oxygen sublattice, while the cation ordering is retained with the lattice parameter decreased slightly. Oxidation at higher temperatures causes disordering of the cation sublattice too and the pyrochlore structure transforms into the defect fluorite structure. In addition, it should be noted, that ^{99}Tc ($t_{1/2}$ = 210,000 years), a long-living FP, can be incorporated in pyrochlores as it also forms pyrochlores with REEs like $Nd_2Tc_2O_7$ [74]. The solubility of Sr in $(Gd_{2-x}Sr_x)\ Zr_2O_{7-x/2}$ is limited to x < 0.05, but can be slightly enhanced by adding equal amounts of Ce(IV) [75].

3.2. Actinide Oxidation State and Oxygen Potential

As mentioned above, the majority of the actinide (An) elements are stable in more than one oxidation state. The oxygen stoichiometry in their solid solutions is usually expressed either by the oxygen to actinide (O / An) or by the oxygen to metal (O / M) ratio, the latter corresponding to 2-x in (Zr, REE, An)O_{2-x}. The actinide oxidation state and thereby the O / M ratio depends on the applied oxygen potential and temperature during the sample production, or during a post production treatment. The oxygen potential $\Delta\overline{G}_{O_2}$ is related to the oxygen partial pressure $p^*_{O_2}$ normalized to standard state pressure (1010 hPa) according to;

$$\Delta\overline{G}_{O_2} = RT \cdot \ln(p^*{}_{O_2})$$

where R is the gas constant and T is the temperature in Kelvin (K). The relationship between the O / M ratio and the oxygen potential of the system is strongly temperature dependent. With increasing temperature the equilibrium O / M ratio of the system generally decreases, i.e., when a solid solution is heated under the same gas atmosphere, the actinide is reduced and oxygen released. This effect is of significant importance for operational safety of nuclear fuels and transmutation targets. During their irradiation, heat is generated and transported from the pellet through the pellet / cladding gap, and then from the cladding to the coolant, resulting in a temperature gradient from pellet center to the rim of the pellet. To avoid the release of oxygen causing a redistribution of fission products and unfavorable fuel-cladding interaction [76], nuclear fuels are generally sintered under reducing conditions.

The O / M ratio can be determined by heating a sintered sample at a given temperature and in a standard atmosphere for which the oxidation state of the actinide is known. The change of O / M ratio relative to the reference state can be measured by weighing the material before and after the reference heat treatment. The more challenging measurement of the oxygen potential itself can be made by thermogravimetric or electromotive force measurements. Additional information about the actinide oxidation state can be obtained from the lattice parameter behavior or by element–specific spectroscopic methods such as X-ray absorption near edge structure (XANES) and Mössbauer spectroscopy.

Table 1. Possible oxidation states of actinides in stabilized zirconia

Th	Pa	U	Np	Pu	Am	Cm
				3^a	3^a	3
4	$4^{a?}$	4^a	4^a	4^b	4^b	$4^{c?}$
	$5^{b?}$	5^b	5^c			
		6^c	$(6^{c?})$			

^{a} reducing conditions (e.g., Ar/H_2 at high temperature)
^{b} oxidative conditions (e.g., air at moderate temperature)
^{c} oxidative conditions in REE doped solid solutions with excess oxygen vacancies
$^?$ feasible, no experimental data

Sintering under reducing conditions, e.g., Ar/H_2, N_2/H_2 or CO at high temperature (1500–1750°C), promotes the formation of a lower O / M system with the lower actinide oxidation states. In contrast, oxidative treatment at moderate temperatures (< 1000°C) in air favors higher actinide oxidation states. Heat treatment under vacuum often results in a less well defined, but slightly reducing condition [33,64]. The possible oxidation states of actinides in stabilized zirconia are summarized in Table 1 [65]. The oxidation state +3 increases its stability for Pu, Am, and Cm. all of which form $An_2Zr_2O_7$ pyrochlores [23,77,78]. Namely, the oxidation state +4 becomes less stable from Np to Cm. Thorium in stabilized zirconia exists only as Th(IV). Although no experimental data are available for Pa in zirconia based compounds, Pa(IV) and Pa(V) should exist. The higher oxidation states, U(VI), Np(V) and Cm(IV), can be stabilized due to the presence of the lower–valent REE dopant cation in stabilized zirconia.

The lowest U oxidation state in stabilized zirconia is U(IV), which requires reducing conditions. Similar to the ternary $(U,REE)O_{2-x}$ oxides [79], the stability of U(IV) decreases with increasing REE dopant concentration in $(Zr,REE,U)O_{2-x}$ solid solutions. For example, $(Zr,Y,U)O_{2-x}$ reduced in Ar/H_2 at 800 and 1600°C revealed a systematic increase of O/U ratio with increasing Y content in the solid solution [65]. This effect is caused by the oxygen vacancies introduced by the REE dopants, which modify the oxygen potential of $(Zr,REE,U)O_{2-x}$ promoting stability of higher U oxidation states. The oxidation of $(Zr,Ca,U)O_{2-x}$ in air occurs mainly between 200 and 400°C, resulting in an O/U ratio of 2.3 at 900°C [62]. Furthermore, the $(Zr,Ca,U)O_{2-x}$ defect fluorite structure is stable at 800°C in air if the O/M is less tan 2.05 [61]. For $(Zr,Y,U)O_{2-x}$ the O/U ratio after oxidation at 800°C in air was monitored for different Y / Zr and Y / U stoichiometries [65] and the O / U ratio depends on the Y / U ratio, rising from ~2.5 to ~3.0 with increasing Y / U ratio (see Fig. 6 left). At higher Y / U ratios, no further O / M increase was observed as O / U = 3 corresponds to the highest U(VI) state (Table 1). As the Y / U ratio determines the concentration of oxygen vacancies available to be filled on oxidizing the sample, it can be concluded that both oxidizing conditions and excess oxygen vacancies (due to the presence of REEs) are needed to fully stabilize U(VI). Diffuse reflectance measurements of $(Zr_{0.85}Y_{0.14}U_{0.01})O_{2-x}$ indicated mainly U(V) after heat treatment at 1450°C in air or N_2 / H_2 [80]. Based on X-ray diffraction (XRD) and XANES measurements an average oxidation state of +5.5 was found for U in a $(Nd_{1.8}U_{0.2})Zr_2O_{7+x}$ pyrochlore oxidized at 800°C in air [69]. Oxygen potential measurements are reported only for U-rich ($U \geq 50$ mol%) solid solutions with Zr and REEs [81].

The dominant oxidation state of Np in stabilized zirconia with low content of oxygen vacancies (i.e., low Y content) is Np(IV). An analysis of the fluorite lattice parameter finds Np(IV) in a $(Zr,Y,Np)O_{2-x}$ solid solution (Y / Zr = 0.17), independent of whether the material was sintered under Ar/H_2 or air [45,82]. The predominance of Np(IV) in $(Zr_{0.69}Y_{0.11}Np_{0.2})O_{2-x}$ sintered under Ar/H_2 was verified by ^{237}Np Mossbauer spectroscopy, which also reveals the presence of Np(V) at a trace level [82]. The presence of small Np(III) contents, feasible according to the Np–O phase diagram [83] was not observed. Neptunium in a $(Nd_{1.8}Np_{0.2})Zr_2O_{7+x}$ pyrochlore is mainly in the Np(IV) oxidation state after sintering in Ar/H_2 and can be oxidized to Np(V) by heating at 800°C in air [69]. A systematic study of the Np oxidation dependence on the oxygen vacancy concentration is reported by Walter et al. [65]. A plot of O / Np versus Y / Np ratio in $(Zr,Y,Np)O_{2-x}$ does not lead to a definitive conclusion, but the O / Np ratio certainly increases almost linearly with Y / (Zr+Np) (Fig. 6 right), implying that the Np oxidation state is influenced by the amount of both Np and Zr, i.e., Np has to compete with Zr for oxygen vacancies introduced by Y(III). Furthermore, the highest O / Np ratio of 2.63 suggested the presence of minor amount of Np(VI).

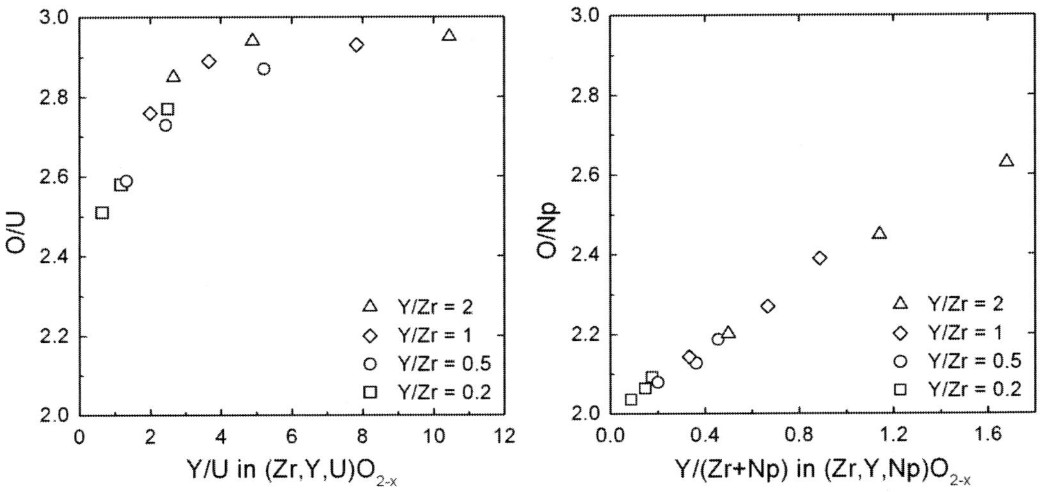

Figure 6. O/U (left) and O/Np (right) ratios in $(Zr,Y,U)O_{2-x}$ and $(Zr,Y,Np)O_{2-x}$ versus the Y/U and Y/(Zr+Np) ratios in the solid solutions [65]. Different oxidation mechanism exists for U and Np. A progressive U oxidation is observed with increasing Y/U ratio (i.e., the concentration of free oxygen vacancies) until U(VI) dominates (O/U=3.0). The correlation of O/Np and Y/(Zr+Np) indicate that the Np oxidation has to compete with Zr affinity to oxygen vacancies introduced by Y(III).

Plutonium in stabilized zirconia can be in both Pu(III) and Pu(IV) oxidation states. The stabilization of the higher Pu oxidation states (e.g., V and VI) by REEs is not reported. Although a fully reduced $Pu_2Zr_2O_7$ pyrochlore (a = 10.692 Å) is reported by Raison et al. [77], the reduction to Pu(III) in stabilized zirconia is often not complete. For example, an O/Pu ratio of 1.61 was estimated for $Pu_2Zr_2O_{7+x}$ sintered at 1700°C under Ar/H_2 [33]. Yamashita et al. [64] sintered $(Zr_{0.8}Y_{0.2})O_{1.9}$ and PuO_2 at 1500°C under Ar/H_2 and reported a mixture state of 2/3 Pu(III) and 1/3 Pu(IV) for solid solutions ranging from 4.5 to 47.4 mol% Pu, independent of the existence of the solid solution in either the fluorite or pyrochlore structure. The more stable oxidation state Pu(IV) can be reached by sintering in air [64] or in Ar/H_2 saturated with several thousand ppm H_2O [37]. Oxygen potential measurements are reported for $Pu_2Zr_2O_{7+x}$ (pyrochlore [84]) and $(Pu_{0.5}Zr_{0.5})O_{2-x}$ (defect fluorite [85]) and are shown in **Fig. 7** for 1078 K [85]. The oxygen potential of $Pu_2Zr_2O_{7+x}$ is 50–70 kJ/mol higher than that of $(Pu_{0.5}Zr_{0.5})O_{2-x}$, indicating that pyrochlore ordering stabilizes Pu(III) in comparison with the disordered defect fluorite structure. The oxygen potential of the Ce analogue $Ce_2Zr_2O_{7+x}$ is ~150kJ/mol higher than that of $Pu_2Zr_2O_{7+x}$, similarly to the CeO_{2-x} – PuO_{2-x} case [84].

Am(III) in stabilized zirconia [69,86] and $Am_2Zr_2O_7$ [78] can be obtained under reducing conditions. A mixture state of Am(III) and Am(IV) is reported when stabilized $(Zr,Am,Cm)O_{2-x}$ was sintered at 1500°C in air [60]. This is in agreement with oxygen potential measurements on AmO_2, indicating that oxygen is released from AmO_2 at temperatures above 1000°C [87]. Am(IV) is stable in stabilized zirconia when the heat treatment in air did not exceed 1000°C [69,86]. The oxygen potential of $Am_2Zr_2O_{7+x}$ (pyrochlore) was measured at 1333 K [88] and reveals a sharp increase of the oxygen potential from -516 kJ/mol (at x = 0) to ~ -100 kJ/mol (at x = 0.12) followed by a smooth increase until 0 kJ/mol (at x = 0.56). The oxygen potential of $Am_2Zr_2O_{7+x}$ is ~200 kJ/mol higher than that of a fluorite–type $(Am_{0.5}Np_{0.5})O_{2-x}$ solid solution at the same O / M ratio [88]. This can be attributed to structural ordering in the pyrochlore structure, including Zr's affinity for oxygen vacancies. Curium incorporation into stabilized zirconia occurs as Cm(III), even if the material is sintered in air [60]. The formation of Cm(IV) might be feasible during low temperature oxidation treatment in the presence of excess oxygen vacancies (e.g., achieved by high Y/(Y+Zr) ratios) [65].

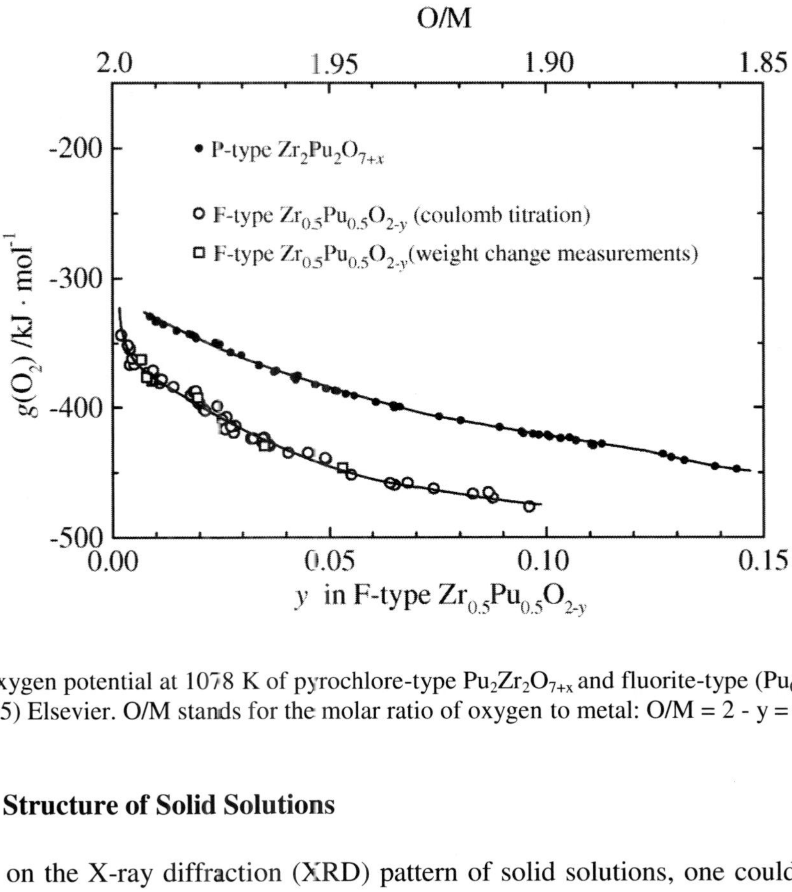

Figure 7. Oxygen potential at 1078 K of pyrochlore-type $Pu_2Zr_2O_{7+x}$ and fluorite-type $(Pu_{0.5}Zr_{0.5})O_{2-y}$ [85] © (2005) Elsevier. O/M stands for the molar ratio of oxygen to metal: O/M = 2 - y = (7 + x)/4.

3.3. Local Structure of Solid Solutions

Based on the X-ray diffraction (XRD) pattern of solid solutions, one could assume that cations and oxygen are randomly distributed on each sublattice in the defect fluorite structure and that the metal-oxygen bond lengths follow Vegard's law. However, X-ray diffraction (XRD) yields average structure information over several unit cells and is not sensitive to the actual local structure of the system when different elements are placed at the same crystallographic site. The latter information can be obtained by extended X-ray absorption fine structure (EXAFS) spectroscopy. EXAFS study of stabilized zirconia allows the individual determination of the first nearest neighbor cation (e.g., Zr, REE, or actinide)-oxygen and the second nearest neighbor cation–cation distances in the two closest coordination shells around the specified central cation. Although the oxygen coordination number is refined during EXAFS analysis, the low reliability of this parameter (typically 10% at best) restricts the discussion of the obtained results mainly to the interatomic distances. Complementary results on local structure can be obtained by nuclear magnetic resonance (NMR) method, but it is limited to NMR-active nuclides like ^{89}Y (and ^{17}O, etc.) and has not yet been applied to stabilized zirconia containing actinides. Both methods are element specific, and thus permit the detection of local structure order around a selected element brought about by, e.g., oxygen vacancies and the incorporation of actinides in different oxidation states etc.

Indeed, early EXAFS measurements on cubic YSZ indicated that the Zr–O bond lengths are shorter and Y–O bond lengths are longer, respectively, than those expected from Vegard's law [20,21,89]. In addition, the Zr–O and Y–O bond lengths remained nearly constant with increasing Y content instead of following the measured increase of lattice parameter. A similar behavior is reported for cubic $(Zr,Yb)O_{2-x}$ [90]. For both solid solutions a preferential association of oxygen vacancies with the Zr atoms was concluded [20,21,90], and was recently validated for YSZ in an ^{89}Y MAS-NMR study; the latter showed that Y remains eight-fold coordinated up to a 20 mol% Y, while the average coordination number of Zr was always lower than that of Y between 20 and 60 mol% Y [22]. According to Li et al. [91] oxygen vacancies in stabilized zirconia are preferentially associated with Zr atoms when the dopant (e.g., Y, Gd, Yb) has a larger ionic radius than Zr. In contrast, smaller cations (e.g., Fe, Ga and Sc etc) compete with Zr for oxygen vacancies leading to a significant local structural disturbance. Additional insight is obtained by an ^{89}Y MAS-NMR spectroscopic study on $(Zr_{0.8-x}Ce(IV)_xY_{0.2})O_{2-x}$, where Ce(IV) can be considered as surrogate for tetravalent actinides, especially for Pu(IV). Based on the quantitative evaluation of the ^{89}Y MAS-NMR peaks corresponding to Y in oxygen 6-, 7-, and 8-fold coordination, a preferential association of oxygen vacancy to Zr(IV), i.e., its order of Zr > Y > Ce(IV), was found [92].

EXAFS studies on stabilized zirconia with defect fluorite structure containing U, Np, and Am have been performed [54,65,86,93,94]. Probing of Pu by EXAFS is difficult as the Zr K-edge (17998 eV) and Pu L_3-edge (18053 eV) overlap. For this reason, besides licensing problems, Ce(IV) was used often as a surrogate for EXAFS analysis [95-97]. Although most of the EXAFS studies reported a single metal-oxygen shell for cubic zirconia, some authors addressed the distortion within the local structure by fitting oxygen atoms at more than one distance [93,95,96], thus complicating any comparison of results obtained.

A structural model proposed for $(Zr_{0.67}Y_{0.14}Am_{0.19})O_{2-x}$ with Am present either as Am(III) or Am(IV) is shown in Fig. 8 [86]. The interatomic distances were obtained from analysis of the Zr K-edge, Y K-edge, and Am L_3-edge EXAFS. The Zr–O and Y–O distances are similar to those determined for YSZ [20,21,89]. Also the Am(IV)–O distance is close to U(IV)–O and Np(IV)–O distances in $(Zr,Lu,U)O_{2-x}$ [65], $(Zr,Mg,Er,U)O_{2-x}$ [54], and $(Zr,Y,Np)O_{2-x}$ [65]. This has led to the assumption that the Pu(IV)–O distance in $(Zr,Y,Pu)O_{2-x}$ with 20 mol% Pu will also be approximately 2.27Å [94]. The Am(III)–O distance of 2.37Å is slightly shorter than the Nd(III)–O distance reported for $(Zr,Y,Nd)O_{2-x}$ with identical stoichiometry [97]. The interatomic distances depend not only on the oxidation state of the element but also on the stoichiometry of the system, as observed for other isovalent solid solutions of ionic character [98-100]. An EXAFS study covering a stoichiometry range from 6 to 94 mol% U(IV) in $(Zr,Lu,U)O_{2-x}$ and 6 to 80 mol% Np(IV) in $(Zr,Y,Np)O_{2-x}$ shows interatomic distances following Vegard's law to a certain degree [94]. Nevertheless, Fig. 8 enables the understanding of the interaction of local structure and oxidation behavior. On changing the oxidation state from Am(IV) to Am(III), the Am–O bond length increases as expected from the ionic radii of the cations. In addition, the local structure around Zr changes with both Zr–O and Zr–Zr distances decreasing from Am (IV) to Am(III). Once Am(IV) is reduced to Am(III), additional oxygen vacancies are introduced and their preferential association with the Zr cations reduces the oxygen coordination number of Zr and therefore the average Zr-O bond length. The same mechanism underpins the oxidation of U(IV) and Np(IV) in stabilized zirconia (see Fig. 6). Uranium in stabilized zirconia can be easily oxidized and oxygen vacancies are eliminated yielding a near stoichiometric compound with O / M ~ 2 [61]. As a consequence, the oxygen coordination number of Zr increases to ~8, which disables the stabilization mechanism for a cubic solid solution. Indeed, this effect was observed for $(Zr,Y,U)O_{2-x}$ with Y/Zr = 0.2, which changed from fully cubic stabilized to a partly stabilized tetragonal structure upon oxidation [65]. Neptunium is much less susceptible to oxidation than U and can only be oxidized in the presence of excess oxygen vacancies.

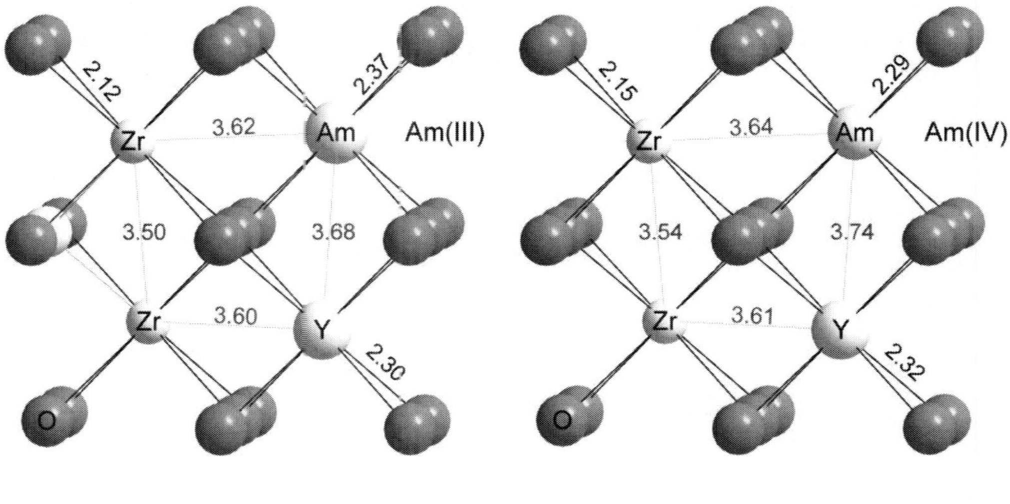

Figure 8. Structural model of the $(Zr_{0.67}Y_{0.14}Am_{0.19})O_{2-x}$ local structure based on interatomic distances determined from Y and Zr K-edge, and Am L_3-edge EXAFS spectra. The Am oxidation state was set to predominantly Am(III) by thermal treatment in Ar/H_2 (at 1600°C) and to Am(IV) by oxidation in air (at 1000°C) [86] © (2009) Elsevier.

The transition from the defect fluorite to the ordered pyrochlore structure was studied on the local scale for $Gd_2(Ti_{1-y}Zr_y)_2O_7$ [101]. Such studies are sparse for actinide doped pyrochlores, but EXAFS has been achieved for $^{241}Am_2Zr_2O_7$ pyrochlore [102,103]. The Zr K-edge and Am L_2-edge EXAFS evaluation indicated a single Zr–O bond distance of 2.13 Å and two Am–O bond distances of 2.31 Å and 2.53 Å, matching those calculated from the XRD data. The metal–metal distances (Zr–Zr, Zr–Am, and Am–Am) determined by EXAFS were identical, and close to 3.75 Å again in agreement with the pyrochlore structure. Due to the ^{241}Am self-irradiation, the $Am_2Zr_2O_7$ pyrochlore disordered within one year to yield a defect fluorite structure. The Zr–O distance became 2.15 Å and the split Am–O configuration merged to a single Am–O distance of 2.36 Å [102,103]. In contrast, the identical metal–metal distances of the pyrochlore changed to give Zr–Zr = 3.51 Å, Zr–Am = 3.64 Å, and Am–Am = 3.75 Å. Continued storage and concomitant damage did not change the structure further [103]. The interatomic distances of the disordered $Am_2Zr_2O_7$ pyrochlore are in fair agreement with those of the $(Zr_{0.67}Y_{0.14}Am(III)_{0.19})O_{2-x}$ defect fluorite structure [86] shown in **Fig. 8**. Another EXAFS study [69] focused on a $Nd_2Zr_2O_7$ pyrochlore, where 10 mol% of the Nd was substituted by Th, U, Np, Pu, or Am. The material was first sintered at 1600°C under

Ar/H_2 to obtain the ordered pyrochlore structure with Th(IV), U(IV), Np(IV), and Am(III). In addition, part of material was annealed at 800°C in air, thus oxidizing the actinides to U(V-VI), Np(V), Pu(IV), and Am(IV). At this moderate temperature, cation ordering in the pyrochlore is preserved, with a pyrochlore lattice parameter smaller than for the reduced samples. All actinide EXAFS spectra, no matter the oxidation state, indicated a split oxygen shell as typical for the Nd position in $Nd_2Zr_2O_7$. The determined actinide–oxygen bond distances correlated well with the ionic radii of the actinides in their appropriate oxidation states.

As a complementary method atomistic simulations were performed to investigate the incorporation of Pu(III) and Pu(IV) in $Gd_2Zr_2O_7$ pyrochlore [104] and of Pu(III), Pu(IV), and U(IV) in a suite of $A(III)_2B(IV)_2O_7$ pyrochlores [73]. As expected, Pu (III) incorporation was found to be energetically favored on the A site in $A_2Zr_2O_7$ (A=La–Gd). For Pu(IV), accommodation on A and B site results in a similar solution energy [73], while a preferential substitution of Zr is also proposed [104]. Due to its larger ionic radius U(IV) was slightly less soluble than Pu(IV). Finally, co-doping of Pu(IV) with Ca(II) yields no further energetic stabilization on the A site in $A_2Zr_2O_7$ [73].

3.4. Thermal Behavior

3.4.1. Thermal Expansion

The linear thermal expansion of a crystalline solid can be measured using high-temperature X-ray diffraction and is calculated according to

$$\alpha(T) = \frac{1}{a_{298K}} \cdot \frac{da_T}{dT}$$

where a_{298K} is the lattice parameter at room temperature and a_T the lattice parameter at the measured temperature, T. The linear thermal expansion coefficient of $(Zr,Y,Pu(IV))O_{2-x}$ sintered in air increases from $7 \cdot 10^{-6}$ K^{-1} at room temperature to $1.2 \cdot 10^{-5}$ K^{-1} at 1000°C, and is nearly independent of the Pu content [64]. Similar measurements were performed on $(La_{1-x}Pu_x)Zr_2O_{7+x}$ sintered in air and Ar/H_2 [33]. Here, the linear thermal expansion coefficient increased from $8 \cdot 10^{-6}$ K^{-1} to $9 \cdot 10^{-6}$ K^{-1} at 1000°C. The thermal expansion of the samples with x = 0.1 was slightly higher than that of pure $La_2Zr_2O_7$.

3.4.2. Thermal Conductivity

The thermal conductivity of stabilized zirconia is low and values close to, or below, 2 $W \cdot m^{-1}K^{-1}$ are observed from room temperature until 2000 K [14,105-107]. A somewhat lower value of 1.3 $W \cdot m^{-1}K^{-1}$ is reported for $Nd_2Zr_2O_7$ [108]. The thermal conductivity of an $(Er_{0.05}Y_{0.10}Pu_{0.10}Zr_{0.75})O_{1.925}$ is nearly temperature independent at around 2 $W \cdot m^{-1}K^{-1}$ [106]. Calculations indicate that doping, e.g., of Y, Er and Pu, reduces the thermal conductivity at lower temperatures [106]. Experimentally, $(Zr_{0.78}Y_{0.16}Am_{0.06})O_{1.89}$ and $(Zr_{0.66}Y_{0.14}Am_{0.20})O_{1.93}$ yielded thermal conductivities close to 2 $W \cdot m^{-1}K^{-1}$ and 1.5 $W \cdot m^{-1}K^{-1}$, respectively [42,107]. A significant increase of the thermal conductivity, especially at lower temperature, can be achieved by diluting the stabilized zirconia of either fluorite or pyrochlore structure, in thermally higher conducting matrices like $MgAl_2O_4$ [105,109] or MgO [108,110,111]. Values even higher than that for UO_2 can be obtained as is shown in Fig. 9 [109]. The thermal conductivity of CERMET composites is even higher and values of 50–70 W·m^{-1}K^{-1} are reported for $(Zr_{0.53}Am_{0.24}Pu_{0.23})O_{2-x}$ (39 vol% metal phase) [112] and $(Zr_{0.67}Y_{0.11}Am_{0.22})O_{2-x}$ (29 vol% metal phase) [111] spheres embedded in a Mo matrix.

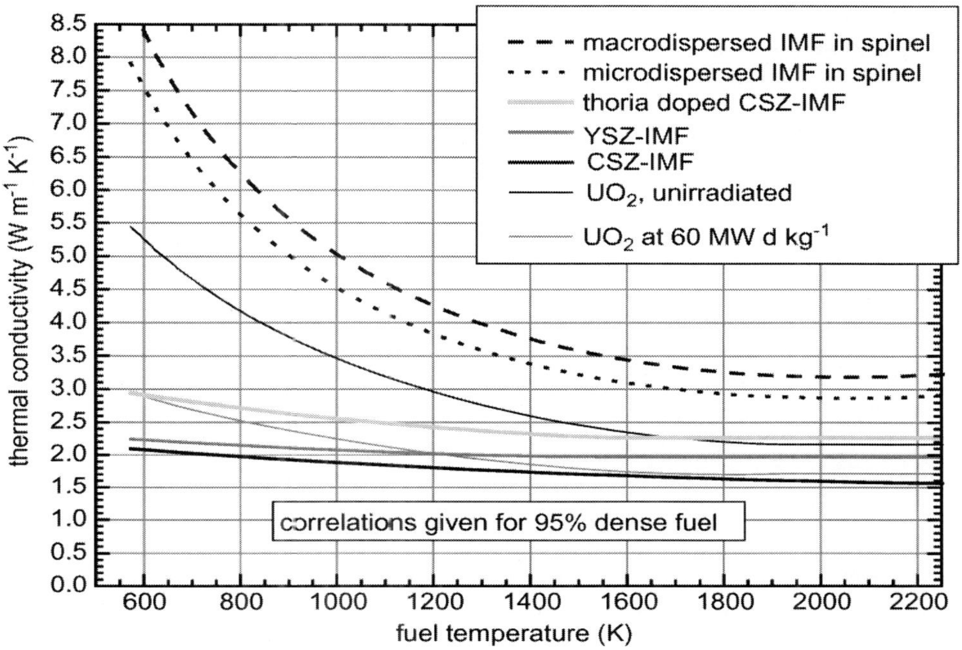

Figure 9. Estimated thermal conductivity of different fuel types [109] © (2006) Elsevier.

3.4.3. Vapor Pressure

The vaporization behavior of both homogenous and composite targets has been investigated by heating samples under vacuum in a Knudsen cell coupled to a mass spectrometer for gaseous phase analysis [113]. Although vaporization and subsequent relocalization of elements from the target is not permitted under normal nuclear reactor conditions, this information is essential to assess safety in severe accidents. The total Pu and Am vapor pressure of the homogeneous $(Zr,Y,Am)O_{2-x}$ and $(Zr,Pu,Am)O_{2-x}$ targets is shown in Fig. 10. Compared to the PuO_{2-x} and AmO_{2-x} reference compounds, the vapor pressure of Pu and Am over the inert matrix compounds is shifted by ~250 K to higher temperatures. This is not true for $(Zr_{0.784}Y_{0.159}Am_{0.057})O_{2-x}$ sintered under Ar. Obviously the non-reductive sintering of this sample lead to a higher O/Am ratio than for the other samples sintered under Ar/H_2 and an Am vapor pressure similar to AmO_{2-x} is observed.

MgO and Mo based composite materials show different behavior. Although MgO is thermally stable under air, it decomposes under vacuum to Mg and oxygen vapor species. Therefore, an enrichment of the actinide bearing zirconia phase at the target surface was observed after Knudsen cell measurements. In addition, the liberated oxygen can increase the O/M ratio in the residual $(Zr,Y,Am)O_{2-x}$. In contrast, the vaporization of Mo is not significant at these temperatures. The Pu and Am vapor pressure in $(Zr,Pu,Am)O_{2-x}$-Mo composites is mainly from the outer $(Zr,Pu,Am)O_{2-x}$ particles at the surface of the material, for that from the inner $(Zr,Pu,Am)O_{2-x}$ particles is effectively suppressed by the Mo matrix [113].

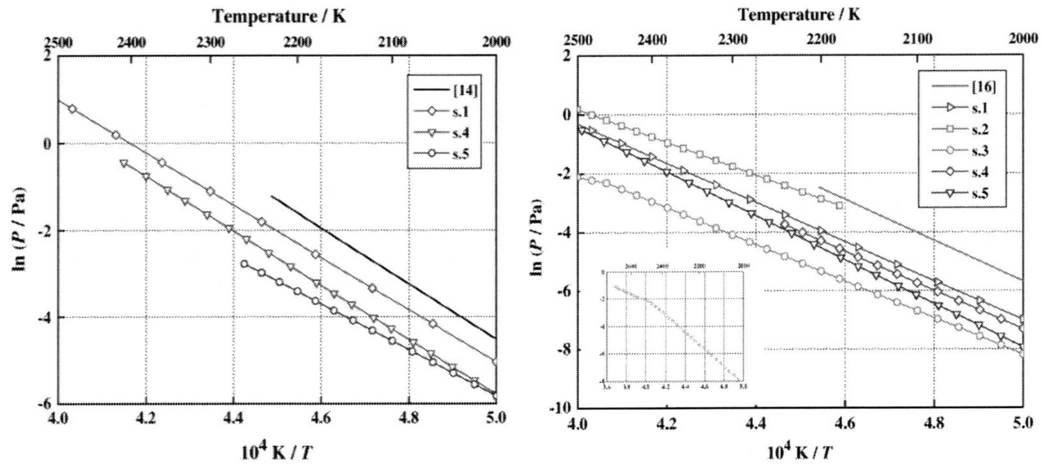

Figure 10. Total Pu (left) and Am (right) vapor pressure under vacuum: s.1 $(Pu_{0.797}Am_{0.198})O_{2-x}$, s.2 $(Zr_{0.784}Y_{0.159}Am_{0.057})O_{2-x}$ sintered under Ar, s.3 $(Zr_{0.800}Y_{0.134}Am_{0.066})O_{2-x}$, s.4 $(Zr_{0.666}Pu_{0.285}Am_{0.049})O_{2-x}$, s.5 $(Zr_{0.579}Pu_{0.248}Am_{0.173})O_{2-x}$ [113] © (2011) Elsevier. With the exception of s.2 all samples were sintered under Ar/H_2. The straight lines noted as [14] and [16] correspond to the reference compounds PuO_{2-x} and AmO_{2-x}.The insertion presents the total AmO_y vaporisation from 300 up to 2700 K.

3.4.4. Helium and Fission Gas Release

During irradiation and storage, significant amounts of noble gases are generated either by fission (Kr, Xe) or by α–decay (He). As their immobilization within the crystal structure is limited, their behavior with increasing temperature is of special safety concern. Helium release from $Nd_2Zr_2O_7$ and $Gd_2Zr_2O_7$ pyrochlores was studied by Knudsen cell measurements on 1 MeV He implanted samples [114]. A first He release peak centered at approximately 700 K was attributed to diffusion to the surface, which was enhanced by the small grain size of the sample. Further He release at 1200 K (only in $Gd_2Zr_2O_7$) and 1600 K were attributed to He present in trapping sites or bubbles [114]. The thermal behavior of 540 keV implanted Xe in YSZ was investigated by Rutherford backscattering, and no change in the Xe depth profile was found upon annealing up to 1500°C [14], while implanted Cs is released sharply above 1173 K [115]. Compared with UO_2 a stronger retention, at similar temperatures, of fission products can be expected for actinide doped stabilized zirconia, as the lattice parameter of $(Zr,Y,Pu,MA)O_{2-x}$ is smaller than that of UO_2, thus slowing the diffusion process [12].

3.5. Radiation Damage Effects

Radiation damage (in the absence of fission) in actinide bearing materials is mainly caused by α-particles (^4He) and the α–recoil nucleus. Although the α–particle has a much higher energy (~5 MeV), most damage results from the slower (70–100 keV), but larger and heavier recoil nucleus. Taking the α–decay of ^{239}Pu as an example, each α–particle causes about 220 atomic displacements (dpa) whereas the α-recoil ^{235}U causes 1180 dpa [116]. In transmutation targets, fission products and neutrons cause additional damage. Such radiation damage alters the material, e.g., by swelling and amorphization. The destruction of the lattice could be accompanied by an increase in aqueous solubility and is a special concern for the conditioning of Pu and MAs for long term storage. Radiation damage effects can be investigated by ion beam irradiation, neutron irradiation and self-irradiation studies using actinides with high specific activity. The radiation tolerance of stabilized zirconia in the defect fluorite structure is high. Ion beam irradiation of stabilized zirconia using 340–400 keV Xe^{+2} (170–300 K) and 72 MeV I^{+} (300–1170 K) ions provided no evidence for amorphization [117], similar indeed to UO_2 [118]. Also $(Zr_{0.5}Er_{0.5})O_{1.75}$ defect fluorite remained crystalline under 350 keV Xe^{+2} ion irradiation at 120 K [119]. The implantation of 400 keV Sr^{+} into YSZ resulted in no amorphization up to 200 dpa. However, a $SrZrO_3$ perovskite phase was formed during thermal annealing at 1000°C [120]. The segregation of Sr, Ba, Zr, and other elements is also observed in conventional fuels, where a separate complex perovskite-type phase $(Ba,Sr,Cs)(U,Pu,Zr,Mo)O_3$ is observed at high burnup [76]. Among the $A_2Zr_2O_7$ pyrochlores (A=La–Gd), $La_2Zr_2O_7$ is the only pyrochlore which has become fully amorphous after a 1.5 MeV Xe+ beam irradiation of 5.5 dpa at room temperature. Its critical amorphization temperature is about 310 K [121]. In other pyrochlores, such as $Nd_2Zr_2O_7$, $Sm_2Zr_2O_7$, and $Gd_2Zr_2O_7$, a gradual pyrochlore to defect fluorite transformation was observed following ion beam irradiation [121,122]. For $Gd_2Zr_2O_7$ the disordering at low temperature is completed at 0.4 dpa [122], whereas the oxygen sublattice disorders prior to the cation sublattice [123]. In the case of $La_2Zr_2O_7$ the transition occurred simultaneously with its amorphization [121]. For the conditioning of Pu and MAs, Ti based pyrochlores or mixed Ti + Zr pyrochlores are also under consideration [24] and ion beam irradiation thereof indicated a substantial tendency for amorphization with increasing Ti content, accompanied with an increase of the critical amorphization temperature [122,124].

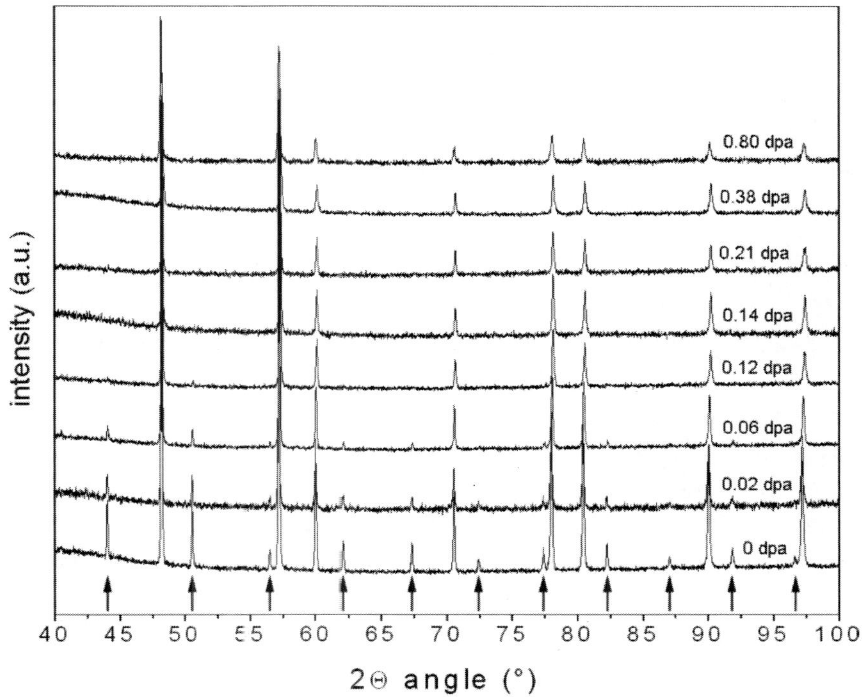

Figure 11. X-ray diffraction patterns of $^{241}Am_2Zr_2O_7$ with different accumulated α–decay dose (dpa) [103] © (2009) The American Chemical Society. Progressive loss of the pyrochlore superstructure starts at 0.02 dpa and is completed after 0.21 dpa (arrows indicate peaks of the pyrochlore superstructure).

The results of self–irradiation studies of actinide doped stabilized zirconia largely substantiate the findings from the ion beam irradiation experiments. The defect fluorite structure of ^{238}Pu doped $(Zr_{0.79}Gd_{0.14}Pu_{0.07})O_{2-x}$ remained without swelling and cracking after a cumulative dose of $6.4 \cdot 10^{19}$ α–decays·g^{-1} [125]. An $Am_2Zr_2O_7$ pyrochlore gradually transformed into a defect fluorite structure due to the ^{241}Am α–decay after a dose of 0.21 dpa, corresponding to $2.5 \cdot 10^{18}$ α–decays·g^{-1} (see Fig. 11 [103]). No further amorphization of the defect fluorite structure was found up to 0.80 dpa. For $^{249}Cf_2Zr_2O_7$ the transformation of the pyrochlore to defect fluorite structure was completed after $1.17 \cdot 10^{18}$ α–decays·g^{-1} [126]. Similar findings are reported for a ^{238}Pu doped $(Nd_{1.8}Pu_{0.2})Zr_2O_{7+x}$ pyrochlore with both Pu(III) and Pu(IV) incorporation. Based on the intensity ratio of the (331) and (222) X-ray diffraction peaks, a first order kinetic behavior was found with a transformation half-life of $6.4 \cdot 10^{17}$ α–decays·g^{-1} for Pu(III) and $6.8 \cdot 10^{17}$ α–decays·g^{-1} for Pu(IV) [69]. In addition, swelling due to irradiation damage is partly compensated by the density increase upon the pyrochlore to fluorite transition [69,103]. Due to the high Ti content, the Ti/Zr pyrochlore phase of a ^{244}Cm doped $Gd_{1.935}Cm_{0.065}TiZrO_7$ ceramic became amorphous following $4.6 \cdot 10^{18}$ α–decays·g^{-1}, corresponding to 0.60 dpa [127]. Further information on radiation damage effects in pyrochlore can be obtained in the review of Ewing et al. [24].

3.6. Irradiation Programs

Test irradiation programs can provide information on the overall safety of transmutation targets under reactor operation conditions. The post-irradiation examination comprises mainly target swelling, densification, fission gas release, achieved burnup, fuel restructuring, and element redistribution. Most of these properties are related to more than one parameter (e.g., target composition, microstructure, irradiation conditions, etc). Due to its complexity, only some characteristic features of irradiation behavior will be highlighted in this chapter. Further information can be obtained from Ref. [4,128].

The general irradiation behavior of homogeneous targets consisting of stabilized zirconia based solid solutions seems acceptable, but hardly yet proven. The low thermal conductivity of the material results in much higher target temperatures compared with UO_2 or MOX fuel operated at the same linear power rate and thus also in significantly higher fission gas release. Depending on the actinide content and linear heat rate in the reactor, target temperatures of approximately 1500–1900°C were estimated and measured for homogenous targets based on Pu and U dissolved in stabilized zirconia [109,129,130]. This is still significantly below the melting point, but a limitation of the linear heat rate is necessary to decrease the target temperature to a more acceptable value [130]. The high target temperature caused partial target densification at the beginning of irradiation (fresh fuel pellets usually have a relative density of less than 95 %) [129]. The target temperature did not increase during the densification, thus the gap between target material and cladding was retained and it is likely that sintering occurred from the center to the edge of the pellet [109]. In addition, a homogenization of the elemental constituents was observed for $(Zr,Y,Er,Pu)O_{2-x}$ transmutation targets prepared by powder blending [131]. The fission gas release from homogeneous target is significant and increases with operating temperature and is close to UO_2 irradiated at similar fuel temperature [130]. The swelling of homogenous targets is low and comparable to MOX fuel [130].

The irradiation behavior of heterogeneous targets is reported for stabilized zirconia based solid solutions embedded in Al_2O_3, $MgAl_2O_4$ (spinel), and Al_2O_3-$MgAl_2O_4$ mixtures [109, 130]. Due to the higher thermal conductivity, the target temperature was lower than in homogenous targets at comparable linear heat rates. High swelling was observed for microdispersion in an $MgAl_2O_4$ matrix, probably due to the low $MgAl_2O_4$ resistance against fission fragment damage. In the case of a Pu containing target consisting of a stabilized zirconia, Al_2O_3, and $MgAl_2O_4$ mixture, the formation of a Pu hibonite phase might have increased the release of fission gases to bubbles, causing high swelling [132]. The fission gas release from $(Zr,Y,Er,Pu)O_{2-x}$ microdispersed in $MgAl_2O_4$ was low [109]. The corresponding macrodisperse $(Zr,Y,Er,Pu)O_{2-x}$–$MgAl_2O_4$ targets showed somewhat opposite behavior, i.e., low swelling and high fission gas release. The lower swelling is in line with the substantial localization of fission fragment damage in the shell around the stabilized zirconia phase. The higher fission gas release is probably caused by the mismatch in their thermal expansion, resulting in microcracks providing pathways for fission gas. Also the local U or Pu content in the stabilized zirconia based solid solutions in composites is higher than in homogeneous targets, causing different local irradiation behavior [130]. The target temperature is also important for the behavior of the actinide-free matrix material. At temperatures >1700 K, $MgAl_2O_4$ was thermally unstable: Due to the low oxygen potential in the pin, it decomposed leaving stabilized zirconia and Al_2O_3 in the target center, forming MgO at the inner surface of the cladding [82,130]. No $MgAl_2O_4$ decomposition was found with an average center line temperature of 1060 K [130]. For Al_2O_3 based macrodisperse composites, very high fission gas release was observed during irradiation at temperatures below 1400 K. Surprisingly, irradiation at higher target temperatures yielded a lower fission gas release. This behavior might be caused by improved damage annealing at higher temperature, whereas the slower annealing of the Al_2O_3 structure at lower temperatures might enable fission gas release from the embedded stabilized zirconia [130]. In conclusion, despite their improved thermal conductivity the Al_2O_3 and $MgAl_2O_4$ based composite targets showed a less satisfactory irradiation behavior than the homogeneous targets. It must also be stressed that given their cost, an insufficient number of tests have been performed, limiting a full evaluation.

Other irradiation campaigns, possibly with less complex materials and still ongoing, focus on the irradiation behavior of $(Zr,Y,Am)O_{2-x}$. Within the CAMIX-COCHIX experiment a homogeneous $(Zr,Y,Am)O_{2-x}$ target was irradiated for comparison with microdisperse and macrodisperse $(Zr,Y,Am)O_{2-x}$–MgO composites [44]. The HELIOS program includes two homogeneous $(Zr,Y,Am)O_{2-x}$ and $(Zr,Y,Am,Pu)O_{2-x}$ target pins [6] to establish the effect of the higher linear heat rate in the Pu bearing material with the higher target temperature and He release. This should result in lower swelling compared with the Pu free target. In addition, $Am_2Zr_2O_7$–MgO and $(Zr,Y,Am)O_{2-x}$–Mo composite targets were irradiated in the HELIOS programme. A $(Zr,Y,Am)O_{2-x}$–Mo composite target is also part of the FUTURIX irradiation program [6]. Post-irradiation examination of these materials is pending and will shed new light on the operational safety of such targets.

3.7. Aqueous Durability

Although stabilized zirconia is considered for conditioning of Pu and MAs, and also transmutation targets disposal after irradiation, knowledge about the dissolution of such compounds in aqueous solutions is limited. Stabilized zirconia is highly durable in the pH range 3–9 causing experimental problems in determining solubility and dissolution rates. For a heterovalent solid solution of stabilized zirconia, actinides and their fission and decay products, an incongruent dissolution at the solid-water interface is expected. This would result in a layer of less soluble tetravalent cations like (Zr, actinides), whereas more soluble ions (e.g., Y, UO_2^{+2}, NpO_2^{+}) are depleted at the surface. Therefore the durability of the disposed material will mainly be governed by the solubility of Zr and tetravalent actinides. The solubility of pure ZrO_2 and cubic YSZ in deionized water is both about 10^{-9} M Zr at pH 9 [133–135]. The solubility of Zr hydrous oxides at near neutral pH-range is approximately 10^{-8} to 10^{-7} M Zr [136,137]. For a bentonite-filled repository, the solubility of stabilized zirconia will be approximately 10^{-9} to 10^{-8} M. Due to the high uncertainty and limited data a conservative solubility limit of 10^{-6} M Zr has been recommended [133]. The solubility of tetravalent actinide oxides and hydrous oxides at near neutral pH is close to the Zr solubility [138]. As no solubility difference between crystalline dioxides and hydrous oxides is observed at near neutral conditions, the same surface structure controls the bulk material solubility both for tetravalent actinides and Zr [135,138]. The high durability of stabilized zirconia is limited by complex formation with carbonate and in the alkaline range, conditions expected in the presence of concrete or in rock-salt repositories. The addition of $NaHCO_3$ increases the Zr solubility to 10^{-7}–10^{-5} M at pH 9 [134,135], indicating the formation of Zr carbonate complexes similar to other tetravalent cations e.g., Th [139,140]. Under alkaline conditions, the solubility of Zr hydrous oxide is greatly increased in the presence of 0.2–1.0 M $CaCl_2$ (see Fig. 12), and this is attributed to the formation of a $Ca_3[Zr(OH)_6]^{+4}$ complex [141,142]. A similar complex also increases the solubility of tetravalent actinides under such conditions [141]. Finally, the solubility of stabilized zirconia under acidic conditions is low, even in case of an irradiated $(Zr,Y,Pu,Er)O_{2-x}$ target [131].

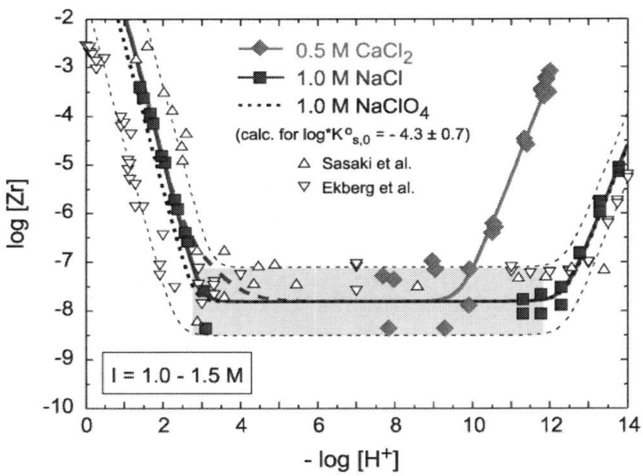

Figure 12. Solubility of $ZrO_2{\cdot}xH_2O$ in NaCl, $NaClO_4$, $CaCl_2$, and $Ca(ClO_4)_2$ at comparable ionic strength [141] © 2008 Oldenbourg Wissenschafts verlag GmbH. Open triangles correspond to values from Ekberg et al. [136] and Sasaki et al. [137].

Another aspect in the assessment of the durability is the dissolution rate of actinide doped stabilized zirconia. The dissolution rate for 32 days contact of $(Zr_{0.82}Y_{0.18})O_{1.91}$ and $La_2Zr_2O_7$ powders in deionized water at 90°C is about 10^{-6} to 10^{-5} $g{\cdot}m^{-2}d^{-1}$, whereas the rates for Y and La are slightly higher than for Zr [143]. Furthermore, the dissolution rates of $(La,Nd)_2Zr_2O_7$ and $La_2Zr_2O_7$-$(CeO_2$-$ZrO_2)$ solid solutions in deionized water at 90°C are Zr 10^{-8} to 10^{-7} $g{\cdot}m^{-2}d^{-1}$, Ce 10^{-7} to 10^{-6} $g{\cdot}m^{-2}d^{-1}$, and La, Nd 10^{-5} to 10^{-4} $g{\cdot}m^{-2}d^{-1}$. It should be noted, however, that the material was pretreated under the same conditions to eliminate higher initial dissolution of smaller particles from powder preparation [144]. Kuramoto et al. report dissolution rates for $(Zr,Y,Nd)O_{2-x}$ and $(Zr,Y,Ce)O_{2-x}$ in deionized water at 150°C in the range of 10^{-9} to 10^{-7} $g{\cdot}m^{-2}d^{-1}$ [145]. Dissolution rates of YSZ single crystal pieces in a saline ground water from the Canadian Shield were determined and the average leach rate after 56 days at 25 °C was $1.2{\cdot}10^{-4}$ $g{\cdot}m^{-2}d^{-1}$ and increased to $6.0{\cdot}10^{-4}$ $g{\cdot}m^{-2}d^{-1}$ at 85 °C. If a granite coupon was added, the leach rate increased by a factor of three [146]. The normalized Pu mass loss from $(Zr_{0.79}Gd_{0.14}Pu_{0.07})O_{1.99}$ following a dose of 6.4 10^{18} α–decays$\cdot g^{-1}$ in deionized water at 90°C averaged over 28 days contact is $1.4{\cdot}10^{-3}$ $g{\cdot}m^{-2}d^{-1}$. The Pu mass loss to the aqueous phase increases with dose [125]. However, the dissolution rate given in [125] refers to the geometric dimensions of the used pellet without porosity correction and care has to be taken when comparing with the dissolution rate determined for powders or single crystals. In general, the dissolution rates determined for stabilized zirconia, present either as defect fluorite or pyrochlore, indicate a high durability of such waste forms. The complexation of Zr and actinides with ground water constituents (e.g., carbonate, Ca) could increase the dissolution rate. In contrast, other ground water constituents, like Si, might result in the formation of surface precipitates as shown for the YSZ alteration in saline ground water in the presence of bentonite [146]. Another effect, not yet considered, is redox reactions during dissolution. This concerns for example U, Np, Am, which are most soluble as Am(III), Np(V), and U(VI). Finally, it is known that the dissolution rate is fast at the beginning of the experiments and becomes lower once equilibrium is reached [135,143]. Therefore additional long-term dissolution studies should be made against the industrial standard – borosilicate glass [133].

Conclusion

Despite its low thermal conductivity, stabilized zirconia seems to be a promising inert matrix material for the transmutation of Pu and MAs in nuclear reactors or accelerator driven systems (ADS). Even if high transmutation rates might not be reached, the fissile Pu isotopes are drastically reduced, thus guarantying proliferation resistance of the spent transmutation target. Stabilized zirconia is able to incorporate sufficient amounts of both Pu and MAs as a solid solution, either in defect fluorite or pyrochlore structures. The pyrochlore structure is neither advantageous nor necessary, as radiation damage will disorder the pyrochlore into the (radiation resistant) defect fluorite structure anyway. Target synthesis routes have been developed satisfactorily, addressing the problems arising from the high α–activity of MAs. Powder blending is possible for the production of Pu targets, whereas the infiltration of stabilized zirconia spheres obtained by sol-gel methods is favorable for Am and Cm. The low thermal conductivity limits the amount of Pu and MAs in transmutation targets to ensure a sufficient safety margin between target operation temperature and the melting point. Low thermal conductivity can be improved by adding thermally better conducting materials such as MgO (CERCER) or Mo (CERMET). Despite the lower target temperature, such targets show a more complex irradiation behavior, for which further data and understanding are needed. In view of the fact that stabilized zirconia based transmutation targets will be sent for disposal after irradiation and such materials are also considered for Pu and MA conditioning, the existing knowledge on their dissolution behavior in aqueous solution is not sufficient and further systematic research on their aqueous durability is mandatory to address effects of porosity, incongruent dissolution, ground water constituents, and actinide valence state.

REFERENCES

[1] J. Magill, V. Berthou, D. Haas, J. Galy, R. Schenkel, H.-W. Wiese, G. Heusener, J. Tommasi, G. Youinou, *Nuclear Energy* (2003) 42, 263.
[2] C. Degueldre, J.M. Paratte, *J. Nucl. Mater.* (1999) 274, 1.
[3] R.J.M. Konings, J.L. Kloosterman, *Progress Nucl. Energy* (2001) 38, 331.
[4] IAEA, *Viability of inert matrix fuel in reducing plutonium amounts in reactors, IAEA-TECDOC*-1516, Wien (2006).
[5] SCK-CEN, Partioning and transmutation European roadmap for sustainable nuclear energy (PATEROS*), http://sckcen.be/pateros/.*
[6] F. Delage, R. Belin, X.-N. Chen, E. D'Agate, F. Klaassen, S. Knol, W. Maschek, J.-P. Ottaviani, A. Rineiski, V. Sobolev, J. Somers, D. Staicu, R. Thetford, J. Wallenius, B. Wernli, *Energy Proc.* (2011) 7, 303.
[7] J. Somers, *Energy Proc.* (2011) 7, 169.
[8] N. Chauvin, R.J.M. Konings, Hj. Matzke, *J. Nucl. Mater.* (1999) 274, 105.
[9] R.J.M. Konings, R. Conrad, G. Dassel, B.J. Pijlgroms, J. Somers, E. Toscano, *J. Nucl. Mater.* (2000) 282, 159.
[10] V. Berthou, C. Degueldre, J. Magill, *J. Nucl.Mater.* (2003) 320, 156.
[11] U. Kasemeyer, J.-M. Paratte, P. Grimm, R. Chawla, *Nucl. Technol.* (1998) 122, 52.
[12] C. Degueldre, *J. Alloys Compounds* (2007) 444–445, 36.
[13] H. Kleykamp, *J. Nucl. Mater.* (1999) 275, 1.
[14] G. Ledergerber, C. Degueldre, P. Heimgartner, M.A. Pouchon, U. Kasemeyer, *Progress Nucl. Energy* (2001) 38, 301.
[15] G.R. Lumpkin, *Elements* (2006) 2, 365.
[16] W.L. Gong, W. Lutze, R.C. Ewing, *J. Nucl. Mater.* (2000) 277, 239.
[17] S. Pillon, J. Somers, S. Grandjean, J. Lacquement, *J. Nucl. Mater.* (2003) 320, 36.
[18] M. Yashima, M. Kakihana, M. Yoshimura, *Solid State Ionics* (1996) 86-88, 1131.
[19] D.-J. Kim, *J. Am. Ceram. Soc.* (1989) 72, 1415.
[20] C.R.A. Catlow, A.V. Chadwick, G.N. Greaves, L.M. Moroney, *J. Am. Ceram. Soc.* (1986) 69, 272.
[21] P. Li, I.-W. Chen, J.E. Penner-Hahn, *Phys. Rev.* (1993) B 48, 10074.
[22] K. Kawata, H. Maekawa, T. Nemoto, T. Yamamura, *Solid State Ionics* (2006) 177, 1687.
[23] R.G. Haire, P.E. Raison, Z. Assefa, *J. Nucl. Sci. Technol.* (2002) *Supp. 3*, 616.
[24] R.C. Ewing, W.J. Weber, J. Lian, *J. Appl. Phys.* (2004) 95, 5949.

[25] E.J. Harvey, K.R. Whittle, G.R. Lumpkin, R.I. Smith, S.A.T. Redfern, *J. Solid State Chem.* (2005) 178, 800.
[26] R.G. Haire, L. Eyring, in: K.A. Gschneidner, L. Eyring, G.R. Choppin, G.R. Lander eds., *Handbook on the Physiscs and Chemistry of Rare Earths*, Vol. 18, Lanthanides / Actinides: Chemistry, Elsevier Science B. V., Amsterdam (1994) pp.413–505.
[27] J. Fuger, R.G. Haire, J.R. Peterson, *J. Alloys Compounds* (1993) 200, 181.
[28] C. Keller, *Über die Festkörperchemie der Actiniden-Oxide. KFK-225, Kernforschungs zentrum Karlsruhe* (1964).
[29] T. Shiratori, T. Yamashita, T. Ohmichi, A. Yasuda, K. Watarumi, *J. Nucl. Mater.* (1999) 274, 40.
[30] S.K. Sali, N.K. Kulkarni, D.G. Phal, S.K. Aggarwal, V. Venugopal, *J. Am. Ceram. Soc.* (2010) 93, 3437.
[31] C. Hellwig, M. Pouchon, R. Restani, F. Ingold, G. Bart, *J. Nucl. Mater.* (2005) 340, 163.
[32] M. Burghartz, G. Ledergerber, F. Ingold, P. Heimgartner, C. Degueldre, *Progress Nucl. Energy* (2001) 38, 247.
[33] S. Yamazaki, T. Yamashita, T. Matsui, T. Nagasaki, *J. Nucl. Mater.* (2001) 294, 183.
[34] B. Burakov, E. Anderson, M. Yagovkina, M. Zamoryanskaya, E. Nikolaeva, *J. Nucl. Sci. Technol.* (2002) *Suppl.* 3, 733.
[35] S. Grandjean, B. Arab-Chapelet, A.C. Robisson, F. Abraham, P. Martin, J.-P. Dancausse, N. Herlet, C. Léorier, *J. Nucl. Mater.* (2009) 385, 204.
[36] K. Holliday, T. Hartmann, S.R. Mulcahy, K. Czerwinski, *J. Nucl. Mater.* (2010) 402, 81.
[37] A. Fernández, D. Haas, R.J.M. Konings, J. Somers, *J. Am. Ceram. Soc.* (2002) 85, 694.
[38] G. Ledergerber, F. Ingold, R.W. Stratton, H.P. Alder, C. Prunier, D. Warin, M. Bauer, *Nucl. Technol.* (1996) 114, 194.
[39] K. Idemitsu, T. Arima, Y. Inagaki, S. Torikai, M.A. Pouchon, *J. Nucl. Mater.* (2003) 319, 31.
[40] M. Couland, S. Fourcaudot, R.J. Abril, A. Fernandez-Carretero, J. Somers, *J. Amer. Ceram. Soc.* (2012) 95, 133.
[41] J. Somers, A. Fernández, *J. Am. Ceram. Soc.* (2005) 88, 827.
[42] C. Nästren, D. Staicu, J. Somers, A. Fernandez, *J. Nucl. Mater.* (2013) 433, 314,.
[43] J. Somers, A. Fernandez, *Progress Nucl. Energy* (2006) 48, 259.
[44] Y. Croixmarie, E. Abonneau, A. Fernández, R.J.M. Konings, F. Desmoulière, L. Donnet, *J. Nucl. Mater.* (2003) 320, 11-17..
[45] H. Kinoshita, K. Kuramoto, M. Uno, T. Yanagi, S. Yamanaka, H. Mitamura, T. Banba, *J. Am. Ceram. Soc.* (2000) 83, 391.
[46] C. Nästren, A. Fernandez, D. Haas, J. Somers, M. Walter, *J. Nucl. Mater.* (2007) 362, 350.
[47] V. Menvie Benkale, C. Legros, C. Haut, G. Sattonnay, A.M. Huntz, *Solid State Ionics* (2006) 177, 3339.
[48] P.G. Medvedev, S.M. Frank, T.P. O'Holleran, M.K. Meyer, *J. Nucl. Mater.* (2005) 342, 48.
[49] K. Richter, A. Fernandez, J. Somers, *J. Nucl. Mater.* (1997) 249, 121.
[50] M. Walter, J. Somers, A. Fernandez, D. Haas, K. Dardenne, M.A. Denecke, *J. Nucl. Mater.* (2007) 362, 343.

[51] R.P.C. Schram, R.R. van der Laan, F.C. Klaassen, K. Bakker, T. Yamashita, F. Ingold, *J. Nucl. Mater.* (2003) 319, 118.
[52] P.G. Medvedev, J.F. Jue, S.M. Frank, M.K. Meyer, *J. Nucl. Mater.* (2006) 352, 318.
[53] K. Holliday, T. Hartmann, K. Czerwinski, *J. Nucl. Mater.* (2009) 392, 487.
[54] K. Holliday, T. Hartmann, F. Poineau, J.R. Kennedy, K. Czerwinski, *J. Nucl. Mater.* (2009) 393, 224.
[55] A. Fernández, R.J.M. Konings, J. Somers, *J. Nucl. Mater.* (2003) 319, 44.
[56] H.M. Ondik, H.F. McMurdie, *Phase diagrams for zirconium and zirconia systems*, The American Ceramic Society, Westerville (1998).
[57] D.F. Caroll, *J. Am. Ceram. Soc.* (1963) 46, 194.
[58] H. Radzewitz, *Festkörperchemische Untersuchungen über die Systeme* $SEO_{1,5}$-ZrO_2 (HfO_2),$AmO_{1,5}$-ZrO_2(HfO_2,ThO_2)-O_2 *und* TiO_2-NpO_2(PuO_2), *KFK-433, Kernforschungs zentrum Karlsruhe* (1964).
[59] P.E. Raison, R.G. Haire, Z. Assefa, *J. Nucl. Sci. Technol.* (2002) *Suppl.* 3, 725.
[60] P.E. Raison, R.G. Haire, *J. Nucl. Mater.* (2003) 320, 31.
[61] J.H. Handwerk, G.D. White, D.C. Hill, *J. Am. Ceram. Soc.* (1963) 46, 29.
[62] N.K. Kulkarni, S. Sampath, V. Venugopal, *Ceram. Int.* (2001) 27, 839.
[63] P.E. Raison, R.G. Haire, *Progress Nucl. Energy* (2001) 38, 251.
[64] T. Yamashita, S. Yamazaki, T. Sato, T. Matsui, *J. Nucl. Sci. Technol.* (2002) *Suppl.* 3, 656.
[65] M. Walter, J. Somers, D. Bouëxière, P. Gaczynski, B. Brendebach, *J. Solid State Chem.* (2009) 182, 3305.
[66] F. Schleifer, A. Naoumidis, H. Nickel, *J. Nucl. Mater.* (1981) 101, 150.
[67] Y. Hinatsu and T. Muromura, *Mater. Res. Bull.* (1986) 21, 1343.
[68] N. Sakai, T. Hashimoto, T. Katsube, K. Yamaji, H. Negishi, T. Horita, H. Yokokawa, Y. P. Xiong, M. Nakagawa, Y. Takahashi, *Solid State Ionics* (2001) 143, 151.
[69] C. Nästren, R. Jardin, J. Somers, M. Walter, B. Brendebach, *J. Solid State Chem.* (2009) 182, 1.
[70] B.P. Mandal, Nandini Garg, S.M. Sharma, A.K. Tyagi, *J. Nucl. Mater.* (2009) 392, 95.
[71] B.P. Mandal, M. Pandey, A.K. Tyagi, *J. Nucl.Mater.* (2010) 406, 238.
[72] N.K. Kulkarni, S. Sampath, V. Venugopal, *J. Nucl. Mater.* (2000) 281, 248.
[73] A. Cleave, R.W. Grimes, K. Sickafus, *Philos. Mag.* (2005) 85, 967.
[74] T. Hartmann, A. Alaniz, F. Poineau, P.F. Weck, J.A. Valdez, M. Tang, G.D. Jarvinen, K.R. Czerwinski, K.E. Sickafus, *J. Nucl. Mater.* (2011) 411, 60.
[75] S.J. Patwe, A.K. Tyagi, *Ceram. Int.* (2006) 32, 545.
[76] H. Kleykamp, *J. Nucl. Mater.* (1985) 131, 221.
[77] P.E. Raison, R.G. Haire, T. Sato, T. Ogawa, *Mat. Res. Soc. Sypm. Proc.* (1999) 556, 3.
[78] R.C. Belin, P.J. Valenza, P.E. Raison, M. Tillard, *J. Alloys Compounds* (2008) 448, 321.
[79] T. Fujino, C. Miyake, in: A. J. Freeman, C. Keller eds., *Handbook on the Physics and Chemistry of the Actinides*, Elsevier, Amsterdam (1991) pp.155-240.
[80] E.R. Vance, Y. Zhang, Z. Zhang, *J. Nucl. Mater.* (2010) 400, 8.
[81] F. Schleifer, A. Naoumidis, H. Nickel, *J. Nucl. Mater.* (1983) 115, 143.
[82] T. Yamashita, K. Kuramoto, M. Nakada, S. Yamazaki, T. Sato, T. Matsui, *J. Nucl. Sci. Technol.* (2002) *Suppl.* 3, 585.

[83] H. Kinoshita, D. Setoyama, Y. Saito, M. Hirota, K. Kurosaki, M. Uno, S. Yamanaka, *J. Chem. Thermodynamics* (2003) 35, 719.

[84] H. Otobe, A. Nakamura, T. Yamashita, *J. Nucl. Sci. Technol.* (2002) *Suppl.* 3, 652.

[85] H. Otobe, A. Nakamura, T. Yamashita, K. Minato, *J. Nucl. Mater.* (2005) 344, 219.

[86] M. Walter, C. Nästren, J. Somers, R. Jardin, M.A. Denecke, B. Brendebach, *J. Solid State Chem.* (2007) 180, 3130.

[87] T.D. Chikalla, L. Eyring, *J. Inorg. Nucl. Chem.* (1967) 29, 2281.

[88] H. Otobe, M. Takano, H. Hayashi, Y. Arai, *J. Am. Ceram. Soc.* (2011) 94, 3596.

[89] M.H. Tuilier, J. Dexpert-Ghys, H. Dexpert, P. Lagarde, *J. Solid State Chem.* (1987) 69, 153.

[90] D. Komyoji, A. Yoshiasa, T. Moriga, S. Emura, F. Kanamaru, K. Koto, *Solid State Ionics* (1992) 50, 291.

[91] P. Li, I.-W. Chen, J.E. Penner-Hahn, *J. Am. Ceram. Soc.* (1994) 77, 118.

[92] H. Maekawa, K. Kawata, Y. P. Xiong, N. Sakai, H. Yokokawa, *Solid State Ionics* (2009) 180, 314.

[93] P. Villella, S.D. Conradson, F.J. Espinosa-Faller, S.R. Foltyn, K.E. Sickafus, J.A. Valdez , *Phys. Rev.* (2001) B 64, 104101.

[94] M. Walter, J, Somers, D. Bouëxière, J. Rothe, *J.Solid State Chem.* (2011) 184, 911.

[95] S.D. Conradson, C.A. Degueldre, F.J. Espinosa-Faller, S.R. Foltyn, K.E. Sickafus, J.A. Valdez, P.M. Villella, *Progress Nucl. Energy* (2001) 38, 221.

[96] C. Degueldre, S.D. Conradson, *Appl. Phys.* (2001) A73, 489.

[97] C. Nästren, M. Walter. in: *ANKA Annual Report 2008, Angströmquelle Karlsruhe, Forschungszentrum Karlsruhe* (2008) pp.168–169.

[98] J.B. Boyce, J.C. Mikkelsen Jr., *Phys. Rev.* (1985) B 31,6903.

[99] J. Purans, G. Heisbourg, N. Dacheux, Ph. Moisy, S. Hubert, *Phys. Scr.* (2005) T115, 925.

[100] S. Hubert, J. Purans, G. Heisbourg, P. Moisy, N. Dacheux, *Inorg. Chem.* (2006) 45, 3887.

[101] N.J. Hess, B.D. Begg, S.D. Conradson, D.E. McCready, P.L. Gassman, W.J. Weber, *J. Phys. Chem.* (2002) B 106, 4663.

[102] P.M. Martin, R.C. Belin, P.J. Valenza, A.C. Scheinost, *J. Nucl. Mater.* (2009) 385, 126.

[103] R.C. Belin, P.M. Martin, P.J. Valenza, A.C. Scheinost, *Inorg. Chem.* (2009) 48, 5376.

[104] R.E. Williford, W.J. Weber, *J. Nucl. Mater.* (2001) 299, 140.

[105] N. Nitani, T. Yamashita, T. Matsuda, S.I. Kobayashi, T. Ohmichi, *J. Nucl. Mater.* (1999) 274, 15.

[106] C. Degueldre, T. Arima, Y.W. Lee, *J. Nucl. Mater.* (2003) 319, 6.

[107] C. Ronchi, J.P. Ottaviani, C. Degueldre, R. Calabrese, *J. Nucl. Mater.* (2003) 320, 54.

[108] S. Lutique, R.J.M. Konings, V.V. Rondinella, J. Somers, T. Wiss, *J. Alloys Compounds* (2003) 352, 1.

[109] Ch. Hellwig, M. Streit, P. Blair, T. Tverberg, F.C. Klaassen, R.P.C. Schram, F. Vettraino, T. Yamashita, *J. Nucl. Mater.* (2006) 352, 291.

[110] P.G. Medvedev, M.J. Lambregts, M.K. Meyer, *J. Nucl. Mater.* (2006) 349, 167.

[111] F. Delage, F. Audubert, R. Belin, X.-N. Chen, E. D'Agata, A. Fernandez-Carretero, E. Gavilan, J.P. Hiernaut, F. Klaassen, W. Maschek, J.P. Ottaviani, A. Rineiski, V. Sobolev, J. Somers, D. Staicu, R. Thetford, J. Wallenius, *Proc. GLOBAL* 2009, Paper 9068, Paris, 2009.

[112] D. Haas, A. Fernandez, D. Staicu, J. Somers, W. Maschek, P. Liu, X. Chen, *Energy Convers. Manage.* (2008) 49, 1928.
[113] P. Gotcu-Freis, J.-P. Hiernaut, J.-Y. Colle, C. Nästrén, A. Fernandez Carretero, R.J.M. Konings, *J. Nucl. Mater.* (2011) 411, 119.
[114] T.A.G. Wiss, J.-P. Hiernaut, P.M.G. Damen, S. Lutique, R. Fromknecht, W.J. Weber, *J. Nucl. Mater.* (2006) 352, 202.
[115] M.A. Pouchon, M. Döbeli, C. Degueldre, M. Burghartz, *J. Nucl. Mater.* (1999) 274, 61.
[116] W.J. Weber, R.C. Ewing, C.R.A. Catlow, T. Diaz de la Rubia, L.W. Hobbs, C. Kinoshita, Hj. Matzke, A.T. Motta, M. Natasi, E.K.H. Salje, E.R. Vance, S.J. Zinkle, *J. Mater. Res.* (1998) 13, 1434.
[117] K.E. Sickafus, Hj. Matzke, T. Hartmann, K. Yasuda, J.A. Valdez, P. Chodak III, M. Nastasi, R.A. Verrall, J. *Nucl. Mater.* (1999) 274, 66.
[118] Hj. Matzke, L.M. Wang, *J. Nucl. Mater.* (1996) 231, 155.
[119] K.E. Sickafus, L. Minervini, R.W. Grimes, J.A. Valdez, M. Ishimaru, F. Li, K.J. McClellan, T. Hartmann, *Science* (2000) 289, 748.
[120] S. Zhu, S.X. Wang, L.M. Wang, R.C. Ewing, X.T. Zu, *Appl. Phys. Lett.* (2007) 90, 171915.
[121] J. Lian, X.T. Zu, K.V.G. Kutty, J. Chen, L.M. Wang, R.C. Ewing, *Phys. Rev.* (2002) B 66, 054108.
[122] S.X. Wang, B.D. Begg, L.M. Wang, R.C. Ewing, W.J. Weber, K.V.G. Kutty, *J. Mater. Res.* (1999) 14, 4470.
[123] J. Lian, L. Wang, J. Chen, K. Sun, R.C. Ewing, J.M. Farmer, L.A. Boatner, *Acta Mater.* (2003) 51, 1493.
[124] J. Lian, S.V. Yudintsev, S.V. Stefanovsky, L.M. Wang, R.C. Ewing, *J. Alloys Compounds* (2007) 444–445, 429.
[125] B. Burakov, E. Anderson, M. Yagovkina, M. Zamoryanskaya, E. Nikolaeva, *J. Nucl. Sci. Technol.* (2002) Suppl. 3, 733.
[126] R.E. Sykora, P.E. Raison, R.G. Haire, *J. Solid State Chem.* (2005) 178, 578.
[127] S.V. Yudinsev, A.N. Lukinykh, S.V. Tomilin, A.A. Lizin, S.V. Stefanovsky, *J. Nucl. Mater.* (2009) 385, 200.
[128] P. Pöml, R.J.M. Konings, J. Somers, T. Wiss, G.J.L.M. de Haas, F.C. Klaassen, Chapter 57 in R.J.M. Konings ed., *Comprehensive Nuclear Materials*, Springer (2011).
[129] Ch. Hellwig, U. Kasemeyer, *J. Nucl. Mater.* (2003) 319, 87.
[130] N. Nitani, K. Kuramoto, Y. Nakano, T. Yamashita, Y. Kimura, Y. Nihei, T. Ogawa, *J. Nucl. Mater.* (2008) 376, 88.
[131] R. Restani, M. Martin, N. Kivel, D. Gavillet, *J. Nucl. Mater.* (2009) 385, 435.
[132] T. Yamashita, K. Kuramoto, H. Akie, Y. Nakano, N. Nitani, T. Nakamura, K. Kusagaya, T. Ohmichi, *J. Nucl. Sci. Technol.* (2002) 39, 865.
[133] E. Curti, W. Hummel, *J. Nucl. Mater.* (1999) 274, 189.
[134] M.A. Pouchon, E. Curti, C. Degueldre, L. Tobler, *Progress Nucl. Energy* (2001) 38, 443.
[135] E. Curti, C. Degueldre, *Radiochim. Acta* (2002) 90, 801.
[136] C. Ekberg, G. Källvenius,Y. Albinsson, P.L. Brown, *J. Sol. Chem.* (2004) 33, 47.
[137] T. Sasaki, T. Kobayashi, I. Takagi, H. Moriyama, *Radiochim. Acta* (2006) 94, 489.
[138] V. Neck, J.I. Kim, *Radiochim. Acta* (2001) 89, 1216.
[139] M. Altmaier, V. Neck, R. Müller, T. Fanghänel, *Radiochim. Acta* (2005) 93, 83.

[140] M. Altmaier, V. Neck, M.A. Denecke, R. Yin, T. Fanghänel, *Radiochim. Acta* (2006) 94, 495.
[141] M. Altmaier, V. Neck, T. Fanghänel, *Radiochim. Acta* (2008) 96, 541.
[142] B. Brendebach, M. Altmaier, J. Rothe, V. Neck, and M.A. Denecke, *Inorg. Chem.* (2007) 46, 6804.
[143] H. Kamizono, I. Hayakawa, S. Muraoka, *J. Am. Ceram. Soc.* (1991) 74, 863.
[144] I. Hayakawa, H. Kamizono, *J. Nucl. Mater.* (1993) 202, 163.
[145] K. Kuramoto, H. Mitamura, T. Banba, S. Muraoka, *Progress Nucl. Energy* (1998) 32, 509.
[146] R.B. Heimann, T.T. Vandergraaf, *J. Mater. Sci. Lett.* (1988) 7, 583.

CONTRIBUTORS

Prof. J. Mizusaki (Co-editor) (Tohoku Univ. Inst. Interdisciplinary Res. for Advanced Materials, Japan) (mizusaki@tagen.tohoku.ac.jp)

Prof. A. K. Tyagi and V. Grover (Div. of Chemistry, BARC, India) (aktyagi@barc.gov.in) & (vinita@barc.gov.in)

Prof. M. Yashima (Tokyo Inst. Technology, Dept. Materials Science and Engineering, Japan) (yashima@cms.titech.ac.jp)

Dr. A. Nakamura (Editor) (Advanced Science Research Center, Japan Atomic Energy Agency, Japan) (nakamura.akio@jaea.go.jp)

Drs. H. Kishimoto & H. Yokokawa (Advanced Insititute for Science and Technology, Tsukuba, Japan) (haruo-kishimoto@aist.go.jp) & (h-yokokawa@aist.go.jp)

Prof. A. Atkinson (Imperial College London, UK) [alan.atkinson@imperial.ac.uk]

Profs. Y. Ikuhara & N. Shibata (Univ. Tokyo, Graduate School of Engineering, Japan) (ikuhara@sigma.t.u-tokyo.ac.jp) & (Shibata@sigma.t.u-tokyo.ac.jp)

Prof. M. Uo (Tokyo Medical & Dental Univ., Graduate School of Medical and Dental Sciences, Japan) (mail@m-uo.com)

Dr. Arima (Kyushu Univ. Graduate School of Engineering, Japan) (arima@nucl.kyushu-u.ac.jp)

Dr. H. Serizawa (Japan Atomic Energy Agency, Nuclear Science and Engineering directorate, Japan) (serizawa.hiroyuki@jaea.go.jp)

Drs. M. Walter & J. Somers (JRC, ITU, Germany) (Marcus Walter [1052076@web.de]) & (joseph.somers@ec.europa.eu)

INDEX

A

B

C

D

E

F

G

H

I

K

L

M

N

O

P

Q

R

S

T

U

V

W

X

Y

Z